建造设计手册
医院和医疗建筑

Construction and Design Manual
Hospitals and Medical Facilities

建造设计手册
医院和医疗建筑

Edited By Philipp Meuser(菲利普·莫伊泽)

Scientific Advisor:
Franz Labryga(弗朗茨·纳布瑞加)

Further Contributions
Klaus Bergdolt, Klaus R. Bürger, Linus Hofrichter, Hartmut Nickel,
Lekshmy Parameswaran, Peter Pawlik, Jeroen Raijmakers, Wolf Dirk Rauh,
Christoph Schirmer, and Álvaro Valera Sosa

中国建筑工业出版社

	Examination / Treatment	┼	Mixed Type
	Care	—	Horizontal Type
	Administration	│	Vertical Type
	Social Services	●	New Construction
	Provision and Removal	○	Reconstruction
	Research and Teaching	∞	Extension
	Other		
	Operational-Technical Installation		

Coloured diagrams on the scale 1:2,000

Contents

INTRODUCTION

009 Hospitals, Medical Practices, Pharmacies Philipp Meuser

HISTORY AND THEORY OF HEALTHCARE ARCHITECTURE

015 From *House for the Sick* to Hospital Philipp Meuser | Christoph Schirmer

027 Healing Architecture and Evidence-based Design Álvaro Valera Sosa

051 People-centered Innovation in Healthcare Lekshmy Parameswaran | Jeroen Raijmakers

PLANNING AND PRACTICE OF HOSPITAL ARCHITECTURE

063 The Out-Patient Department Peter Pawlik | Linus Hofrichter

071 The Operating Theatre Wolf Dirk Rauh

077 The Area of Nursing and Care Franz Labryga

095 Obstetrics Hartmut Nickel

EXAMPLES OF HOSPITAL ARCHITECTURE

106 St. Marienwörth Hospital Bad Kreuznach sander hofrichter architekten

110 Johannes Wesling Clinic Minden TMK Architekten Ingenieure

118 Helios Clinic Berlin-Buch TMK Architekten Ingenieure

124 City Clinic Wolfsburg Architekten BDA RDS Partner, with Koller Heitmann Schütz

130 New Clinic Forchheim RRP Architekten + Ingenieure GbR

136 Johanniter-Krankenhaus Fläming Treuenbrietzen

Planungsring Dr. Pawlik + Co. Generalplanungsgesellschaft mbH

140 Emil-von-Behring-Clinic Berlin-Zehlendorf

Planungsring Dr. Pawlik + Co. Generalplanungsgesellschaft mbH

144 Bundeswehrkrankenhaus, Specialised Medical Examination Posts Berlin

Heinle, Wischer und Partner Freie Architekten

148 Städtisches Klinikum Brandenburg/Havel Heinle, Wischer und Partner Freie Architekten

154 Kreiskrankenhaus Bitterfeld/Wolfen Berg Planungsgesellschaft mbH & Co. KG

158	Müritzklinik Waren/Müritz Thomas Schindler Architekt BDA
162	Städtiches Klinikum Magdeburg Steffen + Peter, Dr. Ribbert Saalmann
166	Medical Service Centre, Ostalb-Klinikum Aalen IAP Isin Architekten Generalplaner GmbH
172	Katholisches Krankenhaus St. Johann Nepomuk Erfurt TMK Architekten Ingenieure, Thiede Messthaler Klösges
178	Diakonie-Klinikum Stuttgart Arcass Freie Architekten BDA
184	University Medical Centre Heidelberg Arcass Planungsgesellschaft mbH
190	Robert-Bosch-Krankenhaus Stuttgart Arcass Freie Architekten BDA, Prof. Joachim Schürmann

PLANNING AND DESIGNING MEDICAL PRACTICES

199	Architecture as a Factor of Quality Philipp Meuser
211	Planning Medical Practices: the German Principle Franz Labryga

EXAMPLES OF MEDICAL PRACTICES

254	Orthopaedics at Rosenberg St. Gallen bhend.klammer architekten
258	Centre for Radio-Oncology Cologne brandherm + krumrey
262	Dental Clinic Marktoberdorf Klaus R. Bürger
266	ENT and Psychotherapy Clinic Düsseldorf Cossmann_de Bruyn
270	Gastroenterology Remscheid Regina Dahmen-Ingenhoven
274	Dental Practice Bremen Gruppe für Gestaltung
278	Children's Dental Practice Berlin GRAFT
282	KU64 Dental Practice Berlin GRAFT
288	MKG-Surgery Airport Clinic Freising holzrausch
292	Radiology Schorndorf Ippolito Fleitz Group
298	Orthodontic Clinic Mindelheim landau + kindelbacher
302	SPORTHOPAEDICUM Berlin Meuser Architekten
308	Orthopaedic Clinic in Adlershof Health Centre Berlin Mateja Mikulandra-Mackat
312	Matrei Health Centre Matrei Gerhard Mitterberger

318	Medical + Dental Suite at Cologne Bonn Airport Cologne pd raumplan
322	Centre d'Endodontie Paris pd raumplan
326	Municipal Clinic, Outpatient Department Frankenthal sander.hofrichter
332	Dental Practice Frankfurt am Main Stengele + cie.
336	OMF Surgery, Lindenarcaden Lübeck Wagenknecht Architekten

PLANNING AND DESIGNING PHARMACIES

343	Pharmacies, Between Tradition and the Modern Market Philipp Meuser
349	From Herb Garden to Mail-Order Pharmacy Klaus Bergdolt
355	Corporate Identity and Design Klaus R. Bürger
359	Planning Pharmacies: the German Principle Franz Labryga

EXAMPLES OF PHARMACIES

388	Zum Löwen von Aspern Vienna ARTEC Architekten
394	Adler Pharmacy Kamen Jörn Bathke
398	Wilhelm Pharmacy Berlin Jörn Bathke
402	Alpin Pharmacy Kempten Klaus R. Bürger
406	Klemensplatz Pharmacy Düsseldorf Klaus R. Bürger
410	OHM Pharmacy Erlangen Klaus R. Bürger
416	TRI-Haus Pharmacy Arnsberg-Neheim Renate Hawig
420	St. Anna Pharmacy[1828] Munich Huber Rössler
424	Linden Pharmacy Ludwigsburg Ippolito Fleitz Group
428	Stadtklinik Pharmacy Frankenthal sander.hofrichter
434	Index of Architects and Designers
436	Authors and Co-Authors
438	Further Readings

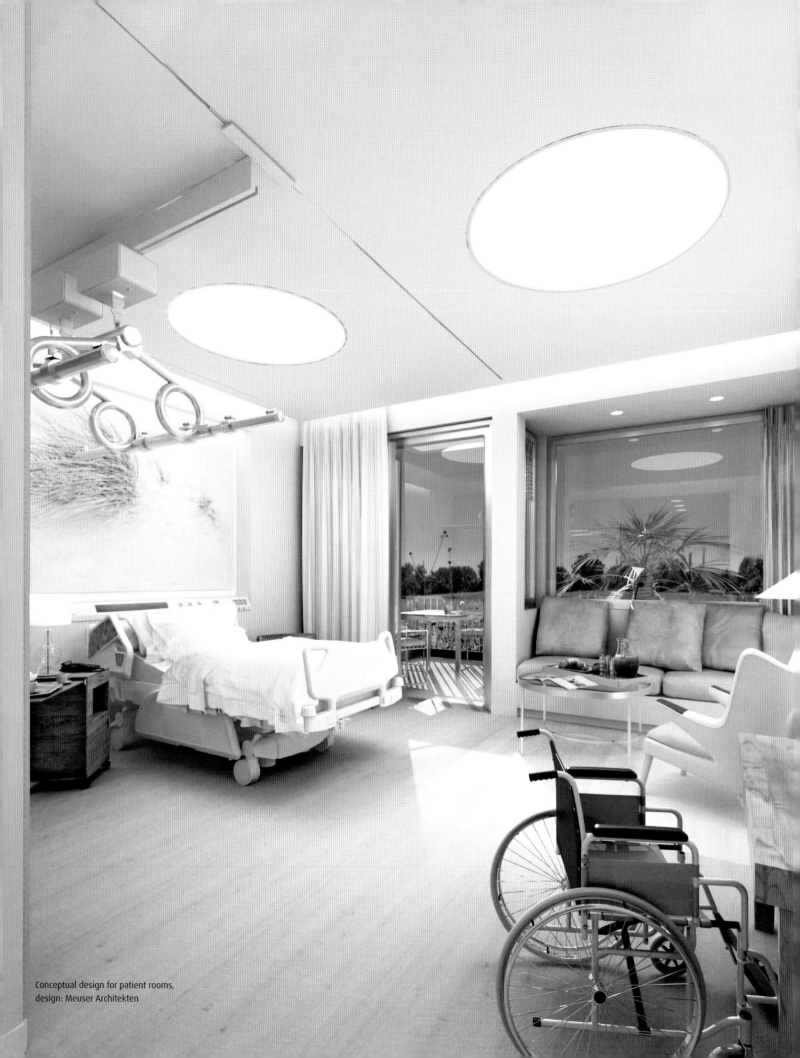

Conceptual design for patient rooms,
design: Meuser Architekten

Introduction

Philipp Meuser

Hospitals, Medical Practices, Pharmacies

This manual is intended as a compendium for questions about healthcare buildings. It is designed as a communication medium that can be used by teachers and trainees, as well as planners and their clients. The basic idea is to render the sometimes very complex and abstract use requirements for medical and pharmaceutical buildings easy to understand. As different as the planning parameters for hospitals, medical practices, and pharmacies may be, all three building tasks are connected by the idea that architecture and interior design, even the design of furniture and other equipment, can lead to a sense of well-being amongst the users and thus support the recovery process.

In his principal text on art theory[1], the sociologist and social theorist Niklas Luhmann (1927–1998) appositely pointed out that the environment must keep changing if people are to continue noticing it. He derived this view from the world of art, which must be different from other art if it is to attract attention. Clearly, this assertion does not just apply to the visual arts. If we regard architecture as the mother of all art forms, it is also true of the designed environment. Particularly in the twentieth century, architectural differentness was celebrated as innovation – so intensely and uncritically that today, a certain embarrassment creeps in when we speak of modernism and its drive to reinvent both architecture and society.

From today's viewpoint, architectural innovation is a milestone only if it represents a significant change brought about by new technology or by economic or political events. Hospitals, classified for a number of years (quite rightly, from a political perspective) as healthcare buildings, are particularly dependent on new and inventive technology. This relatively young type of construction[2], dating back some 300 years, has been shaped by a revolution in medicine and public health, with every new technological discovery being incorporated into the fabric of the building. Like almost no other type of construction, the history of hospital architecture is also a history of technological progress in medicine and health research since the early eighteenth century.

As doctors, pharmacists, researchers, and therapists develop new treatment methods, accordingly, designers and architects have had to incorporate new materials and layouts into their creations. Much of the renewal is fuelled by the construction industry, and by the countless medical equipment manufacturers who measure their success by the numbers of products they sell, rather than on whether the products are needed. Even antiseptic materials with nanosurfaces, for example, appear on closer viewing not to be wholly mature technology, so there is no reliable information concerning their safety. Designing a hospital is within the reach of only a small number of experienced architectural firms, thanks to the large number of regulatory requirements and the long period from initial sketch to turnkey handover. Outsiders seeking design contracts face much greater barriers to entry than with other types of projects, often because planning a hospital – or even a small part of a hospital – is such a complex undertaking. It is therefore understandable that hospital architects are expected to design buildings that will evolve with future medical progress if possible. To succeed in this area, you also need to combine the talents of a futurologist and a Science-Fiction author.

Learning from the German example

Demographic change, changing values, and the debate on the principles of healthcare make it particularly important to examine the future of hospital construction. A few years ago, the German government set up a high-level committee to discuss the hospitals of the future. The committee is also required to come up with ideas for an improved health sector that offers a higher quality of care at lower cost. In Germany, this debate is partly about affluence, an issue which is difficult for people outside Europe to understand.
At the end of 2017, there were 1,942 hospitals and clinics in Germany, employing 895,000 people and treating 19.4 million inpatients per year. With a total cost of 105 billion euros, they are a key factor in the health

1 Luhmann, Niklas: Die Kunst der Gesellschaft (The Art of Society), Frankfurt am Main, 1997, p. 77
2 The first European hospital was Berlin's Charité, founded as a plague hospital in 1710. The first modern hospitals were built in the mid-nineteenth century.

Hôpital du Valais in Martigny, Switzerland. The area of sterilisation today has become an attractive workplace. From 2014 to 2017, the new spaces were designed by bauzeit architekten. (photo: Yves André)

economy, and the biggest employer in many regions.[3] Even though the number of hospitals in Germany has decreased by more than 20 per cent since 1991, it is one of only two European countries (the other being Austria) with more than six inpatient beds per 1,000 population. The number of beds will need to keep growing as the population ages, so hospitals are both an investment in the future and a major source of income for the construction industry. In 2017, the investment backlog in German public hospitals was quantified at around 10 billion euros, half of it earmarked for buildings and the other half for medical technology.[4]

The continued need for new hospitals is a big opportunity for the construction industry to draw on new medical technology and hospital management know-how in the planning process, with traditional floor plans now being accompanied by flowcharts to make routes and layouts more transparent. As well as these complex functional dependencies, patients are increasingly well-informed and emancipated. They research hospitals, doctors, and illnesses online, have a stronger sense of involvement with their care, and sometimes even claim superior knowledge to the medical staff looking after them. What is more, patients no longer necessarily expect their hospitals to be like five-star hotels with their own VIP entrances, or to see an emergency admissions department right beside the main entrance.

Clearly, hospitals cannot be totally reinvented, but thinking about things that were previously taboo creates a new impetus for the planning and construction of hospitals.

These include comfortable patient rooms where relatives can stay. If this is impossible for space or cultural reasons, such as the increase in medical tourism, they should be accommodated in nearby hotels. Hospitals could also have their own leisure facilities that are open to the public – for example, a fitness centre could also be used for post-operative physiotherapy. Clearly, no one wants to turn hospitals into bustling urban entertainment centres with travel agencies, cinemas, museums, and service providers targeting both patients and external customers. But they must be part of the fabric of our cities, and not banished to greenbelt land or the suburbs.

New requirements for healthcare buildings

There is no doubt that the health market is undergoing revolutionary change. Future hospitals should be planned in accordance with the *hospitable hospital*[5] paradigm, focusing on financial, technological, and architectural efficiency and, most importantly, on human wellbeing. After all, a broken leg will always be a broken leg, and a heart operation will always be a heart operation. What will change is the way in which patients are classified. The patients with private insurance will book single rooms, those with compulsory insurance will sleep three to a room, and those without will be hastily and sometimes prematurely discharged after treatment. If we accept this classification system, which is already a reality, general hospitals will also offer 350 square-metres suites for wealthy medical tourists and their retinues.

3   Source: Bundesamt für Statistik, 2018
4   Deutsches Ärzteblatt, issue 10/2018
5   Panel "Creating healing environments", led by the author as part of *Light + Building*, 13 April 2011, Frankfurt am Main

Introduction

Inselspital in Bern, Switzerland. The entrance hall (designed by bauzeit architekten, 2012–2016) welcomes both patients and visitors with a high standard of design. (photo: Yves André)

After the emergence of the idea of a five-star standard patient room in the late 1990s, the demands on healthcare facilities have continued to increase in the recent past. Today, the registration with a representative lobby character or the hospital room in the living room style no longer apply to the exceptions in the interior of the hospitals. The owners of medical practices have long recognised that an exceptional interior design not only optimises the patient care workflow but also contributes to a commitment of patients to the attending physician. Today it is important to consider special patient groups. Patients with dementia or their sequelae are increasingly being treated.[6] The needs of these clients – be it in the hospital or in the doctor's office – must be met in the same way as an increasing number of patients from different cultures. This not only applies to immigrant patients following behaviour patterns from their respective homelands in host hospitals, but also to an increasing number of so-called health tourists visiting a hospital with their families for several days of treatment or surgery with several weeks of rehabilitation.

The planning of treatment rooms is only marginally changed by these new patient groups. The need for adaptation is based on medical progress and psychological findings, as has been the case since the beginning of hospital architecture. Spatial changes mainly affect the bed tract of a hospital. Today, the qualitative classification of patient care between single, double, and shared rooms is supplemented by the possibility of overnight accommodation for additional family members in close proximity to the patient. At the same time, the number of childcare facilities in large hospitals has increased; while a sick family member is personally cared for, they want to know their offspring is being well looked after for a certain period of time. And what at first only seemed to be a problem in the USA is now also part of everyday life in Europe: seriously overweight patients who can barely be lifted from a couch into a wheelchair by two nurses. It is hardly surprising, therefore, when both hospitals and medical practices have begun to treat not only diseases but to take preventative action through information, exercise, and nutritional counselling.

6 In reference to the research study "Dementia-sensitive hospital architecture" at Dresden Technical University 2016–2018

Hospitals

Philipp Meuser | Christoph Schirmer
From *House for the Sick* to Hospital

Álvaro Valera Sosa
Healing Architecture and Evidence-based Design

Lekshmy Parameswaran | Jeroen Raijmakers
People-centered Innovation in Healthcare

Peter Pawlik | Linus Hofrichter
The Out-Patient Department

Wolf Dirk Rauh
The Operating Theatre

Franz Labryga
The Area of Nursing and Care

Hartmut Nickel
Obstetrics

Philipp Meuser | Christoph Schirmer

From *House for the Sick* to Hospital

Hospitals offer round the clock care by doctors specialising in different areas, from diagnosis to monitoring to intensive care. In short, the entire spectrum of medical care is to be found here under one roof. But this essentially positive characteristic contains both pros and cons. A hospital must also guarantee so-called hotel service encompassing total care and nursing of the patient, and for this reason it has high running costs. Often the patient is transported to the hospital where he or she usually finds themselves in an impersonal, perhaps even alienating atmosphere. The patient is separated from the family and mostly cannot develop a personal relationship with the medical team treating them. The problem of hospitalisation syndrome, the appearance of mental changes caused by a long hospital stay, especially evident in children and the elderly, has been known for a long time.

As early as 1859 Florence Nightingale observed that the death rate in city hospitals was much higher than among patients with the same illnesses who were treated outside these institutions. Even today, at least 1,500 people die every year in German hospitals of infections that they contracted there.[1] No precise numbers are available because mortality figures for clinics are governed by statistics and one statistical rule is that a place claiming to heal people and keep them alive will, whenever possible, deny anything that might run counter to this. In June 2005 alone, twelve patients in England died of an infection called clostridium difficile. This intestinal bacterium is also a recognised danger in Germany; it can be fatal, especially for the elderly and those with severely compromised immune systems. But what hospital worthy of the name would want to admit that it makes people sick? In the place where they are supposed to be restored to health, there are 800,000 additional infections that a patient can catch resulting from inadequate hygiene and the excessive use of antibiotics.[2] What we thought had been left far behind, has made a sudden, unwelcome return into our consciousness in our a bright and modern age. At a time when people have an ever higher life expectancy, thanks to the apparently limitless advances of medicine, it must seem like a mockery and sacrilege that this omnipotence is so deceptive. "It may seem a strange principle to enunciate as the very first requirement of a hospital, that it should do the sick no harm."[3]

But the opposite is the case. A technically highly sophisticated medical science is powerless when confronted by old and new germs of an untold aggressive nature which everyone had thought were a think of the past. Aspects of our modern society that make people sick, such as stress and psychological problems represent another threat. All this flies in the face of the meaning of the hospital, but according to an old Chinese proverb, the treatment should not be worse than the illness itself. Then there is the fact that most people don't really want to go into hospitals at all, not even as visitors. Patients have differing expectations as to how the hospital of their choice should look: some want a high-tech centre with internet connection, others prefer a comfortable, cosy atmosphere with hotel service. That is why the public image of a hospital – apart from medical competence – is, above all, dependent on its architecture and interior design.

In hospital planning, what was emphasised for a long time, especially during the 1970s, were the physical needs of patients. Their psychological and social needs were addressed; functionality and efficiency were the cornerstones of treatment. In German hospital financing law, hospitals are defined as "institutions in which illnesses, suffering or physical damage are established through medical and nursing services, or in which help is given in childbirth, and in which the persons to be cared for can be accommodated and nursed". Two characteristics essential to the hospital building are covered in this definition: on the one hand, the duration of a patient's stay in the building which can last several days or longer; on the other, the medical dimension

Florence Nightingale
(1820–1910)

1 Stolze, Cornelia: Erst waschen. In deutschen Kliniken grassieren gefährliche Keime. In: DIE ZEIT 17/2005
2 see Der Tagesspiegel, 11 June 2005
3 Nightingale, Florence: Notes on Hospitals. London 1859

Infant hospital patients, London (1870)

A nurse tending a soldier during the Franco-Prussian War. Etching by L. Huard, published on Christmas Eve 1870.

Insurgents' hospital, Grahovo, Herzegovina (1876) (source: Illustrated London News)

of the specialised orientation of the building towards the goal of reducing and curing illness. The "helping services" of the building in this process and the character of its architecture have played a supporting role for a long time. But what was long dismissed as banal by the pharmaceutical industry has now taken centre-stage: the basic rules of life, the healing powers of nature.

The present-day German word Krankenhaus ("house for the sick") has its origins in the Middle High German *siechen-hûs,* which designated a hospital for lepers. Only later did the krancenhûs (Krankenhaus) develop out of the siechenhûs. Contrastingly, the root of the word "clinic" originated in the Greek term kliniè téchnè, the art of healing ill people confined to their beds.[5] The English word "hospital" is usually translated as Krankenhaus, although the English term is used occasionally today, mostly in Austria and Switzerland. "Hospital" contains the word Spital, which formerly designated a poorhouse or home for the aged, but its etymology is different: similarly to the French word hôtel, the word goes back to the Latin adjective hospitalis, which has to do with a welcoming friendliness towards guests, i. e. hospitality. The German word Herberge developed out of this.[6] The idea of a development from "sick-house" to hospital is now more relevant than ever. Apparently there is once again a general consensus that the quality of their environment has a profound influence on people's behaviour and well-being. One of the basic tenets is that beyond all technology, organic health requires a holistic approach to a person's well being. Recovery and healing are no longer to be attained through medical competence and state-of-the-art medical equipment alone, but are also decisively influenced by the environment. Concrete deserts vs. beautiful landscape; industrial healing factory on the one hand, *Schwarzwaldklinik* ("Black Forest Clinic", a German television series) on the other. Nothing illustrates the popular conception of the relationship between physical health and "feeling at home" better than that evening TV series with its emphasis on human interest. But the one cannot be had without the other, not even in a hospital set in a rural idyll on the edge of the forest. Is the *Schwarzwaldklinik* our ideal hospital?

From Hospital City to Garden City
The idea of healing people in a beautiful setting strongly influenced an entire generation of architects. It was the time when cities, literally overrun and bowed down by industrialisation, simultaneously became centres for diverse epidemics of civilisation. In 1892 in Hamburg, for example, on the threshold of the twentieth century, a cholera epidemic broke out infecting 16,956 people out of a total population of 640,000 and killing 8,605.[7] Since 1830 approximately half a million people have died of cholera. Cities, bursting at the seams, were forced to take measures against the germs that lived and multiplied in their drinking water. The answer was sewage and water purification, the rigorous separation of waste water and drinking water.[8] The existence of the cities exacted their tribute for civilisation in other ways as well. "The overall morbidity of population through frequent outbreaks of tuberculosis as well as the neuro-psychiatric consequences of alcohol abuse"[9] was enormous. In the contest of a population growth of 13.2 per cent in the years 1889 to 1895, the "number of psychically ill in municipal care during the same period rose 55.1 per cent", according to a report by the Berlin authorities at that time.[10] The new, highly mechanised mode of civilisation in the densely populated concrete desert simply drove people mad. In order to cure these early urban neurotics, Berlin discovered isolation: "A little way beyond the village, in open country on completely flat land",[11] as the psychiatrist and author Alfred Döblin wrote in 1900. Döblin noted this when construction of the Third Berliner Irrenanstalt (Lunatic Asylum) had begun, in accordance with the plans of the Berlin municipal building advisor Ludwig Hoffmann

4 see section 2 Begriffsbestimmungen, nummer 1, Gesetz zur wirtschaftlichen Sicherung der Krankenhäuser und zur Regelung der Krankenhauspflegesätze, Krankenhausfinanzierungsgesetz (KHG). 1972
5 Lexer, Matthias: Mittelhochdeutsches Handwörterbuch, Stuttgart 1992
6 Herkunftswörterbuch (Duden, vol. 7). Mannheim 2001
7 see Erik: Das Abenteuer, das Hamburg heißt. Hamburg 2003

Monastery in St. Gallen, ground plan (about 820)

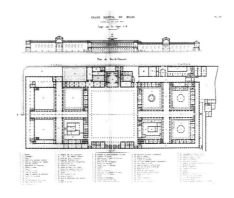

Ospedale Maggiore, Milan, ground plan, design: Filarete (1456)

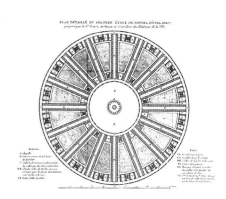

Hôtel-Dieu, Paris, ground plan, design: Bernard Poyet (1785)

(1852–1932), on the north-eastern periphery of Berlin in the village of Buch. It was opened in 1906, the first ever entire hospital complex with 45 buildings on an area of almost 68 hectares. Keeping the mentally ill away from ordinary society, more or less excluding them, had a social dimension which explains why psychiatric institutions were built outside the city. Building a hospital outside the city gates was however nothing new. "Sick-houses" and hospitals, as well as Spitäler were to be found outside the city during the Middle Ages and following the urban growth during the 13th and 14th centuries, at least on the city wall and thus on the fringe of the main settlement, so as to avoid infection. But this was also because care of the sick, according to the Christian precept "love thy neighbour" was largely left to monasteries and convents[12] if the buildings were located beyond the bustle of the city or behind their own enclosure. The spatial proximity of the church to the treatment of illness was the expression of a striving sense of community combined with a comforting, spiritual-religious type of care. The St.-Gallen monastery plan of AD 820, with its enclosed setting and spatial arrangement around the square cross cloister, was the ideal ground plan for a Benedictine monastery. Over the centuries the proximity of the church to the social areas arranged around it thus became a model for hospital buildings everywhere. Here, there was not only accommodation, but also nursing facilities for sick travellers and pilgrims. Nursing was considered so important in the Benedictine monasteries, that the monasteries became bastions of medical knowledge. From the late 19th century, nursing facilities were often the seed-beds for the first large hospitals, such as the Berlin Charité, which evolved from the so-called Quarantine House founded by the Prussian king Friedrich I in 1710. Since treatment there was free, in 1727 Friedrich Wilhelm I gave the institution the name Charité from the French word for love of one's neighbour. With its general and military hospitals and six specialised departments, the Charité developed into the Allgemeines Krankenhaus (General Hospital) where sick citizens and soldiers were treated and doctors trained. The unrealised 1774 design for the Hôtel-Dieu in Paris by Bernard Poyet shows the extent to which the requirements of a hospital had changed since the Middle Ages, following the growth of the cities. On the one hand, the 5,000-bed hospital had grown to an enormous size; on the other, the previously recognised connection between hospital and church no longer existed. The isolated location of the chapel in the centre of the building complex had become the expression of the transformation from a spiritual and religious type of nursing care to medical care based on scientific knowledge.

The organisation of the Third Irrenanstalt in Berlin-Buch, the later Hufelandkrankenhaus, in today's terms a combination psychiatric clinic and hospital, embodies a hospital city with all its functional buildings – from the large nursing houses and management buildings to the country houses, admitting houses, observation houses and offices – as well as accommodation for doctors, nursing staff and management. Not even the church and morgue are missing. This hospital established the norm in those days and is today a unique constructional monument. The gracious-seeming buildings, in a "style close to English architecture" with their baroque features, are embedded in a cultivated landscaped park. As a building complex, the hospital presents itself as right-angled facilities on three axes, enclosed by a white wall. The Public Health buildings are arranged in a row set in the greenery, while other buildings are organised in a cruciform plan on the central axis round the fountain. In principle, the facilities have much in common with the monastery plan for St. Gallen with their clear, modular order, containing all the facilities required to allow them to operate independently. Since its ideals formed the basis for the founding of further hospitals, the Third Irrenanstalt in Berlin-Buch became the masterplan for other hospital projects by Hoffmann.

8 Vasold, Manfred: Es ist eine böse Zeit. In: DIE ZEIT 19/2003
9 Wolff, Horst-Peter et al.: Zur Geschichte der Krankenanstalten in Berlin-Buch. Berlin 1996
10 ibid.
11 Döblin, Alfred: Berlin Alexanderplatz. Berlin 1929
12 Knefelkamp, Ulrich: Die Heilig-Geist-Spitäler in den Reichsstädten. In: Müller, Rainer A. [Ed.]: Reichsstädte in Franken. Munich 1987

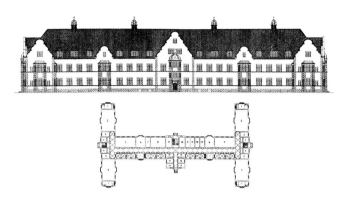

Klinikum Berlin-Buch, elevation and ground plan

Klinikum Berlin-Buch, aerial view,
architect: Ludwig Hoffmann (1899–1906)

Klinikum Berlin-Buch,
administration building

The Fourth Berliner Irrenanstalt in Buch was already opened in 1914. A municipal psychiatric institution across from the Schlosspark, Buch was built on a square ground plan. The similarity to an aristocratic park with the typical view axes is clear, such that the main buildings, like the management and communal house, are built in the neo-classical manner. There are colonnades – similar to the Old National Gallery on the Berlin Museum Island – that enclose the space, conjuring up an image of Muses under shady trees for the unbiased visitor. The nursing buildings, partially adorned with round towers and high roofs, are vaguely reminiscent of East Prussian castle Romanticism. The houses form an enclosed space with a garden on the inside in an open horseshoe arrangement with ample green inner courtyards, giving an impression of quiet and comfort. This Fourth Irrenanstalt was built literally at the beginning of the First World War and never used for its original purpose; it first served as a military hospital and was primarily used as a children's hospital after the war. A Berlin city spokesman enthusiastically announced in 1917 "that our municipal military hospital in Buch is a model institution in every sense of the word". He further enthused: "The military hospital is not only an ornament for Berlin, but a classic example of Germany's perfect humanitarian and sanitary installations in this war."[14]

For the Hospital Buch-Ost, originally a home for the aged built in 1908 (before the Fourth Irrenanstalt), Hoffmann compensated for the building mass of 21 houses on the relatively small area of barely ten hectares with a country-house style with garret roofs. The *Berliner Tageblatt* printed the following on 19 June 1909 about "The City of the Elderly": "This home does not look at all like a hospital or even a 'sick-house'. The twenty pretty houses which make up this city have bright walls and high, red brick roofs; they are separated from each other by areas of green lawn and yellow gravel walkways. Door openings and aligned walls lend the whole a unified and organic appearance. And in the background a tower overlooks the cheerful bustle of roofs, round and thick, like an old watchtower."[15] In order to arrange the buildings in a square around the garden adorned with fountains, thereby flanking a representative central axis, Hoffmann was able to develop the construction of the Fourth Irrenanstalt in a much more expansive way, on an area of 41.2 hectares with 32 buildings. A decisive factor in terms of municipal building regulation was that Ludwig Hoffmann oriented himself on the influences of social reform apartment-building originating in the private apartment-building societies, both in his architecture and his conception of space.

These influences had led to the following development: at the time, dense block structures were broken up from within in favour of large, green inner courtyards and communal buildings; at times, the space-forming quality of house and street were negated and turned inwards. Examples are Albert Gessner's social reform rental apartments, Paul Geldner's and Andreas Voigt's Goethepark, Paul Mebes' apartment group on Horstweg in Berlin-Charlottenburg, his residential street in Berlin-Steglitz and the facilities for companies like those on the Berliner Hohenzollerndamm built to Max Welsch's designs. All these buildings were constructed during the first decade of the twentieth century with the philosophy of making inner city living conditions healthier. They represent the first phase in this direction from which, from a historical perspective, it was only a small step to the abandonment of the traditional block-edge building in favour of the city built in the middle of a landscape, as the urban architectural strategies of the "new architecture" or "new objectivity" have made clear as a direct consequence of the Garden Cities Movement. Ludwig Hoffmann was also influenced by the social reform movement of the turn of the century, declaring light, sunshine and nature not to be the sole preserve of the elite, but the basic existential needs of all, in the context of social health. Hoffmann's rehabilitation communes were seen from the perspective of the reformers as the first garden cities. However,

13 quoted in: Ludwig Hoffmann: Die Wiederentdeckung eines Architekten. Ausstellungskatalog (source: Landesarchiv). Berlin 1986
14 Wolff, Horst-Peter et al.: Zur Geschichte der Krankenanstalten in Berlin-Buch. Berlin 1996
15 quoted in: ibid.
16 Reinhard, Hans J. und Schäche, Wolfgang: Ludwig Hoffmann in Berlin. Die Wiederentdeckung eines Architekten. Berlin 1996
17 Lampugnani, Vittorio Magnago: Moderne, Lebensreform, Stadt und Grün. Urbanistische Experimente in Berlin 1900–1914. In: Scheer, Thorsten et al. [Ed.]: Stadt der Architektur. Architektur der Stadt. Berlin 1900–2000. Berlin 2000

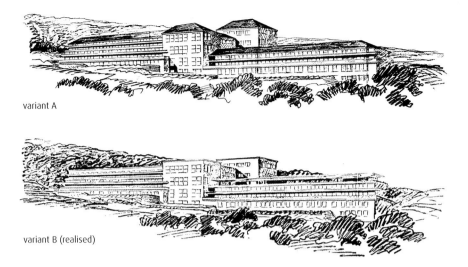

variant A

variant B (realised)

Krankenhaus Maulbronn,
architect: Richard Döcker (1930–1932)

in their eyes they remained only a half-hearted step, because the healing places, in which the patients only stayed for a limited period of time, did not lead to a change in their circumstances but offered only temporary respite. Hoffmann was therefore perceived by the aforementioned reform architects as an "architect between historicism and modernism".[16] Life-style reform ultimately means breaking out of the swamp of morally circumscribed and patronising Wilhelmine culture into a contrasting design of naturalness – feeding oneself from one's own soil, knowing that one's body is in harmony with creation and living accordingly: "Economically self-supporting and culturally autonomous."[17] The life reform movement, Germany's first mass socio-cultural protest and ecology movement were that society's re-action to the rapid industrialisation which had radically changed its cities and landscapes. Life reform meant more in those days in Germany than it did later, e. g. between the 1960s and 1980s, in the Federal Republic of Germany. Life reform, above all, meant getting away from the customary living conditions: "flight from the bourgeoisie and flight from the city became one".[18]

The architecture of the New Objectivity was the artistic expression of this communal movement towards sunshine, air, naturalness and freedom. In 1920 the influential architect for new building in southern Germany, Richard Döcker (1894–1968), formulated the evolutionary leap thought to have been undertaken by *Homo sapiens* as follows: "People of today and of the future are self-determining, healthy and free, life-affirming and without rules."[19]

From Scaled Building to the Brutalist Architecture

In Kaiser Wilhelm's Germany, the natural healing properties of air, sunshine and landscape were already considered parameters of a humane architecture; they were already architectonically applied in staggered and stacked buildings and in those with terraces. This architectural form took the place of pavilion hospitals for well-off tuberculosis patients; the good healing results achieved by the terraced sanatoriums also subsequently came to influence the architecture of general hospitals. A good example of this is the main building of the Düsseldorfer Krankenanstalten (medical establishment) built between 1903 and 1907, provided with broad verandas in front of the patient rooms. But the ultimate departure from the stuffiness of society and the absolutist ruling systems of Europe only took place after the First World War. A naturalistic idea of society, conveyed by the idea of self-determination for the individual, began consistently to establish itself in a new art free of limitations. This development was not confined to Germany but occurred throughout Europe, out of the period of the economic crisis and relatively late during the second half of the 1920s. The building of hospitals as institutions of municipal welfare delivery received an additional stimulus during the Weimar Republic in the midst of a drive for social reform that placed almost all social services under municipal control. During this process, the architecture of individual hospitals developed as part of a public service. A new type of tall building came into being with the so-called terrace hospitals. "Our time imperiously demands a hygienic life for all levels of human society. The breaking out of the old block of the full, closed building has taken place. Closed-off life within the building has ceased; it presses out towards light and sunshine, searching for connectedness with nature and landscape. Other housing bodies and forms are coming into being from this demand alone" wrote Richard Döcker in 1929 of the style of terrace buildings on which he had a marked influence.[19] In single-family houses, shared family houses and rented apartments alike, Döcker applied terracing regardless of building style, e. g. in the style of acute-angled rooms with bay-windows, both horizontal and vertical. He also transferred this construction strategy to residential and municipal buildings in order to do justice to the principles of a healthy living. What is good for the healthy person can only be cheap for the sick person;

18 ibid.
19 quoted in: Mehlau-Wiebking, Friederike: Richard Döcker. Ein Architekt im Aufbruch zur Moderne. Braunschweig 1989
20 see Vogler, Paul and Gustav Hassenpflug [Ed.]: Handbuch für den Neuen Krankenhausbau. Munich / Berlin 1951

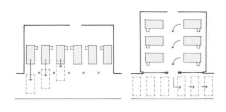
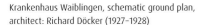

Krankenhaus Waiblingen, schematic ground plan, architect: Richard Döcker (1927–1928)

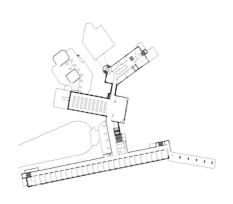

Tuberculosis Sanatorium, Paimio/Finland, ground plan, architect: Alvar Aalto (1928–1933)

Tuberkulosis Sanatorium, Paimio/Finland, terrace (photo: Suomen Rakennustaiteen Museo)

and thus Döcker set new standards in southwest Germany with his terraced hospitals in Waiblingen (1927–1928) and Maulbronn (1930–1932).[20] Hospitals were now no longer built in the city but outside it, in remote green areas according to the model of the sanatoriums. Döcker knew that the medical profession supported him in his designs, for contemporary scientific investigations had revealed that infection could be greatly limited by a good amount of sunshine radiation. The result was hospitals with broad terraces in front of the patient rooms, where the patients could lay in their beds like holiday-makers, contentedly enjoying the view of the idyllic landscape. These were hospitals free of the typically cold, stationary atmosphere, intended for a maximum of 120 patients and resembling a holiday hotel. The hospital in Waiblingen is a single-wing, two-storey, elongated building which contains all the rooms. In order to attain the same depth of space on all storeys, Döcker set the upper storey back along the breadth of the veranda to the north. The space resulting from the protruding storey on the north side of the building was closed off and used as a colonnade. Döcker attempted to develop a prototype for a future-orientated architecture from the planning of the Waiblinger Krankenhaus. This was transferable onto three- and four-storey buildings as well, but was not successful during his time. The reason for this was that Alvar Aalto (1898–1976), with his Tuberculosis Sanatorium Paimio, almost simultaneously set a precedent that became a landmark in architectural history. The sanatorium, financed by over 50 communities, was built in the remote isolation of the forests of southwest Finland. Subdivided into three wings, the building clearly separates the functional areas for patients and staff. Nurses, medical orderlies and doctors, as well as employees in the areas of management and building maintenance, received spatially separated living and sitting areas from the hospital. The bed-house as centrepiece of the recovery institution was designed by Aalto as a long, extended six-storey, slender block ending up in the terrace wing. The main formative motif is the clear arrangement of interior life in the form of patient rooms next to each other and oriented towards the sun, air and landscape. The roofs are in dark colours, the artificial light only indirectly brightening the room. The elegant functional aesthetic of the buildings and their spatial concept became a model to emulate for many hospital buildings after 1945. What the building forms of Richard Döcker and Alvar Aalto had in common was that the concept of the tuberculosis sanatorium not only influenced hospital architecture, but also developments in the area of residential buildings. In terms of architectural history, the modern period developed a new relationship with the landscape; individual buildings were now part of the landscape. Contrary to this, nature was still part of the overall architectural concept in Hoffmann's traditional form of building. A city in the hospital developed out of the green hospital city. The fact that high-quality buildings were created may have been thanks to the geographically open space that had had a lasting influence on municipal architecture and to the locations where healing took place. The trend of concentrating all the functions of a hospital together emanated from America, Scandinavia and Switzerland. If the architecture of Modernism in the early post-war years was definitely simply, especially in western Germany, but also fresh, light and transparent, this had changed by the 1960s. The catchword now was "New Concentration". The horizontal type of building such as in Borna, Wolgast and Belzig (built before the Berlin Wall was constructed in the GDR) still dominated the scene, partially in a homeland-style variant of the 1930s with a maximum of four storeys, bay-windows and hole-façades. At the same time, enormous buildings of unprecedented size were constructed in western Germany, such as the Klinikum Berlin-Steglitz. The architecture of the clinic was in keeping with the sociological and architectonic philosophy of building for the masses and the idea of the machine for living, as demonstrated by Le Corbusier's 1947 *Unité d'Habitation*

Hospitals

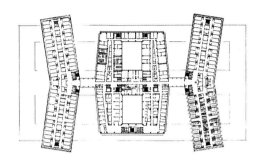

Klinikum Steglitz, ground plan,
architects: Curtis & Davis and Franz Mocken (1959–1969)

Klinikum Steglitz, aerial view, 2011
(photo: Philipp Meuser)

in Marseille. The tendency towards high-density living soon found its parallel in hospital architecture. According to the concept of the new concentration, almost all institutes and clinics were brought together into one building in Berlin-Steglitz from 1959 to 1969. The result was a "bed-castle" with 1,450 beds. It is subdivided into three parts above which rises an equally high element with all the examination and treatment rooms between two five-storey wings housing the wards. All the nursing, care and disposal logistics were also spatially brought together and automated. With the Klinikum Steglitz, a modern "health machine" had been created that would serve as a model for numerous university clinics of the late 1960s and 1970s. The highly technological healing machines give the impression that the physical wreck of a patient more or less went in "at the front" and came out "at the back" in full working order. Monstrous structures were created, such as the Universitätsklinikum Münster and the Großklinikum Aachen, modelled on McMaster University in Hamilton/Canada by architects Weber, Brand & Partner starting in 1969 and completed in 1984. Although operational costs were under close scrutiny in these compact hospital units, with the consequent abandonment of the trend towards horizontal construction in favour of the more economical vertical one, there was already criticism of these rationalistic solutions at the beginning of the 1970s. In the specialist journal *Baumeister* in 1972, editor-in-chief Paulhans Peters asked whether the individuality of the human being was any longer a topic up for discussion, demanding that patients not only be restored to biological health but that they should also be considered and cared for as a unity of body, mind, and soul.[21] But it took almost another whole generation before the philosophy behind the building of the huge hospitals was seriously questioned and criticised by the boards of directors of hospital management. Only at the end of the 1980s the idea of complementing huge, cumbersome and expensive structures with more flexible and therefore more economical units gained currency.

From Hospital to "Health House"

Due to the long period needed for decision-making and planning in hospital architecture, which can take several years just for the operational location decision in respect of state spatial regulation processes and until official building permission is granted, medical insights and architectonic improvements in their entirety can only have a time-delayed influence on the planning process. Often, essential decisions for a project are only taken when the innovative ideas have been approved and accepted by the professionals. The planning therefore drags behind the current state of discussion. Extensive changes in planning during the planning process are frequently unavoidable. Since the 1990s the following insight from the USA has begun to make itself felt: to improve the image of the hospital through provision of service and leisure facilities, and to make the building capable of competition on the health market. Along with architectonic appearance, the question of corporate identity comes to the fore. The new trend is international and is called "Healing Hospitals," as *The New York Times* referred to it in 2004.[22] This turning point is different from the hospital buildings commissioned by the authorities since the 1960s; it is not a diktat, but has more or less developed out of the ecology movement, analogous to the movement the turn of the century around 1900. The critical confrontation with a materialistic reception of Descartes's attitude towards nature as a machine and the consequent perception of man as a part of nature led back to a new consideration of old insights. These included the fact that a person's physical condition reflects the spiritual and mental state of consciousness, and that an illness can definitely be psychosomatically based. According to this insight, organic "repair" on the operating table and treatment with pharmaceutical drugs is certainly not everything and, at worst, can even be wrong. This also explains the (re-) discovery of healthy nutrition and ecological farming, natural healing practices and the popularity of Far Eastern and holistic medicine.

21 Baumeister, 5/1972
22 see Alvarez, Lisette: Where the Healing Touch starts with Hospital Design. In: The New York Times, 7 September 2004

Klinikum Aachen, façade detail, architects: Wolfgang Weber, Peter Brand & Partner, 1969–1984 (photo: Holger Bischoff)

In short, this includes everything apart from drugs produced in chemical laboratories and commercial methods for the treatment of illness and which, still somewhat ironically, are declared to be "alternative". Nonetheless, these hitherto unorthodox practices are slowly gaining a place on the deduction lists of social insurances thanks to their lasting effectiveness. These are healing processes that in the long run prevent an infarct in the human-medical care-organism in the face of exploding health costs because they follow a medical ethic. The American inventor Thomas Edison (1847–1931) already described this at the beginning of the twentieth century: "The doctor of the future will give no medicine, but interest his patients in the care of the human frame, diet and in the cause and prevention of disease."[23] The physician should function as a health advisor, as the person who protects people from illness, in this way preventing them from becoming patients. Edison's statement could hardly be more relevant today.

With this background in mind, it is an irony of history that the Waiblinger Krankenhaus was torn down after the Second World War – during the lifetime of Richard Döcker and in spite of resistance by the head doctor, who called attention to the importance of the terraces for the patients' speedy recovery. Paradoxically, however, it was precisely this argument that was used to justify the demolition. The terraces had become superfluous because the stationary accommodation of patients had to be kept as short as possible for financial reasons. Academic medical sanatoriums had more or less abolished themselves, and precisely that could also happen to large clinics today for the very reasons. The literally massive reduction in the number of beds provides supporting evidence; ambulatory day treatment is on the rise, stationary in-treatment is on the wane and kept to a minimum, as in earlier times.

In Germany in 2004 the Federal Bureau of Statistics noted a further decrease in hospital admissions: 3.4 per cent out of 16.7 million patients. The duration of the average hospital stay was also shortened by 0.2 days, or down to 8.7 days. Since 1993 the duration of stay has been reduced by nearly a third, not to mention the further increase in privatisation. The proportion of private hospitals is constantly growing. Correspondingly, the proportion of publicly funded hospitals sank to only 26 per cent in 2004.[24] The very thing that represents a positive development for the overall population is at the same time confronting the medical and hospital sector with a new challenge. Although capacities are less burdened thanks to shorter stays and fewer patients due to medical advances, inter-hospital competition for patients is becoming tougher. The so-called Health Care Building has been under discussion for several years as an architectonic response to these health-based and societal changes; health, illness prevention and treatment overlap in this system.

The pioneers in this area are the private hospitals. In the USA these Health Care Buildings must compete for their patients, and as a result they often resemble luxurious hotels or holiday camps; the hospital presents itself as a "health house". These establishments frequently offer seminars on subjects from illness prevention to nutritional counselling. Their new self-determined image is comprised of external factors such as value for money, landscape and architecture. One example is the 210 bed Parrish Medical Center in Titusville on the Florida coast, designed by architect Earl Swensson and completed in 2002. It enjoys an exclusive view of the Cape Canaveral Space Center. There is a fitness area directly in the foyer for patients; from the outside, it communicates the image of life-affirming people exercising their way back from the "sick-camp" to normal daily life. Many of the healthcare buildings are correspondingly oriented to accommodate ambulatory patients, in line with their philosophy of health. This form of care, in which the patient does not stay in the hospital very long, requires a completely new distribution of rooms. Flagler Hospital in St. Augustine/Florida, designed by the architecture

23 Tames, Richard: Thomas Edison. London 2003
24 Statistisches Bundesamt Deutschland, 29 August 2005
25 Philipp Meuser: Genesung durch Architektur. Neue Krankenhausbauten im Süden der USA. In: Neue Zürcher Zeitung, 12 April 1996

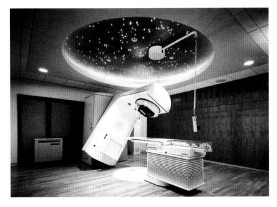

University of California, Mt. Zion Comprehensive Cancer Center, architects: Smith Group, 2000 (photo: Michael O'Callahan)

practice Nix Mann Perkins & Will, was structured so that the medical facilities are accessible by both internal and external patients, thus avoiding a costly double facility. The common care of long-term and day patients is not in line with the financial realities, however, but also helps the former on their road to good health. The architect's task of room distribution and design therefore contributes to this process of regaining health. Even formerly dull care machines such as Emory University Hospital in Atlanta show that an atmosphere conducive to healing can only be brought about by a systematic architectural remodelling of the rooms. In 1995 the new centrepiece of Emery University, Cox Hall, was presented, with a renovated clock tower as a "campanile" and a sun-terrace surrounded by trees in an atmosphere of southern friendliness, like a holiday hotel on the Riviera. The focal point of this conference and event hall is the former restaurant courtyard, where both the healthy and the sick meet together today. The project is representative of developments in the USA in the 1990s and was discussed in the international press.[25] The work of California practice Smith Group belongs to those examples of contemporary healthcare design esteemed by professionals. Numerous projects show their successful attempt to give the medically advanced yet intimidating place called "hospital" the familiar feeling of a living room with a connection to nature through interior design. The examination room of the University of California Cancer Center in San Francisco, for example, is encased in high-quality wood panelling. One can view the stars through the roof; that roof, together with rotating tomographs, also gives patients the impression of being in an extraterrestrial world. The contrast between home comfort and medical examination could help distract from an impending, less-than-welcome test outcome. This is also true of the other examination and nursing areas, which resemble nothing so much as a five-star hotel because of the high quality of their design. The insight that medical technology alone cannot put patients back on their feet can meanwhile be observed in Europe, and therefore in Germany as well, not least in terms of the achievements of outstanding design and execution in the area of architecture. One example is the Hessingpark-Klinik in Augsburg built by Jürgen Schröder and Sebastian Meissler, in which a service facility based on the American model is planned.[26] The hotel-like character of the new building, completed in 2002, has rooms on the top floor with terrace access for patients requiring a longer hospital stay due to a treatment or rehabilitation. The hospital seems, in this respect, to gradually be developing into a place of recovery, thus keeping step with the *Zeitgeist* which is an expression of a leisure society that no longer wants to be ill. In this context, the hotel also takes on further functions, slowly developing from a place to sleep into a health centre; a combination and assimilation of both types of building is underway.

The catchphrase is medical well-being and the destination Hotel Clinic. A changing pattern in holiday behaviour is revealed in the tendency to want and ask for more services and additional facilities during shorter stays. The changes in Public Health have thus led to a flood of new offerings. The new area of medical tourism promises patients that they can do something for their health or beauty during the course of their holiday. The imagination knows no bounds in this area; offerings range from simple care to rehabilitation to so-called "Surgeon and Safari" packages combining surgery with adventure holidays in a paradise-like atmosphere. This market has also been tapped into in Germany; specialised medical procedures carried out in private clinics are combined with stays in first class hotels.[27] It may soon be possible to exchange a multibedded room in a hospital for a single room in a hotel, for hospitals are also accommodating the new demands made by their patients. At several university clinics, such as those in Münster, Kiel, Lübeck and Dresden, the idea of housing patients in three- or four-star hotels for post-operative observation and treatment is under consideration. The hospital management

Emory University Hospital, Atlanta, architects: Nix Mann Perkins & Will, 1992–1995

26 Project documentation in: Meuser, Philipp; Schirmer, Christoph: New Hospital Buildings in Germany. Berlin 2006, p. 268 (vol. 1)
27 see Storcks, Holger: Hospital Branding. In: Meuser, Philipp; Schirmer, Christoph: New Hospital Buildings in Germany Berlin 2006 (vol. 2)

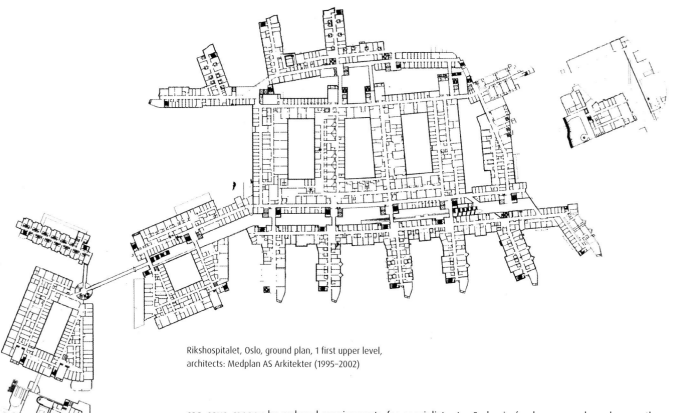

Rikshospitalet, Oslo, ground plan, 1 first upper level, architects: Medplan AS Arkitekter (1995–2002)

can save money by reduced requirements for specialist staff and technologies, while patients hope for a beautiful environment and more rapid recovery process. All these developments influence the new hospital architecture.

From Human Being to Architecture

A model project exemplifying the new hospital architecture can be found in Norway. The Rikshospitalet in Oslo, built from 1995 until 2002, is permeated by the old insight that the atmosphere and environment of a hospital, i.e. the way it has been constructed, are as important for well-being as are light and air. The Rikshospitalet functions like a city with streets and squares, with its imposing entrance hall as centrepiece and its continuation as a backbone. It was important for the architects of Medplan AS to choose materials and colours corresponding to human needs, and to design the buildings, embedded in the landscape, in as friendly and bright a manner as possible. Along with Norway's largest public collection of art outside a museum, the communal rooms and balconies give the patients the incentive to leave their rooms.

Le Corbusier had already developed a similar idea in 1965 with his unrealised design for a hospital in Venice conceived as a city within the city. An important characteristic of his "Clinic for Life" was the single rooms for patients; Le Corbusier called them an intimate and protective withdrawal possibility. The minimal rooms, developed from the proportional system modulor (a mathematical order orientated on the human scale), were to have been windowless and only illuminated by overhead natural lighting; the idea was to motivate patients to discover and take over the corridor. Social life was to have taken place in the rooms designed as Venetian streets and alleyways, offering a view onto the city.

Le Corbusier's plans were based upon three stages in the patients "career": the initial fear and uncertainty is transformed into a stage of initiative, concluding with the recovery of one's independence when discharged from the hospital.[28] Architecture is intended to support the patient in this process, since being ill forces people into a passive role and is mostly a process of waiting. "One waits until one is awakened, then treated, then until one is healthy again, all inactive activities."[29] The room design should therefore encourage patients physically and mentally, for example by positioning switches and television so as to induce a decision to move, so that patients will not fall into a state of lethargy and inertia. It has been confirmed that the healing process can be positively influenced by a change in room occupancy arrangements, thus, in psychiatric institutions in Germany, single rooms are dispensed with whenever possible. In other facilities, disadvantages were revealed in both multi-bedded rooms (through a greatly reduced private sphere) and in single rooms (through isolation and the feeling of being at someone else's mercy). Although a mutually beneficial effect does exist in double rooms, the incidence of single rooms in hospitals will continue to rise due to the growing number of privately insured patients.[30]

If health reform and the rejection of "bed-castles" in Germany have not yet taken hold, an increase in hospital competition in this area is also evident analogous to the situation in the USA and Scandinavia. The patient is no longer transported and delivered for repair like a defective motor, but instead he or she carefully selects the hospital like a holiday-maker choosing a hotel. It is the private and religious hospitals, including several outstanding public institutions, which reluctantly turn away from dogmatic functionalism. An example is the Kreiskrankenhaus in

28 see Sarkis, Hashim: Le Corbusier's Venice Hospital and the Mat Building Revival. Munich/London/New York 2001
29 Schricker, Rudolf: Grundlagen der Gestaltung von medizinisch-therapeutischen Räumen. In: DBZ, 2/2001
30 Rauh, Wolf Dirk: Gesundheitsfördernde Architektur. In: Krankenhaus Umschau, 1/2002
31 Project description. In: Meuser, Philipp; Schirmer, Christoph: New Hospital Buildings in Germany. Berlin 2006 (vol. 1), p. 156
32 Siedler, Wolf Jobst: Phoenix im Sand. Glanz und Elend der Hauptstadt. Berlin 2000

Agatharied in Upper Bavaria, built like a small city by the architects Nickl & Partner.[31] The up-to-date medical care which has made this one of Germany's most modern hospitals corresponds to an attractive architecture communicating the atmosphere of a comfortable hotel rather than that of a hospital. A stay here due to illness seems almost like a holiday, with much natural light streaming into every corner of this building located on the edge of an idyllic landscape. The patient, taken out of his accustomed environment and also psychologically impressionable due to his illness, finds a homelike atmosphere here. Wooden floors, birch wall panelling and appropriate furniture, a winter-garden sunroom as well as technical connections to the contemporary world of telecommunications allow the patients to feel "at home". Since warmth of the nest is just as subjective a perception as the often overused term "beauty", the patient can arrange the furniture in his or her room as desired, thus creating a completely private and individual space. The personal patient room becomes a temporary living-room, further complemented by hotel-like services, with coffee bars on each floor and breakfast buffets, personally served breakfast, a patient library, hairdresser, kiosk, flower-shop and chapel.

The Deutsche Orden has created a similarly inviting building with its Spital in Friesach. Built according to the insight that architecture reflects a society's cultural milieu and character, the Deutsche Orden decided upon an objective building which plays economically with geometry, the interior of which is all the more convincing for its greater expressionistic power. According to the philosophy behind it, light and rounded forms are the material out of which well-being and comfort emanate. According to the architect Walter Hildebrand, rounded forms create a cosy, warm feeling. With his new buildings he wanted to show that hospital architecture was also capable of uniting function with aesthetic harmony, thereby offering people a feeling of beauty as a component of healing. These recent buildings from the area of hospital architecture provide a powerful antidote to the experience that architecture can be a "weapon with which to murder citizens"[32], be it house-building or municipal design. They confirm the intuition by scientific proof, gathered from resonance research, that architecture can heal. According to a study conducted at the University of Münster, well-rounded spaces set off favourable vibration patterns, which in turn create a "feeling of regeneration and healing in people".[33] Light and colour are, as in all architectural endeavours, two essential components. In order to create the right atmosphere for patients' well-being, it is not only the right colour that counts, but the right dosage as well. The Institute for Light and Colour Dynamics has recommended that colours which are too powerful and intense be dispensed with. Instead of using glaring colours, the right idea is to skilfully apply and combine cheerful pastel tones.[34] The effect of colours on the psyche has not yet been uniformly assessed by either physicians or architects. For example, numerous walls in the Olso Rikshospitalet are painted with powerful colour-fields. In contrast to the predominantly bright walls, they represent a welcome variety, also providing an aspect of design. More important than the choice of colour tones is their composition in interaction with the structure of the materials and the functionality of the furniture. Moreover, coloured markers for departments or patient rooms – as in multi-storey carparks – contribute to a more rapid orientation. Before a patient has had enough of a certain colour, he will probably have been discharged from the hospital anyway, in this age of Diagnosis Related Groups (DRG).

Health Centres with Patient Hotel

With the background of an ever-older society in mind, together with the simultaneous privatisation of large portions of the healthcare system, the entire hospital sector is confronted with a far-reaching transformation process. Since the city continues to step back from its role as building authority, e.g. in social apartment architecture, and private investor groups are only gradually beginning to fill this gap, an investment jam of 30-50 billion euros has accumulated in Germany, according to the Marburg Union and the Main Association of the Building Industry.[35] About two thirds of this falls upon the building sector. The debate over how to allocate these investments will surely come to a head in the near future. Consequently, more highly specialised centres of excellence will be established which will also compete for patients on the international market. However, while the present large hospitals will not disappear overnight, they will decrease in number. One result of public health reform is the fact that medical management is increasingly being taken over by private concerns in Germany, and medical care is returning to a type of familiar clarity. There is also a trend towards the replacement of the traditional hospital ward with so-called patient hotels. The healing process can be continued without interruption in a rehabilitation programme or a holiday setting.

Models for health centres are already found in antiquity. There were private clinics in Athens and Rome, so-called doctors' schools that were partially publicly financed. In ancient Greece these clinics were also holy places where knowledge about health maintenance and healing overlapped with worship of the gods. Hippocrates of Kos (460–377) worked at this kind of doctors' school when he broke with religious-mystical medical teachings in favour of rational-natural knowledge, attempting to activate the self-regenerative of the body with the help of nature's healing powers. He worked preventively and coined the saying, "the doctor helps but nature heals"[36]. Today one might add that architecture can also contribute to this goal.

Hippocrates of Kos
(photo: Uffizi, Florence)

33 www.uni-muenster.de
34 Schilling, Gerd et al.: Symbolsprache Farbe. Munich 1996
35 Deutsches Baublatt, number 291/292. January/February 2002
36 quoted in: Grensemann, Hermann [Ed.]: Der Arzt Polybos als Verfasser hippokratischer Schriften. Mainz 1968

Álvaro Valera Sosa

Healing Architecture and Evidence-based Design

"Is there an architecture that helps you live?" For almost three decades Charles Jencks – co-founder of Maggie's Centres in the UK – has insisted that "architecture does matter for health, as placebo or to evoke hope for those in need".

Despite placebos known to lack of clinical value, many assure they do work for illnesses deriving from emotional or mental stressors[1]. The same principle applies to these centres created for people affected by cancer. Its users forgive the possible functional issues in exchange for quality experience[2].

What Charles Jencks refers to as good architecture for health, other authors call Healing Architecture. Defined by Michael Mullins (Aalborg University) as "the supporting factor in the human healing process" or more extensively, the planning approach that recognises architecture as a variable to support the physical and mental wellbeing of staff, patients, and relatives.

This chapter develops on the premise that Healing Architecture works but cannot explain its curing capacities without support of an Evidence-based Design (E-bD) approach. A field redefined in this essay as the process that ensures architecture develops to enhance human health. As this relation is described, questions arise on the significance of architecture in well-known E-bD recommendations, which for decades have guided designers. Clarification is sought with a background review on how architecture has aimed at care process, followed by three sections that elaborate on: the need to distinguish technical devices from architectural features; medical planning preference over architectural design; and the failure in defining environmental factors for healing as natural, technical, or architectural.

To close the chapter, reflections are shared on how E-bD as an evolving field can not only assist architecture, but also Public Health! An area in need of studying how environment interventions affect and influence health behaviour.

Has architecture been healing us?
Yes, as mentioned, the premise is that architecture heals. The question remains: how?

We've intuitively known that the physical environment (natural and built) affects our health, maybe since the times way before ancient Greece. In recent history, what has seemed to matter most are facts and proof, to the point that science in architecture has overruled its artistic best half. Healing Architecture, during its modern conception, leaned on the side of science in three distinctive occasions: sanitation, environmental risk, and perception.

A pioneering document for buildings, was the patient ward design guidelines from Florence Nightingale's 1859, "Notes on Hospitals". Through statistical records, it alerted architects about the effects healthcare settings were having on human health. Her notes structured a number of measures that significantly improved the deplorable sanitary conditions of the Barracks Hospital (in the Crimean War of 1854). What is commonly referred to as the "Nightingale Ward", became a reference for hospital buildings; a space with a limited amount of beds, three windowed sides, elements designed to trap dust, admit light, fresh air, plus other features that generally enhanced cleanliness and the comfort of patients[3].

Nearly a century later (mid-1940s), the World Health Organisation (WHO) redefined the concept of health, eventually including the environment (social, natural, and built) as one of its determinants. The concept took distance from a merely medical perspective towards a more holistic approach, which included the effects of environmental factors on the health of individuals and societies. Thereon, health research combined multidisciplinary efforts within the scientific community to better understand the environment and develop tools for its assessment.

In the early 1960s, facing the vast and fast-growing scientific knowledge, architecture started considering environmental risk theory and its survey methods[4]. At the time, scientific communities were already concluding that

1 Erle C H Lim and Raymond C S Seet, "What Is the Place for Placebo in the Management of Psychogenic Disease?," *Journal of the Royal Society of Medicine* 100, no. 2 February 2007): 60–61.
2 Dr Fionn Stevenson and Professor Mike Humphris, "A Post Occupancy Evaluation of the Dundee Maggie Centre" (Scotland: Ecological Design Group, School of Architecture, University of Dundee, March 2007), https://www.ads.org.uk/wp-content/uploads/4560_new-maggiecentre1.pdf.
3 Marie T. O'Toole, Mosby's Medical Dictionary, 9th edition (St. Louis, Mo.: Elsevier/Mosby, 2013).

| Architectural Quality | | Psychological recovery (reduced stress) | | Immune system strengthening | → | Disease development reduction/control |

Healing Architecture Rationale
The blue box shows what is scientifically proven: reducing stress also reduces disease. What needs substantial evidence is how architectural quality reduces stress.

socio-physical environments are a medium for disease transmission, a stressor, and a source of danger. Along the evermore duality of disease and health, environments were also starting to be considered a possible enabler for heath behaviour[5], here the importance of Salutogenesis. Introduced by the medical sociologist Aaron Antonowsky in 1979, the theory offered a deeper knowledge and understanding of health and disease. It aimed at identifying factors originating health, contrary to the still ruling pathogenic approach that focuses on those causing disease. This definitely marked a milestone in conceptualising what Healing Architecture would be years later.

In the 1980s, environmental psychology was moving forward in investigating the psychological effects of buildings. Two scientifically proven findings redefined architecture for health with renewed knowledge: that surrounding environments induce a psycho-physiological arousal, and the fact that humans have a limited capacity for processing stimuli and information.

These theoretical grounds encouraged environmental designers to set course on pursuing behaviour adaptation and stress coping through design; a path also known as architectural determinism. The opportunity was given for architects to tackle the underlying causes of stress, linked to the environment such as the lack – or excess – of social contact, access to privacy, and control over environment[6]. As here depicted until now, Healing Architecture's early roots – relating environment and health – stem from a solid scientific background. It leaves the question open to whether architecture alone has been able to heal us or not. A casual conversation a few years ago, gave away that this question might remain unanswered for quite some time to come.

When attending an international conference on urban health (ICUH, Manchester 2014), I had the great opportunity to sit next to Trevor Hancock (WHO, Healthy Cities) for the official get-together dinner. I shared the highlights of my presentation held earlier, titled "Walkability for Health"; a work on possible links between urban streetscapes (street visual structure) and health status in Berlin. Back then, for an ICUH, some kind of architecture intertwined with public health was a rare combination. Anyhow, after much discussion with other colleagues at the table, Hancock graciously came back to the work saying, "if we know for certain that ugly makes us ill, then we should explore more how aesthetics make us healthy."

Architecture aiming the care process
Science laid the initial basis, and continued so with two important events in the 1980s that shaped our early understanding of healing architecture as part of the care process. The first event was a clinical-based research, conducted by environmental psychologist Roger Ulrich in 1984 considered a landmark study in built environment and health outcome. Ulrich, a Professor at Texas A&M University, led a clinical research project that empirically proved that a room with a view onto nature does improve a patient's post-operative recovery. His quasi-experimental study showed a reduction in length of stay and pain medication in patients whose room had a nature view compared to those with a brick wall view. The study provided data on the direct impact of an environmental variable on the patient's outcome[7]. Roger Ulrich´s research boosted the curiosity of architects about the interface between clinical/medical research and design. For healthcare managers, the cost reduction of such recovery processes was eye opening and a motivation to keep exploring.

The second event was the development of a patient-centred care and healing hospital concept by the Planetree Organisation (USA). Despite it being founded in 1978, it was not until the mid-1980s that their research was materialised into a full testable model depicting the relationship between healthcare science and environmental science. They opened a 13-bed medical-surgical unit in San Francisco, which included and evaluated the environment as a variable in patient recovery. It was the first time that a healthcare design was built to structure a case study.

4 John. Zeisel, *Inquiry by Design: Tools for Environment-Behavior Research* (Cambridge: Cambridge Univ. Press, 1984); Wolfgang F. E. Preiser, Harvey Z. Rabinowitz, and Edward T. White, *Post-Occupancy Evaluation* (New York: Van Nostrand Reinhold, 1988).
5 Daniel Stokols, "Establishing and Maintaining Healthy Environments: Toward a Social Ecology of Health Promotion," *American Psychologist* 47, no. 1 (1992): 6–22.
6 Paul A. Bell et al., *Environmental Psychology* (New York: Psychology Press, 2011).
7 R. Ulrich, "View through a Window May Influence Recovery from Surgery," *Science* 224, no. 4647 (27 April 1984): 420–21, https://doi.org/10.1126/science.6143402.

The design principles of the model were developed by Roselyn Lindheim, a professor of architecture at UC Berkeley who worked in collaboration with epidemiologists. The research and findings brought architectural solutions that evoked feelings of home, welcomed the patients' family and friends, valued human beings over technology, enabled patients to fully participate as partners in their own care, provided flexibility to personalise the care of each patient, and encouraged caregivers to be responsive to patients and foster a connection to nature and beauty[8].

The Planetree hospital became an exemplary model across the globe, settling healing architecture as a concept and to be considered for further exploration. In 2007 the Planetree Designation Program was launched to award organisations with the highest level of achievement in patient-centred care and healing environments based on best practice and standards.

The patient-centred approach was early adopted by other organisations such as the Picker Institute, founded in 1986, which focused in assessing the patients' actual experience in hospital settings. As well did the Joint Commission International (JCI) in the early 1990s, develop an accreditation and certification system with emphasis on patient, staff, and visitor safety.

The need of evidence in design
The continuous work of Planetree, the Picker Institute, and similar organisations have caught the eye of building professionals accountable for design solutions that mainly seek hospital cost-effectiveness and return on investment. For them, the economic benefits of designing environments that control patient anxiety and stress is palpable but not as evident as engineering for energy efficiency, for medical error prevention, or to reduce hospital-acquired conditions such as infections, falls, and injuries to staff, patients, and visitors. Strong evidence of the healing capacities of architecture was needed to structure compelling business cases.

Since the seminal study of Roger Ulrich in 1984, the most relevant effort in relating hospital environment design with health-related outcomes belongs to the Center for Health Design (CHD). Founded in 1993, its main purpose has been to launch several research and practice programmes for the healthcare industry gradually defining Evidence-based Design (E-bD) as a discipline. CHD in clear reference to the concept of evidence-based medicine[9] defined E-bD as "the process of basing decisions about the built environment on credible research to achieve the best possible outcomes". In 1995, this centre, with medical researchers from their database, began conducting systematic reviews of clinical literature on facility design and its effects[10].

The first grand review was commissioned to the Johns Hopkins University in 1998. It consisted in revising all published research showing a connection between design interventions and medical outcomes, such as where to place sinks to encourage hand washing, and how to position rooms and windows to reduce length of stay. 78,761 articles were reviewed and only 84 were acceptable from a scientific standpoint[11].

A second systematic review was commissioned in 2004, titled "The Role of the Physical Environment in the Hospital of the 21st Century: A Once-in-a-Lifetime Opportunity". Over 600 studies were found in reputable journals from which 240 were included for analysis linking "a range of hospital environment aspects to: staff stress, patient safety, patient and family stress and healing, and overall healthcare quality and cost"[12].

E-bD then was defined by Ulrich as, a process of creating healthcare buildings, informed by the best evidence available, with the goal of improving health outcomes and continuing to monitor the success of designs for subsequent decision-making.

The third and last CHD review to date was realised in 2008: "A Review of the Research Literature on Evidence-based Healthcare Design". Thirty-two search keywords, referred to health-related issues and physical

8 Laura Gilpin and M. Schweitzer, "Twenty-five Years of Plantree Design," HCD Magazine (blog), 31 August 2003, https://www.healthcaredesignmagazine.com/architecture/twenty-five-years-planetree-design/; B. Arneill and F. Frasca-Beaulieu, "Healing Environments: Architecture and Design Conducive to Health," in *Putting Patients First: Designing and Practicing Patient-Centered Care*, by Susan B. Frampton, Laura Gilpin, and Patrick A. Charmel (San Francisco: Jossey-Bass, 2003).

9 A. L. Cochrane, *Effectiveness and Efficiency: Random Reflection on Health Services* (London: The Nuffield Provincial Hospitals Trust, 1972).

10 Haya R. Rubin et al., *Status Report (1998): An Investigation to Determine Whether the Built Environment Affects Patients' Medical Outcomes* (Martinez, Calif.: Center for Health Design, 1998).

11 Stefan. Lundin, "Healing Architecture: Evidence, Intuition, Dialogue" (Chalmers University of Technology, 2015).

12 Roger Ulrich et al., "The Role of the Physical Environment in the Hospital of the 21st Century: a Once-in-a-Lifetime Opportunity," September 2004.

(Table 1) CHD–Literature growth: the number of studies included for review increased significantly from 84 in 1998 to more than 1,200 in 2008.

Study	Inclusion/exclusion criteria	No. of studies for inclusion
1998 An Investigation to Determine Whether the Built Environment Affects Patient Medical Outcomes.	Articles in English published from 1966 on.	78,761 articles reviewed only 84 were accepted from a scientific standpoint.
2004 The Role of the Physical Environment in the Hospital of the 21st Century.	No information provided.	600 relevant articles, 240 articles were analysed.
2008 A review of the research literature on evidence-based healthcare design.	Studies in English. 32 keywords referred to healthcare-related issues and physical environmental factors.	Ca. 1,200 studies reviewed.

environment factors, were employed to yield over 1,200 studies. After the review, CHD defined E-bD as "the process of basing decisions about the built environment on credible research to achieve the best possible outcomes". Nearly a decade after Roger Ulrich first defined E-bD, director emeritus of the CHD, Kirk Hamilton, and colleague David Watkins[13], extended the definition to multiple building types, by stating, "evidence-based design is a process for the contentious, explicit, and judicious use of current best evidence from research and practice in making critical decisions, together with an informed client, about the design of each individual and unique project."

Expert practitioner Rosalyn Cama elaborated further, indicating the four basic components of this process as: gathering qualitative and quantitative knowledge; map strategic, cultural, and research goals; hypothesise design outcomes and implement translational design; and measure and share outcomes[14].

The three prominent reviews led to other important milestones for the CHD, being the most relevant the launch of the Pebble Project in 2000; an initiative aiming at producing E-bD documents on patient, staff, and economic outcome improvement. Also important, was the creation of EDAC (Evidence-based Design Accreditation and Certification) in 2008, which still today offers architects, hospital executives, healthcare providers and researchers, a certification for introducing an evidence-based process in the design and development of healthcare settings.

Technical devices over architectural features

The three reviews redefined E-bD as a concept, positioned it as a research field, and it rapidly gained the interest of practitioners, as the amount of research exponentially increased. From the first review realised in 1998, finding 84 studies to its last in 2008, with over 1,200 studies included, meant nearly a 1,300 per cent increase of research in just one decade. A growth that Debra J. Levin, president and CEO of the CHD, predicted as sustainable in 2014, "If we were to do the search again today, I have no doubt the number would surpass 2,000."

The amount of research in the field without doubt increased, what today is still questioned, is if the amount of findings has also increased and most important, if there is strong evidence for an architecture that heals. The following comparison of the three CHD reviews (see table 1), finds inspiration in an exercise architect Stefan Lundin included in his 2015 dissertation on healing architecture. Perhaps on this occasion, under the cap of a public health researcher, my search for evidence turns suspiciously more rigorous.

Two tables here presented, contrast two trends: one showing literature growth (table 1) against another pointing evidence growth – or of significant findings – to be applied in practice (table 2). A third table summarises all E-bD recommendations and discriminates hard factors attributable to technical devices from soft factors, proper of architecture.

For this analysis, the definition of hard and soft factors will be borrowed from business management (due to the common economic purposes with E-bD) and conceptualised for architecture as follows. Hard factors, are those features which visibly affect functions and processes with objective (measurable) outcomes such as injuries, errors, infection rates, among many others. Soft factors, are qualities that support human behaviour (individual or collective) influencing subjective outcomes (less easy to measure) such as satisfaction, stress, social cohesion, and others.

From comparing and analysing results from these reviews one can conclude: (1) the volume of evidence finding architectural strategies supportive in care processes has improved but is not abundant, (2)

13 Kirk Hamilton and David H Watkins, *Evidence-Based Design for Multiple Building Types* (Hoboken (N.J.): Wiley, 2009).
14 Rosalyn Cama, *Evidence-Based Healthcare Design* (Hoboken, N.J.): J. Wiley, 2009).

Hospitals

Study	Strategies to apply in practice	Significant new findings
1998 An Investigation to Determine Whether the Built Environment Affects Patient Medical Outcomes.	1. Quiet Coronary Care Unit (unclear if architectural development) 2. Music during Minor Surgery (technical devices, non-architectural) 3. Air Quality (technical devices, non-architectural) 4. Exposure to Daylight and Sunlight	1. Quiet hospital environment 2. Daylight and sunlight exposure is the only strategy Architecture directly relates to In general, no new insight was provided.
2004 The Role of the Physical Environment in the Hospital of the 21st Century.	1. Single-bed rooms 2. Acuity-adaptable rooms 3. Quiet hospital environments (strategy suggested in 1998) 4. Views of nature 5. Other positive distractions 6. Develop way-finding systems 7. Appropriate lighting (technical devices, most cases non-architectural) 8. Design wards and nurses' stations to reduce staff walking and fatigue	1. Single-bed rooms 2. Acuity-adaptable rooms 4. Views of nature 5. Other positive distractions 6. Develop way-finding systems 7. Appropriate lighting 8. Design wards and nurses' stations to reduce staff walking and fatigue
2008 A review of the research literature on evidence-based healthcare design.	1. Single-bed rooms (strategy suggested in 2004) 2. Access to daylight (strategy suggested in 1998) 3. Appropriate lightning (strategy suggested in 2004) 4. Views of nature (strategy suggested in 2004) 5. Noise-reducing finishes (technical devices, non-architectural) 6. Ceiling lifts (technical devices, non-architectural)	None

(Table 2) CDH–Evidence growth: despite the increase of studies for inclusion throughout 10 years, the latest review did not show new findings.

CHD Systematic Reviews 1998 2004 2008	11 E-bD strategies in total suggested	Review analysis, feature classification
	1. Exposure to daylight/sunlight 2. Single-bed rooms 3. Acuity-adaptable rooms 4, 5. Quiet hospital environments/ noise-reducing finishes (technical feature, non-architectural) 6. Views of nature 7. Positive distractions (amenities) 8. Develop way-finding systems 9. Appropriate lighting (technical feature, most cases non-architectural) 10. Design wards and nurses' stations to reduce staff walking and fatigue 11. Ceiling lifts (technical feature, non-architectural)	3 Pertaining to Architecture (soft factors): • Exposure to daylight/sunlight • Acuity-adaptable rooms • Views of nature 7 Non-architectural (technical/hard factors): • Provide single-bed rooms • Positive distractions (amenities) • Develop way-finding systems (signage) • Appropriate lighting (technical devices, most cases non-architectural) • Quiet hospital environments/noise-reducing finishes (technical devices, non-architectural) • Design wards and nurses' stations to reduce staff walking and fatigue • Ceiling lifts (technical devices, non-architectural)

(Table 3) E-bD strategies into architectural and technical features: from the 3 CHD systematic reviews, 11 E-bD strategies were recommended in total; quietness and noise reduction overlap leaving the count in 10 strategies.

growth of new findings has decreased, and (3) the relevance of architectural recommendations raise serious doubts. Doubts as the one architect Stefan Lundin phrases in his dissertation: "Is the research referred to merely confirming what has long been sensed, understood and applied already?" During a recent trip to Barcelona, similar doubts mirrored everywhere in the Hospital de la Santa Creu i Pau, which today is essentially a museum. As walking along its corridors and landscape, the purpose of designing exclusively for healing was called into question. Looking at its rooms mostly stripped of medical equipment, I wondered if Healing Architecture was not more than simply good architecture.

Nature seen from pavilions

Art integrated in architecture

Natural light supporting underground hallways

Hospital de la Santa Creu i Sant Pau, Barcelona.
architect: Lluís Domènech i Montaner (1901–1930)
(photos: Wilfried Humann)

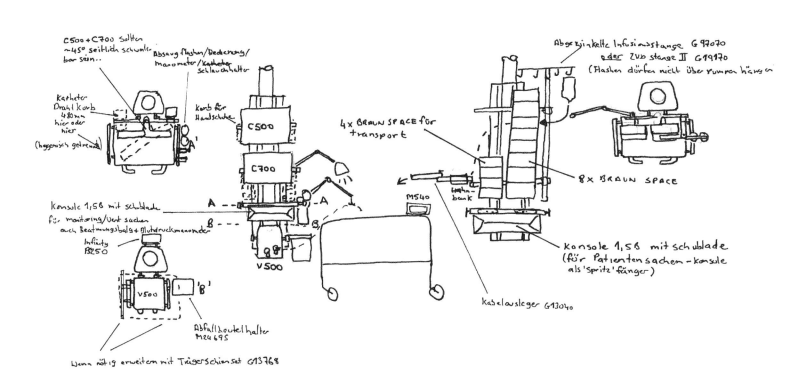

Sketch by Prof. Dr. Walter Schaffartzik,
Unfallkrankenhaus Berlin, and
David Biddel, Dräger

Hospitals

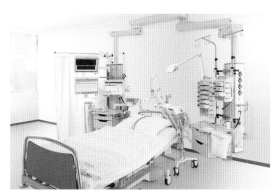

Staff-focused environment by Dräger
(photo: Drägerwerk AG & Co. KGaA)

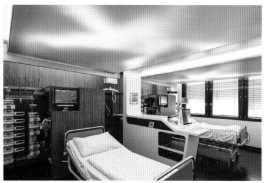

Patient-focused environment by GRAFT Architects,
Virchow Klinikum, Berlin
(photo: Tobias Hein)

Nurse station by GRAFT Architects,
Virchow Klinikum, Berlin
(photo: Tobias Hein)

Architecture or technical-medical plans?
If Healing Architecture is not more that good architecture, then why insist on developing this concept? After analysing the CHD systematic reviews, it seems to be more sensible for E-bD practitioners to implement technical features rather than a rightful development of an architectural design. As experienced in our formation and practice, we also tend to succumb to what hospital functionality and efficiency dictates over creativity and exploration. Medical input, in synergy with technological requirements, often trigger a process of re-drawing in two dimensions, spatial demands over and over in detriment of spatial quality and other architectural factors.

As seen in the CHD reviews, most studies miss distinguishing technical features from architectural quality and its factors. The problem might lie in the evaluation frameworks used to assist surveyors in differing evidence-based designs with an architectural character from those showing extruded medical-technical plans (from 2D to 3D).

Aside from conceptualising terms properly, research activities in general face other common limitations, such as attaining useful results within a limited budget. Analysing small but representative samples of a problem, cuts research times and resources, making studies feasible. This is the case of studying the intensive care unit (ICU) in regards to the hospital. The ICU is arguably the department with the highest impacts on care delivery within hospitals and of greatest concern for healthcare professionals[15]. Its economic, technical, spatial, and staff demands can topple a hospital's budget with services estimated to suffer a higher demand and growth in upcoming years[16]. Studying the ICU environment and its complexities could very well clarify how to tackle larger scale issues concerning architecture and care processes.

In 2013, two architects from the Academy for Design & Health realised an environment evaluation study on ICUs called "Critical Care Design – Trends in Award Winning Designs". It was based on an annual competition organised by the Society of Critical Care medicine between 1992 and 2013. The competition jury used two scoring sheets to assess relevant characteristics of the projects.

Scoring sheet 1 studied environmental qualities and sheet 2, its particular features. Using both sheets, the researchers made a comparative data analysis to 12 winning projects, resulting in the definition of ten design trends[17].

The more I read through this evaluation study, the more arguments I found to establish differences between architectural projects and medical-technical plans. In trail for a future study, both scoring sheets were distinguished into architectural and non-architectural features, using the classification from previous CHD reviews (see table 4). All ten design were classified as follows: technical recommendations, technical-medical planning, and architectural design (see table 5). This merely indicative study, showed the need to develop or improve conceptual frameworks for architecture evaluation in healthcare settings. More differences between medical planning and architecture are emphasised with the following image comparison of two intensive care environments. As architects for health would say, one with a staff-focused design, the other with a more patient-centred one.

Simply explained, a staff-focused design helps medical teams easily navigate the environment with comfort and safety. A patient-centred design ensures the patients and relatives an environment stress-free from care delivery mechanisms. Ideally, these two strategies are not mutually exclusive, on the contrary they should be reciprocal and interdependent. Many are the cases where patient-centred designs trade-off staff satisfaction to ensure patient wellbeing, disregarding the fact that staff is a leading factor for quality of care frequently carrying out long working shifts under harsh environmental conditions.

15 Charles D Cadenhead, "Critical Care Design Twenty Years of Winners and Future Trends: An Investigative Study" (Healthcare Design Conference, Orlando, Florida, 18 November 2013).

16 Jason N. Katz, Aslan T. Turer, and Richard C. Becker, "Cardiology and the Critical Care Crisis: a Perspective," Journal of the American College of Cardiology 49, no. 12 (27 March 2007): 1279–82, https://doi.org/10.1016/j.jacc.2006.11.036.

17 Charles D Cadenhead and Diana C Anderson, "Critical Care Design: Trends in Award Winning Designs," Critical Care Design: Trends in Award Winning Designs, 2013, http://www.worldhealthdesign.com/critical-care-design-trends-in-award-winning-designs.aspx.

(Table 4) Scoring sheets distinguished by architectural and non-architectural features

SSCM Scoring Sheet 1 Environmental Qualities:	CHD Reviews – feature classification	SSCM Scoring Sheet 2 Features	CHD Reviews – feature classification

1. Visual (colour, light)
2. Simplicity (neatness)
3. Organisation (layout)
4. Auditory
   (noise, avoidance, therapeutic sound)
5. Psychological Amenities
   (TV, VCR, plants)
	Architectural
	• Visual
	• Simplicity (neatness)
	• Organisation (layout)
	
	Non-architectural
	(technical factors)
	• Auditory
	• Psychological Amenities
		1. Size
		2. Functionality
		3. Safety/Security
		4. Decor
		5. Amenities
		   (refreshment, toiletry, sleep, seating)
		6. Technology
			Architectural
			• Architectural
			• Size
			• Functionality
			
			Non-architectural
			(technical factors)
			• Safety/Security
			• Decor
			• Amenities
			• Technology

2 Technical recommendations (non-architectural)
- Stabilised patient room size. The standard size will be approximately 23 sqm. Important design considerations derive from patient bed placement and delivery of medical support substitution of headwalls (medical devices placed vertically at the head of the patient) for ceiling-mounted articulating arms called booms (monitoring, outlets, and gasses)
- Remote technology and support systems. In ICU patient rooms, ceiling-mounted booms are preferred over traditional headwall devices.

5 Technical-medical planning solutions
- Larger, consolidated units. As demand for service grows, an increase in number of units, larger units, and space for support areas, will be seen.
- Continued design for interdisciplinary teams. Staff work stations tend to have a combination of centralised and decentralised layouts.
- Integration of diagnosis and treatment facilities. These services are eventually shared with the entire hospital.
- Integration of administration and support spaces within the unit.
- Segregated circulation. Distinction of circulation regarding on-stage (patients with staff) and off-stage (only staff) separations.

2 Architectural design directives
- Defined in-room family space. Most recent units incorporate designated family and visitor space in the unit, or within the patient room itself.
- Visual and Physical Access to Nature. Nature incorporated in the unit for patients, families and staff.

1 No-trend
- Variable unit geometric form. There are no clear trends pointing at a specific ICU geometry

(Table 5) 10 ICU design trends–classified

The first images (p. 032 bottom, p. 033 top left), depict a technical sketch developed by Prof. Dr. Schaffartzik (UKB) with David Biddel (Dräger) and an ICU room as result of a tight research collaboration. This teamwork has led Dräger – a well-known healthcare manufacturing company – to constantly improve its ICU products in the workplace. The other two images (p. 033 top centre and right) were taken at the Charité Medical University Berlin, where Prof. Dr. Claudia Spies and GRAFT Architects also teamed to research, delivering a new treatment concept within a new kind of intensive care unit. This ICU design comprehensively combined factors for stress-reduction such as: room acoustics (reducing noise of alarms and signals), temperature control, and visual structure (from material, light, colour, and media surfaces). It maximised privacy for patients and family members, disguising the technical equipment in the background and buffering alarm sounds.

Hospitals

Staff-focused design integrating light and art, central sterilisation in Martigny Valais Hospital, architects: bauzeit architekten (2017)
(photo: Yves André)

Environmental factors and its healing effects
The CHD reviews and ICU evaluations have helped discern technical devices from architectural features and differ medical-technical planning from architecture. In an effort to keep defining the elements and capacities of Healing Architecture, it is important to look at environmental factors and variables proven to influence human health and well-being.

In late 2012, a review was realised at the Technical University of Berlin about the physiological and psychological influences of environmental features impacting patient recovery and staff performance[18]. The following section of this chapter updates the text and descries which factors are natural, technical, or architectural. Written as a glossary of empirical findings, it stands alone from the rest of the chapter. Here the reader is encouraged to move on to the last section, "E-bD research an evolving field" and come back for facts and references.

Natural factors
1. Light
There is quite a significant amount of clinical and non-clinical evidence showing the effects of light on human health recovery and well-being. Light can have an impact on: pain, sleep, circadian rhythm, hospitalisation period, medical errors, mortality, stress, depression, user satisfaction, mood and orientation, as well as staff effectiveness[19]. Natural daylight is however preferred over electric light as a primary source of illumination in working and living settings[20]. It is not superior to artificial light when it comes to carrying out activities, but it does have clear advantages for all kinds of physiological processes and overall health[21]. Daylight tends to be brighter and have a more balanced spectrum of colours than most artificial light sources. It has effects on health through the visual system, the biological system, as well as the psychological system[22].

1.1 Length of patient hospitalisation and mortality
Beauchemin and Hays[23] show in their research that patients with severe depression and placed in sunny rooms, stay on average 2.6 days less than patients in dull rooms. According to Benedetti et al.[24], patients with bipolar disorder having access to direct sunlight in the morning stay on average 3.67 days less than patients in rooms with sunlight access in the evening[25]. Female patients with myocardial infarction in a cardiac intensive-care unit treated in sunny rooms stayed a shorter time in than those in dull rooms (2.3 days in sunny rooms, 3.3 days in dull rooms). Mortality in both sexes was also higher in dull rooms[26].

1.2 Human biological processes and circadium rythm
According to Aarts and Westerlaken[27], daylight (among other factors) controls the biological clock responsible for body temperature and the sleep-wake rhythm through production of hormones, such as melatonin (sleeping, activity, and energy hormone) and cortisol (stress hormone).

1.3 Pain
According to Walch et al.[28], patients recovering from spinal surgery, placed in a brighter part of the hospital experienced less perceived stress, marginally less pain, and took 22% less analgesic medication per hour than patients on the dim side of the hospital.

1.4 Depression
Wirz-Justice et al.[29] affirm that patients with seasonal affective disorder reduce depressive symptoms and improve daily secretion of melatonin and cortisol after regular morning walks outdoors.

1.5 Mood and perception
Daylight impacts satisfaction, mood, and performance of work through sensory stimulation, changes in daylight (colour, shadow, brightness contrast, position of the sun)[30], and thermal sensations (perceived effect of sunlight, wind, and humidity)[31]. It also offers people a sense of place and time and prevents feelings of disorientation[32].

Nurses who are provided with three hours of exposure to daylight during work shifts reported greater work satisfaction[33].

1.6 Physiological processes
According to McColl and Veitch[34], most of the vitamin D in the blood can only be derived from exposure to light.

2. Nature
The "Biophilia Hypothesis" suggests that there is an instinctive bond between human beings and other living systems[35]. Research on the effect of nature on human health is based on this hypothesis. In healthcare environments, nature is connected to the three main subjects: views of nature, therapeutic gardens, and indoor plants.

2.1 View of nature
Views of nature in buildings are obviously connected to the subject of windows. According to Devlin and Arneill[36], access to windows and views helps patients develop a perceptual and cognitive link with the external environment. Patient satisfaction is achieved when windows occupy 20–30 per cent of the room.

2.2 Pain and human physiological responses
Views of nature or images of nature may provide relief from pain, raise pain tolerance, and reduce post-surgical recovery time. It also provides additional support to reduce pain as "distraction therapy"[37]. Patients with rooms with a view of nature after bladder surgery required fewer strong painkillers and a shorter length of stay, compared to those who were assigned to a room with the view of a brick wall[38]. According to Wilson[39], views of nature in intensive care units lower levels of organic delirium. Natural scene murals was found to reduced pain during bronchoscopy procedures[40].
The blood pressure and pulse of blood donors were lower while watching videos of natural settings (a park and a stream) in waiting rooms[41]. Views of real aquariums and/or ocean scenic images improved the food intake of people with Alzheimer's disease[42].

2.3 Stress alleviation - restoration theory
"Restoration theory" describes the relationship between the view of green areas and improvement in health. It is a stress-recovery mechanism categorised in three types[43]:
– Affective recovery refers to positive emotions and mood improvement.
– Physiological recovery refers to sympathetic-specific mechanisms related to positive change in blood pressure, heart rate, and skin[44].
– Cognitive recovery assumes that nature stimulation and fascination invoke involuntary attention, modestly allowing directed-attention mechanisms a chance to regenerate[45].

According to Van den Berg and Winsum-Westra[46], if natural views were associated with better performance in attention measures, it would hence be plausible to assume that a view of greenery will also have significant positive effects to reduce the chances of medical errors.

Adults and children (in particular females) who live in houses with views of urban nature have a greater ability to concentrate, are less aggressive, and more self-disciplined than individuals who live in houses with views of built environments. The former also reported greater well-being than the latter[47].

2.4 Therapeutic gardens
Stress restoration is the key motivation for patients, family members, and staff to use gardens in healthcare facilities[48]. This idea is supported by two important studies: in their studies, Cooper-Marcus and Barnes[49] and Whitehouse et al[50] found that hospital gardens improved the mood of all hospital users and that many healthcare employees used gardens as an effective means to escape from work stress and aversive conditions. As more evidence is showing that hospital gardens increase staff satisfaction, it may help hire and retain qualified personnel[51]. Also, according to Sadler[52] gardens and nature in hospitals can significantly increase patient satisfaction and perception of the overall quality of care. This increased patient satisfaction can create a positive market identity and thereby improve economic or financial outcomes[53]. Exercising and social support are other mechanisms through which gardens and natural settings may improve people's health and well-being[54]. A study in 1991, Hartig, Mang, and Evans exemplify this association between nature and health. After performing mentally fatiguing tasks, the students who walked through nature as a means to recover showed higher performance in attention tests afterwards in comparison to those who recovered through passive relaxation[55].

2.5 Indoor plants
Research on indoor plants in clinical settings mainly focused on health risks rather than benefits. Transmission of diseases through the soil and water of plants has not been scientifically confirmed. On the contrary, Fjeld[56] (Study 2 in the research) found out that foliage plants and full spectrum lamps reduced sick building syndromes, such as fatigue, headaches, dry throat, and itching, and/or dry hands in a radiology department at a Norwegian hospital. Additionally, an inverse linear relationship was found between performance in productivity tasks and number of plants in the office; lower concentration levels but higher self-reported perceptions of performance improvement[57].

3. Smell
Aromatherapy is applying compounds for improving psychological or physical well-being through inhalation. In a study regarding 40 post-open-heart surgery patients in Iran, lavender essential oil at 2 per cent was placed with a cotton swab in patients' oxygen masks and the patients breathed for 10 minutes. The results show that aromatherapy significantly alleviated stress and improved sleep quality in intensive care unit patients after two days of the experimental treatment[58]. It implies the possibility of applying this method as an independent nursing intervention to stabilise vital signs such as blood pressure, heart rate, and central venous pressure, etc.[59]

Staff-focused design integrating views of nature, central sterilisation in Martigny Valais Hospital, architects: bauzeit architekten (2017)
(photo: Yves André)

Technical Factors
1. Lighting
1.1 Staff performance and medical error
The level of light needed for task performance increases with age due to reduced transmittance of aging eye lenses. Performance on visual tasks increases as light levels increase[60]. Bright light (1,500 lux) improves the performance of duties, which is especially important in reducing errors in medication[61]. High level daylight without glare, shadows, and reflection is superior for tasks involving fine colour discrimination[62]. There is some indication that certain properties of indoor lighting, such as luminance level, lamp colour, and flicker can affect people's mood and performance[63]. Dim lighting in counselling rooms could enhance communication between patients and doctors[64].

1.2 Sleep
Providing cycled lighting (reduced light levels in the night) in neonatal intensive-care units results in improved sleep and weight gain among pre-term infants[65]. Exposure to higher levels of light (1,000 lux) for longer periods during the day increases sleeping efficiency for people with dementia[66].

1.3 Depression
Exposure to artificial high-intensity light (usually ranging between 2,500 lux and 10,000 lux) in the morning has been successfully used in the treatment of patients with seasonal affective disorder[67] and reducing agitation of patients with Alzheimer's disease[68].

1.4 Mood and perception
Intermittent bright light during night shifts is effective in adapting circadian rhythms of nightshift workers, improving well-being, and reducing distress level[69].

1.5 Physiological processes
Exposure to light is an effective treatment for neonatal hyperbilirubinaemia (neonatal jaundice)[70].

2. Acoustics
There are many manifestations of sound in the healthcare setting: noise, music, speech privacy, and speech intelligibility[71]. Peace and quiet are also important for good communication, both with patients and among the staff[72]. There are different sources of noise in hospital environments, such as alarms, equipment, computers, printing, people, staff communication, etc. Besides,

Staff-focused design integrating access to nature, central sterilisation in Martigny Valais Hospital, architects: bauzeit architekten (2017)
(photo: Yves André)

hospital materials are sound-reflecting rather than sound-absorbing[73]. As a result, noise in the hospital setting usually exceeds the values recommended in the guidelines of The World Health Organisation (WHO). These guidelines recommend continuous background noise limits in hospital patient rooms at 35 dB(A) during the day and 30 dB(A) during the night, with peaks in wards not to exceed 40 dB(A) at night. However, many studies indicate that peak hospital noise levels often exceed 85 dB(A) to 90 dB(A)[74]. Poor acoustic environment may well lead to many errors in automatic transcription of doctors' spoken notes, and automatic dispensing of pharmaceuticals, etc.[75] Moreover, speech recognition systems, critical for the functioning of a digital hospital, cannot interpret sound signals in poor acoustic environments[76].

2.1 Noise effects on patients
Noise is a source of awakenings and sleep disruption among patients. Studies by Slevin et al. in 2000[77], Johnson in 2001[78], and Zahr and de Traversay in 1995[79] show that in the NICU unit, loud noise levels decrease oxygen saturation (increasing need for oxygen therapy), elevate blood pressure, increase heart and respiration rate, and worsen sleep.

In 2000, Liu and Tan[80], found that elevated noise levels induce cardiovascular and endocrine effects. Minckley[81] observed that noise levels higher than 60 dB (A) increase the pain medication required by post-surgery patients. In Fife and Rappaport's[82] study in 1976, patients were found to need more recovery time after the cataract surgery when noise level were elevated due to construction.

2.2 Noise effects on staff
Unexpected noises may increase medication errors, perceived work pressure, stress, and annoyance. High levels of noise increases fatigue and emotional exhaustion. In better acoustic conditions, staff experienced less work demands and reported less pressure and strain. A study by Murthy et al.[83] showed under typical noise level in operating rooms (over 77 dB(A)), the threshold level for speech reception increased by 25 per cent, meaning verbal communication was only possible when speaking in a raised voice, while speech discrimination level decreased by 23 per cent. The same study also shows that anesthetists' short-term memory and efficiency declined under such noise conditions[84].
As Joseph and Ulrich cited Parsons and Hartig[85], adequate performance during elevated noise level is maintained

Patient accommodation at the Bispebjerg Pychiatric Center, Copenhagen, Denmark, architects: Henning Larsen Architects (2015)

by increasing effort, as evidenced by heightened cardiovascular response and other physiological mobilisation.

3. Air Quality
3.1 Ventilation and hospital safety
The rate at which the indoor air is renewed per unit of time is called "ventilation rate". It is usually measured in litres per second (L/s). In all building types, a ventilation rate of less than 10 L/s per person is proven to lead to health problems and adversely affect the perception of the air quality[86].
Ventilation can be improved by both natural and artificial routes. Studies on artificial ventilation and its impact on health outcomes are mainly associated with the dissemination of infectious diseases, while studies on natural ventilation are mainly related to window types and sizes[87]. Hospital air quality plays a decisive role in determining the concentration of pathogens in the air, and thereby has major effects on the frequency of airborne infectious diseases. During the SARS outbreak epidemic in Canada, higher ventilation rates resulted in a significantly lower infection rate among healthcare workers[88]. Boswell and Fox's[89] study shows that the use of portable High Efficiency Particulate Air (HEPA) filters in a clinical setting significantly reduces environmental contamination by Methicillin-resistant Staphylococcus aureus (MRSA).
Immune-compromised and other high-acuity patients have a lower incidence of infection when housed in HEPA-filtered isolation. HEPA filters, combined with Laminar Air Flow (LAF) can reduce air contamination to the lowest level; thus it is recommended for operating rooms and areas with ultra clean room requirements. Airflow direction also has an impact on the rate of nosocomial infections. Rooms with infectious patients should have

Unit and work environment at the Norwegian Radium Hospital
for cancer research and treatment, Oslo,
architects: Henning Larsen Architects (2015)
(photo: Adam Mørk)

negative pressure to prevent the spread of contaminated air. The immune-compromised and immune-suppressed accommodation should have positive pressure to protect them from contaminated air[90].

3.2 Temperature and human health
Patients generally find a stable temperature between 21.5°C to 22°C and a humidity rate between 30–70 per cent comfortable[91]. Extreme highs and lows in temperature lead to complaints and dissatisfaction among the staff in office environments and adversely affect their performance of duties[92].
Sick Building Syndrome (SBS) symptoms increase linearly at temperatures exceeding 22°C[93]. Hot temperatures can lead to negative social reactions such as crowding, aggression, and other negative reactions to others[94].

Architectural Factors
1. Stress reduction features
Ulrich, Bogren, and Lundin[95] developed a design theory that could reduce aggression in psychiatric facilities. The architectural features that reduce stress from involuntary admission, thereby reduce aggression are: single patient rooms with own bathrooms; smaller wards for smaller patient group size; moveable seating in spacious dayrooms or lounges; low noise level with good acoustics; views onto nature; art resembling nature; accessible gardens; daylight exposure; staff stations close to patients with good visibility; homelike qualities; and easy wayfinding, etc[96].

2. Elasticity and flexibility
Since the early 2000s, neuroscience and architecture has explored the broad range of human experiences with elements of space and design. Many have been the findings and results on improving disabilities due to brain damage or neurological disorders in general. Strategies emphasise the use of natural light and stimulating spaces to directly impact neuron growth, thereby empower a person's rehabilitation.
For example, in a neuro-rehabilitation facility for people with specific health issues from birth, accidents, and injury (which draw psycho-emotional differences), our purpose is not only to help them re-learn their everyday activities but evermore improve performance beyond expectations with renewed brain capacities. In order to foster this recovery and rehabilitation pathway, the design of healthcare facilities should consider its elasticity (the ability to expand and possibly reduce in size) and flexibility (the possibility to change room functions)[97].

3. Unit and work environment
There is a growing and convincing body of evidence suggesting that improved hospital design can make the jobs of staff easier. As found in studies by Burgio et al.[98] in 1990, walking accounted for 28.9 per cent of nurses working time, followed by patient-care activities that accounted for 56.9 per cent. The time nursing staff spent on walking responds to the type of unit layout (e.g. radial, single corridor, double corridor). Time saved from walking can be translated into patient care activities and interaction with family members[99].
Radial type reduces walking time compared to single corridor and rectangular units because it provides better visual control of the patient from the nursing station. However radial designs might provide less flexibility in managing patient loads[100]. Decentralised nurse stations can reduce staff's walking time only when a decentralised supply is placed near the nurse stations. Central location of supplies could double staff-walking even when nurse stations are decentralised. Decentralised pharmacy systems reduce medication delivery times by over 50 per cent[101]. In 1990, Pierce et.al[102] redesigned an outpatient pharmacy layout to improve workflow, reduce waiting times, and increase patient satisfaction with service.

Pathway finding with natural light, main entrance to the University Hospital Bern,
architects: bauzeit architekten (2017)
(photo: Yves André)

4. Patient accommodation
4.1 Single-bed versus multibed
Single accommodation is recommended for quality of care such as safety, privacy, dignity, confidentiality, and flexibility. National Health Service Estates found out that 52 per cent prefer to stay in a single room while 37 per cent prefer a shared space[103]. Conflicting preferences in hospital accommodation among patients showed a link between the severity of illness and the desire for privacy[104].

4.2 Hospital-acquired infection
Single-bed rooms, cubicles with partitions, and isolation rooms decrease the risk of hospital acquired infection by airborne, contact, and waterborne transmission, compared to multiple-bed rooms. Multibed accommodations increase the probability and speed of outbreaks; for example, the SARS outbreak in Canada where multibed rooms failed in preventing and controlling hospital acquired infections. A study by Farquharson and Baguley[105] shows that ca. 75 per cent of the SARS cases in Canada resulted from exposure to hospital settings.
Single-bed rooms facilitate cleaning and decontamination of rooms. On the contrary, cleaning of multibed patient rooms implies disruption in functionality and costly transportation of patients, i.e. the temporary removal of all patients from these rooms[106].

4.3 Medical errors
Single rooms might decrease the number of the medical errors due to patient transfer between rooms or units. NHS Estates[107] reported that transfers fell by 90 per cent and medication errors by 67 per cent when the US Clarian Hospital changed its coronary intensive care from 2-bed rooms to single acuity-adjustable family-centred rooms[108].

4.4 Sleep quality
Noises from other patients are the most disturbing factor and major cause of sleep loss in multibed rooms, whereas single-bed rooms can reduce noise disturbance from roommates, visitors, and healthcare staff and thereby improve patient sleep[109].

4.5 Care quality
Single-bed rooms increase patient privacy through perception of control and autonomy. This facilitates good communication between patient, staff, and family. This is particularly important because patients are more likely to withhold information when they experience a lack of auditory and visual privacy[110]. This also applies to staff members. In multibed rooms, healthcare staff are reluctant to discuss patients' issues or give information when they are within hearing distance of a roommate, out of respect for patient privacy[111]. Single-bed rooms are thus better than multibed rooms in supporting or accommodating the presence of family and friends.
Patient-family interactions improve patients' physiological outcomes, facilitate progress, and help effectively deal with treatments. The support from interacting with family lowers a patient's levels of stress, fear, anxiety, and depression. A study by Chatham[112] in 1978 shows that specific social interactions with families (such as eye contact, frequent touch, and verbal orientation to time, person, and place) can reduce disorientation, alertness, confusion, anxiety, and improve sleep quality of open-heart surgery patients. Restricted visiting hours in open-plan multibed rooms deter family visit and thereby reduce family members' social support.

5. Orientation and wayfinding
Illegible public buildings might confuse users and create a feeling of incompetence. As topological complexity increases, the overall legibility of the environment decreases, reducing understanding in spatial layout and wayfinding performance. A regular but asymmetrical layout is easier to remember and learn than a regular and symmetrical one. Continuity in paths, i.e. loop-like paths, is preferred over dead ends because the latter cause frustration for people[113].
The lack of differentiation in an environment affects orientation and wayfinding of both newcomers and more experienced users. Creating landmarks and spatial differentiation in appearance are thus essential for users' understanding of a building's spatial organisation. Using colour and shape, art, graphic information as reference

points can improve building interior memory[114]. Good signposting combined with written and verbal information improves people's movements through complex buildings[115]. A clear routing system is especially important in healthcare settings for cognitive impaired patients, such as people with dementia. According to Marquard[116], the following four guidelines could be implemented in all designs to support the way finding abilities of people with dementia: 1. no need for new or higher skills; 2. allow visual access and overviews; 3. reduce decision making; and 4. increase architectural legibility.

6. Interior design
A study with telephone interviews realised on 380 discharged inpatients helped determine that environmental satisfaction was a significant predictor of overall satisfaction with healthcare, ranking only below perceived quality of nursing and clinical care[117]. The study also identified specific environmental factors that were perceived to be pleasing and satisfactory to patients, including: 1. wall colour, artwork, comfortable bed, television working properly, and easy access to anything in the patient room; 2. A window with a nice view, an accessible bathroom in the room, and a room located away from noisier areas of the unit; 3. adequate lighting, quiet surroundings, and a comfortable temperature; 4. A private room, environmental means for privacy (e.g. A closed door); and 5. cleanliness of the room[118].
Redecorating and renovating often lead to positive hospital evaluations. Changing the environment to improve comfort and appeal increases satisfaction in the patient and their families. Appropriate interior design can also impact the patient and staff safety. Non-slippery floors, appropriate door openings, placement of rails and accessories, and appropriate heights of toilet and furniture decrease patient fall accidents in bathroom and bedroom areas. Available and appropriate ceiling lifts reduce the incidence of musculoskeletal injury of staff and the cost of injury claims. However, bedrails are ineffective for reducing falls. Appropriate numbers and locations of hand-washing facilities influence compliance and infection rates[119].

7. Interiors and social interaction
Lounges, day rooms, and waiting rooms with comfortable movable furniture facilitate social interactions and improve eating behaviours, as indicated by the increased food consumption of geriatric patients[120]. A study in 1972 found out that different seating arrangements of hospitalised male psychiatric patients can discourage or encourage social and personal interaction. Chairs in rows along the walls in waiting rooms discourage social interaction[121].

8. Materials
Sound-absorbing ceiling tiles and panels reduce noise levels and sound reverberation time perceptions, improving patient outcome, speech intelligibility, and lowering work pressure among staff[122]. Easily cleanable, nonporous material for floor and furniture coverings decrease the rate of the contact infections[123]. The use of homely material increases social interaction and the feeling of control (carpeted flooring increases the time of visitor stay compared to vinyl flooring)[124].

9. Colours
Colours can manifest themselves in the interior in different ways: in light composition and in the finishings of walls, floors, furniture, as others. There are four properties in colour stimuli: the brightness/intensity (amount of light energy contained in the colour spectrum), luminance (perceived brightness), hue (dominance wavelength), and saturation (determines the vibrancy of the colour)[125].
Colours can affect people's perception and experience in certain environments (e.g. perception of spaciousness is attributed more to the brightness than the hue of a colour) but there are no causal relationships between particular colours and health outcomes[126]. In Jacobs and Hustmyer's[127] study, no significant effects of red, yellow, and blue is found to affect respiration or heart rates. Besides, associations between certain colours and emotions are culturally learned and determined by the physiological and psychological makeup of people, it is ineffective to develop universal guidelines of colour use in healthcare settings[128].

Interiors and social interaction at Herlev Hospital, Helev, Denmark, architects: Henning Larsen Architects (2015)

10. Integrated Art
10.1 Visual art

The effect of visual arts as live and video-recorded performances, drawings and paintings, and traditional and contemporary art on mental health are widely studied. A literature review by Daykin et al.[129], in 2006, suggests that art can have a therapeutic effect on people suffering with mental disorders by mitigating depression, anxiety, and low self-esteem, improving social integration, and alleviating isolation. However, Ulrich[130] revealed that inappropriate visual art styles are related to the disturbance of mental health conditions; Staricoff and Loppert[131] also showed that the psychological effects of being engaged with creative arts, such as dance, drama, music, visual arts, and creative writing in mental health institutions can be too demanding for some patients.

10.2 Contemplative art
10.2.1 Music

Music can induce relaxation and pleasure to the human body. This lowers the activity levels of neuroendocrine and sympathetic nervous systems, creating a decrease in anxiety levels, heart and respiratory rates, and increase in body temperature[132]. Music may also have a calming, relaxing, and even therapeutic effect, as it has been used in different healthcare settings such as oncology, maternity, postoperative, intensive care, pediatric care[133]. Listening to individualised music, based on personal preferences, is effective in decreasing behavioural problems and decreasing stress level significantly. In Gerdner's[134] study, classical music was found to reduce the level of agitation among patients with dementia.

18 Valera Sosa, Álvaro, and Matthys, Stefanie, *From Concepts of Architecture to German Health Economics*, 2012. Review and update of this section by Alvaro Valera Sosa and Weng Ian Au.

19 Ulrich, Roger S. et al., 'A Review of the Research Literature on Evidence-Based Healthcare Design', HERD: *Health Environments Research and Design Journal 1*, no. 3 (April 2008), pp. 61–125, https://doi.org/10.1177/193758670800100306.

20 Van den Berg,A.E., *Health Impacts of Healing Environments; a Review of Evidence for Benefits of Nature, Daylight, Fresh Air, and Quiet in Healthcare Settings* (UMCG, 2005).

21 Boyse, Peter, Hunter, Claudia, and Howlett, Owen, *The Benefits of Daylight through Windows* (Troy, NY: Lighting Research Center, Rensselaer Polytechnic Institute, 2003).

22 Ibid.

23 Beauchemin, Kathleen M., and Hays, Peter, 'Sunny Hospital Rooms Expedite Recovery from Severe and Refractory Depressions', *Journal of Affective Disorders 40*, no. 1 (9 September 1996), pp. 49–51, https://doi.org/10.1016/0165-0327(96)00040-7.

24 Benedetti, Francesco, et al., 'Morning Sunlight Reduces Length of Hospitalisation in Bipolar Depression', *Journal of Affective Disorders 62*, no. 3 (1 February 2001), pp. 221–23, https://doi.org/10.1016/S0165-0327(00)00149-X.

25 Joseph, Anjali, 'The Impact of Light on Outcomes in Healthcare Settings', *The Center for Health Design*, Issue paper no. 2 (August 2006), https://www.healthdesign.org/sites/default/files/CHD_Issue_Paper2.pdf.

26 Ulrich et al., *a Review of the Research Literature on Evidence-Based Healthcare Design.*

27 Aarts, Mariëlle, and Westerlaken, Adriana, 'Licht en gezondheid bij senioren', *Bouwfysica 18*, no. 3 (2007).

28 Walch, Jeffrey M., et al., 'The Effect of Sunlight on Postoperative Analgesic Medication Use: a Prospective Study of Patients Undergoing Spinal Surgery', *Psychosomatic Medicine 67*, no. 1 (January 2005), pp. 156–163, https://doi.org/10.1097/01.psy.0000149258.42508.70.

29 Wirz-Justice, Anna, et al., ''Natural' Light Treatment of Seasonal Affective Disorder', *Journal of Affective Disorders 37*, no. 2 (12 April 1996), pp. 109–120, https://doi.org/10.1016/0165-0327(95)00081-X.

30 Joseph, *The Impact of Light on Outcomes in Healthcare Settings.*

31 Lomas, K.J., and Giridharan,R., 'Thermal Comfort Standards, Measured Internal Temperatures and Thermal Resilience to Climate Change of Free-Running Buildings: a Case-Study of Hospital Wards', *Building and Environment 55* (September 2012), pp. 57–72, https://doi.org/10.1016/j.buildenv.2011.12.006.

32 Devlin, Ann Sloan, and Arneill, Allison B., 'Health Care Environments and Patient Outcomes: a Review of the Literature', *Environment and Behavior 35*, no. 5 (September 2003), pp. 665–694, https://doi.org/10.1177/0013916503255102.

33 Ulrich, et al., *a Review of the Research Literature on Evidence-Based Healthcare Design.*

34 Joseph, *The Impact of Light on Outcomes in Healthcare Settings.*

35 Kellert, Stephen R., and Wilson, Edward O., *The Biophilia Hypothesis* (Island Press, 1995).

36 Devlin and Arneill, *Health Care Environments and Patient Outcomes.*

37 van den Berg, *Health Impacts of Healing Environments; a Review of Evidence for Benefits of Nature, Daylight, Fresh Air, and Quiet in Healthcare Settings.*

38 Ulrich, *View through a Window May Influence Recovery from Surgery.*

39 Wilson, L.M., 'Intensive Care Delirium: The Effect of Outside Deprivation in a Windowless Unit', Archives of Internal Medicine 130, no. 2 (1 August 1972), pp. 225–226, https://doi.org/10.1001/archinte.1972.03650020055010.

40 van den Berg, *Health Impacts of Healing Environments; a Review of Evidence for Benefits of Nature, Daylight, Fresh Air, and Quiet in Healthcare Settings.*

41 Ibid.

42 Ibid.

43 Ibid.

44 Ulrich,R.S., 'Effects of Interior Design on Wellness: Theory and Recent Scientific Research', *Journal of Health Care Interior Design: Proceedings from the Symposium on Health Care Interior Design 3* (1991), pp. 97–109.

45 Kaplan, Stephen, 'The Restorative Benefits of Nature: Toward an Integrative Framework', *Green Psychology 15*, no. 3 (1 September 1995), pp. 169–182, https://doi.org/10.1016/0272-4944(95)90001-2.

46 van den Berg, A.E., and van Winsum-Westra, M., *Ontwerpen met groen voor gezondheid: richtlijnen voor de toepassing van groen in "healing environments"* (Wageningen: Alterra 2006).

47 Kaplan, Rachel, 'The Nature of the View from Home: Psychological Benefits', *Environment and Behavior 33*, no. 4 (July 2001), pp. 507–542, https://doi.org/10.1177/00139160121973115.

48 Ulrich, Roger S., 'Effects of Gardens on Health Outcomes: Theory and Research', in *Healing Gardens: Therapeutic Benefits and Design Recommendations*, by Marni Barnes and Clare Cooper Marcus (New York, NY [u.a.: Wiley, 1999).

49 Barnes, Marni, and Cooper-Marcus, Clare, *Healing Gardens: Therapeutic Benefits and Design Recommendations* (New York, NY [u.a.: Wiley 1999).

50 Whitehouse, Sandra, et al., 'Evaluating a Children's Hospital Garden Environment: Utilisation and Consumer Satisfaction', *Journal of Environmental Psychology 21*, no. 3 (1 September 2001), pp. 301–314, https://doi.org/10.1006/jevp.2001.0224.

51 Ulrich, Roger S., *Health Benefits of Gardens in Hospitals*, in *Plants for People*, 2002.

52 Ibid.

53 Ibid.

54 van den Berg, *Health Impacts of Healing Environments; a Review of Evidence for Benefits of Nature, Daylight, Fresh Air, and Quiet in Healthcare Settings.*

55 Hartig, Terry, Mang, Marlis, and Evans, Gary W., 'Restorative Effects of Natural Environment Experiences', *Environment and Behavior 23*, no. 1 (1 January 1991), pp. 3–26, https://doi.org/10.1177/0013916591231001.

56 Fjeld, Tove, *The Effect of Interior Planting on Health and Discomfort among Workers and School Children*, 2000, p. 7.

57 van den Berg, *Health Impacts of Healing Environments; a Review of Evidence for Benefits of Nature, Daylight, Fresh Air, and Quiet in Healthcare Settings.*

58 Salamati, Armaiti, Mashouf, Soheyla, and Mojab, Faraz, *Effect of Inhalation of Lavender Essential Oil on Vital Signs in Open Heart Surgery ICU*, 2017, p. 6.

59 Ibid.

60 Joseph, *The Impact of Light on Outcomes in Healthcare Settings.*

61 Ulrich et al., *a Review of the Research Literature on Evidence-Based Healthcare Design*.
62 Boyse, Hunter, and Howlett, *The Benefits of Daylight through Windows*.
63 van den Berg, *Health Impacts of Healing Environments; a Review of Evidence for Benefits of Nature, Daylight, Fresh Air, and Quiet in Healthcare Settings*.
64 Ulrich et al., *a Review of the Research Literature on Evidence-Based Healthcare Design*.
65 van den Berg, *Health Impacts of Healing Environments; a Review of Evidence for Benefits of Nature, Daylight, Fresh Air, and Quiet in Healthcare Settings*.
66 Hanford, Nicholas, and Figueiro, Mariana, 'Light Therapy and Alzheimer's Disease and Related Dementia: Past, Present, and Future', *Journal of Alzheimer's Disease : JAD* 33, no. 4 (1 January 2013), pp. 913–922, https://doi.org/10.3233/JAD-2012-121645.
67 Parry, Barbara L., and Maurer, Eva L., 'Light Treatment of Mood Disorders', *Dialogues in Clinical Neuroscience* 5, no. 4 (December 2003), pp. 353–365.
68 Hanford and Figueiro, *Light Therapy and Alzheimer's Disease and Related Dementia: Past, Present, and Future*.
69 Ulrich et al., *a Review of the Research Literature on Evidence-Based Healthcare Design*.
70 Joseph, *The Impact of Light on Outcomes in Healthcare Settings*.
71 Joseph, Anjali, and Ulrich, Roger S., 'Sound Control for Improved Outcomes in Healthcare Settings', *The Center for Health Design Issue Paper*, no. 4 (January 2007), https://www.healthdesign.org/sites/default/files/Sound%20Control.pdf.
72 van den Berg, *Health Impacts of Healing Environments; a Review of Evidence for Benefits of Nature, Daylight, Fresh Air, and Quiet in Healthcare Settings*.
73 Joseph, Anjali, and Rashid, Mahbub, 'The Architecture of Safety: Hospital Design', *Current Opinion in Critical Care* 13, no. 6 (December 2007), pp. 714–719, https://doi.org/10.1097/MCC.0b013e3282f1be6e.
74 Joseph and Rashid.
75 Ulrich et al., *a Review of the Research Literature on Evidence-Based Healthcare Design*.
76 Joseph and Rashid, 'The Architecture of Safety.'
77 Slevin, M., et al., 'Altering the NICU and Measuring Infants' Responses', 2000, p. 7.
78 Johnson, A.N., 'Neonatal Response to Control of Noise inside the Incubator', *Pediatric Nursing* 27, no. 6 (December 2001), pp. 600–605.
79 Zahr, L.K., and de Traversay, J., 'Premature Infant Responses to Noise Reduction by Earmuffs: Effects on Behavioral and Physiologic Measures', *Journal of Perinatology : Official Journal of the California Perinatal Association* 15, no. 6 (December 1995), pp. 448–455.
80 Liu, E.H., and Tan, S., 'Patients' Perception of Sound Levels in the Surgical Suite', *Journal of Clinical Anesthesia* 12, no. 4 (June 2000), pp. 298–302.
81 Minckley, Barbara Blake, 'A Study of Noise and Its Relationship to Patient Discomfort in the Recovery Room', *Nursing Research 17*, no. 3 (1968), https://journals.lww.com/nursingresearchonline/Fulltext/1968/05000/A_STUDY_OF_NOISE_AND_ITS_RELATIONSHIP_TO_PATIENT.18.aspx.
82 Fife, D., and Rappaport, E., 'Noise and Hospital Stay', *American Journal of Public Health* 66, no. 7 (1 July 1976), pp. 680–681, https://doi.org/10.2105/AJPH.66.7.680.
83 Murthy, V.S.S.N., et al., 'Detrimental Effects of Noise on Anaesthetists', *Canadian Journal of Anaesthesia* 42, no. 7 (1 July 1995), p. 608, https://doi.org/10.1007/BF03011878.
84 Ibid.
85 Joseph and Ulrich, *Sound Control for Improved Outcomes in Healthcare Settings*.
86 Rashid, Mahbub, and Zimring, Craig, 'A Review of the Empirical Literature on the Relationships Between Indoor Environment and Stress in Health Care and Office Settings: Problems and Prospects of Sharing Evidence', *Environment and Behavior* 40, no. 2 (March 2008), pp. 151–190, https://doi.org/10.1177/0013916507311550.
87 Codinhoto, Ricardo, et al., 'The Impacts of the Built Environment on Health Outcomes', ed. Daryl May, *Facilities* 27, no. 3/4 (27 February 2009), pp. 138–151, https://doi.org/10.1108/02632770910933152.
88 Ulrich et al., *a Review of the Research Literature on Evidence-Based Healthcare Design*.
89 Boswell, T.C., and Fox, P.C., 'Reduction in MRSA Environmental Contamination with a Portable HEPA-Filtration Unit', *Journal of Hospital Infection* 63, no. 1 (1 May 2006), pp. 47–54, https://doi.org/10.1016/j.jhin.2005.11.011.
90 Ulrich et al., *a Review of the Research Literature on Evidence-Based Healthcare Design*.
91 Rashid and Zimring, *a Review of the Empirical Literature on the Relationships Between Indoor Environment and Stress in Health Care and Office Settings*.
92 Ibid.
93 Ibid.
94 Anderson, Craig A., 'Heat and Violence', *Current Directions in Psychological Science* 10, no. 1 (1 February 2001), pp. 33–38, https://doi.org/10.1111/1467-8721.00109.
95 Ulrich, Roger, Bogren, Lennart, and Lundin, Stefan, 'Towards a Design Theory for Reducing Aggression in Psychiatric Facilities' (ARCH12 Conference, Gothenburg, Sweden, 2012), p. 14, http://vbn.aau.dk/files/71203129/FINAL_pdf_UlrichBogren_Lundin_Toward_a_design_theory_for_reducing_aggression_Oct_2_.pdf.
96 Ibid.
97 Glostrup Hospital, *New Hospital Glostrup Neuro-Rehabiliation Facility Restricted Desgin Competition Brief 2012*, (Glostrup Hospital, 4 December 2012).

98 Burgio, L.D., et al., "'A Descriptive Analysis of Nursing Staff Behaviors in a Teaching Nursing Home: Differences among NAs, LPNs, and RNs', *The Gerontologist* 30, no. 1 (February 1990), pp. 107–112.
99 Ulrich et al., *a Review of the Research Literature on Evidence-Based Healthcare Design*.
100 Ibid.
101 Ibid.
102 Pierce, R.A., et al., 'Outpatient Pharmacy Redesign to Improve Work Flow, Waiting Time, and Patient Satisfaction', *American Journal of Hospital Pharmacy* 47, no. 2 (February 1990), pp. 351–356.
103 Whitehead, Sarah, et al., 'Cost-Effectiveness of Hospital Design: Options to Improve Patient Safety and Wellbeing', *York Health Economics Consortium*, October 2010, p. 148.
104 Ibid.
105 Farquharson, Carolyn, and Baguley, Karen, 'Responding to the Severe Acute Respiratory Syndrome (SARS) Outbreak: Lessons Learned in a Toronto Emergency Department', *Journal of Emergency Nursing* 29, no. 3 (June 2003), pp. 222–228, https://doi.org/10.1067/men.2003.109.
106 Ulrich et al., *a Review of the Research Literature on Evidence-Based Healthcare Design*.
107 NHS Estates, *NHS Estates Annual Report 2004-2005*, (London, 20 July 2005), https://assets.publishing.service.gov.uk/government/uploads/system/uploads/attachment_data/file/273532/0230.pdf.
108 Whitehead et al., 'Cost-Effectiveness of Hospital Design: Options to Improve Patient Safety and Wellbeing.'
109 Ulrich et al., *a Review of the Research Literature on Evidence-Based Healthcare Design*.
110 Joseph and Ulrich, 'Sound Control for Improved Outcomes in Healthcare Settings.'
111 Ulrich et al., *a Review of the Research Literature on Evidence-Based Healthcare Design*.
112 Chatham, M.A., 'The Effect of Family Involvement on Patients' Manifestations of Postcardiotomy Psychosis', *Heart & Lung: The Journal of Critical Care* 7, no. 6 (December 1978), pp. 995–999.
113 Baskaya, Aysu, Wilson, Christopher, and Özcan, Yusuf Ziya, 'Wayfinding in an Unfamiliar Environment: Different Spatial Settings of Two Polyclinics', *Environment and Behavior* 36, no. 6 (November 2004), pp. 839–867, https://doi.org/10.1177/0013916504265445.
114 Ibid.
115 Ulrich et al., *The Role of the Physical Environment in the Hospital of the 21st Century: a Once-in-a-Lifetime Opportunity*.
116 Marquardt, Gesine, 'Wayfinding for People with Dementia: a Review of the Role of Architectural Design', *HERD: Health Environments Research & Design Journal* 4, no. 2 (January 2011), pp. 75–90, https://doi.org/10.1177/193758671100400207.
117 Harris, Paul B., et al., 'A Place to Heal: Environmental Sources of Satisfaction Among Hospital Patients', *Journal of Applied Social Psychology* 32, no. 6 (31 July 2006), pp. 1276–1299, https://doi.org/10.1111/j.1559-1816.2002.tb01436.x.
118 Ibid.
119 Ulrich et al., *a Review of the Research Literature on Evidence-Based Healthcare Design*.
120 Ibid.
121 Holahan, Charles, 'Seating Patterns and Patient Behavior in an Experimental Dayroom.', *Journal of Abnormal Psychology* 80, no. 2 (1972), pp. 115–124, https://doi.org/10.1037/h0033404.
122 Joseph and Ulrich, *Sound Control for Improved Outcomes in Healthcare Settings*.
123 Ulrich et al., *a Review of the Research Literature on Evidence-Based Healthcare Design*.
124 Ibid.
125 Majumder, Aditi, *Chapter 5 Percieving Color*, 2008, http://www.ics.uci.edu/~majumder/vispercep/chap5notes.pdf.
126 Brent Tofle, Ruth, et al., *Color in Healthcare Environments* (United States: Coalition for Health Environments Research, 2004).
127 Jacobs, Keith W., and Hustmyer, Frank E., 'Effects of Four Psychological Primary Colors on GSR, Heart Rate and Respiration Rate', *Perceptual and Motor Skills* 38, no. 3 (June 1974), pp. 763–766, https://doi.org/10.2466/pms.1974.38.3.763.
128 Tofle et al., *Color in Healthcare Environments*.
129 Daykin, Norma, et al., 'Review: The Impact of Art, Design and Environment in Mental Healthcare: a Systematic Review of the Literature', *The Journal of the Royal Society for the Promotion of Health* 128, no. 2 (March 2008), pp. 85–94, https://doi.org/10.1177/1466424007087806.
130 Ulrich, *Effects of Interior Design on Wellness: Theory and Recent Scientific Research*.
131 Staricoff, R., and Loppert, S., 'Integrating the Arts into Health Care: Can We Affect Clinical Outcomes?', in *The Healing Environment: Without and Within*, ed. Deborah Kirklin and Ruth Richardson (London: Royal College of Physicians of London, 2003).
132 Yehuda, Nechama, 'Music and Stress', *Journal of Adult Development* 18, no. 2 (June 2011), pp. 85–94, https://doi.org/10.1007/s10804-010-9117-4.
133 Joseph and Ulrich, *Sound Control for Improved Outcomes in Healthcare Settings*.
134 L. A. Gerdner, 'Effects of Individualised versus Classical 'Relaxation' Music on the Frequency of Agitation in Elderly Persons with Alzheimer's Disease and Related Disorders.', *International Psychogeriatrics* 12, no. 1 (March 2000), pp. 49–65.

Evidence-based design research, an evolving field
Since evidence-based design started offering insights and strategies to facility designers[135], it has received justified criticism for promoting solutions to the detriment of architectural quality. As professor Cor Wagenaar (University of Groningen) recently implied, "Architecture cannot be reduced to E-bD without it being destroyed"[136]. E-bD has insisted in breaking down the robustness of an architectural project into its elements, expecting to find parts that induce a specific effect or impact on individuals' preferences. A task that brings along a very complex multivariable and multidisciplinary problem, escaping the most skilled statisticians. As a result, when evaluating Healing Architecture, studies have attributed the healing process to measurable technical factors instead of spatial design quality.

Therein, E-bD has succeeded in offering a framework for technical solutions. Its rational and scientific approach for evaluation has potential to help architects within transdisciplinary teams, in together assessing problems and embarking in systematic research. This kind of exploration could permit artistic processes be recorded and verify if design as an output complies with the needs and requirements of problems; a viable path for Healing Architecture (see graph).

Both the actual shortcomings and potential of E-bD research could lead to its future development, in very different ways.

The statistical problem – of breaking down architecture into physical environmental factors – can partly be solved with machine learning (ML) technologies. Design processes in general, start with background data containing lists of factors and variables concerning a problem and frenzied sketching finding solutions. In trying to reach the "best" design possible, the sketching attempts are numerous, often restarting from zero when a new problem is commissioned. Apparently not an issue for the architect Renzo Piano who affirms: "one of the great beauties of architecture is that each time, it is like life, starting all over again."

ML systems find solutions using previous knowledge on problems by bridging extensive data bases from various sources. It is able to provide new insights without being explicitly programmed to do so[137].

Today these systems have reached a sufficient multivariate processing power, capable of offering optimal designs to the aerospace industry[138]. It avoids recurrent modeling procedures, which are extremely expensive and time consuming, by storing them for its convenient use when starting new tasks.

As for the field of architecture, Professor Patrick Hebron (New York University) affirms ML cannot replace human thinking or problem solving but sooner than expected will provide evidence to support the human decision-making process[139].

For Healing Architecture, ML could help cipher the multiple health and design-related variables – from complex health delivery processes – and propose initial spatial arrangements for designers to start with.

As mentioned earlier, E-bD can also leap forward in a less algorithmic manner, as a research and practice framework for environmental interventions. In developing architecture for health (as for many transdisciplinary projects), possibly the most demanding implication is to establish a tight scientific and artistic dialogue, free of translation issues.

To start with, what research is for artists, is fundamentally different for natural scientists, leaving architects more or less trapped in the middle. Systematic research is linear and straightforward, while design processes are sometimes ongoing and never-ending. In any case, there is a good chance for both to co-exist if we first recognise their particular differences and how these hamper communication and joint development.

---

[135] The Center for Health Design, *The Center for Health Design*, The Center for Health Design, 2018, https://www.healthdesign.org/.

[136] Wagenaar, Cor, et al., *Hospitals: a Design Manual* (Basel Birkhäuser, 2018).

[137] Sapp, Carlton E., 'Preparing and Architecting for Machine Learning', *Gartner Technical Professional Advice*, no. ID: G00317328 (17 January 2017), p. 37.

[138] Tan Wei Min, Alan, et al., 'Knowledge Transfer Through Machine Learning in Aircraft Design', *IEEE Computational Intelligence Magazine* 12, no. 4 (November 2017): pp. 48–60, https://doi.org/10.1109/MCI.2017.2742781.

[139] O'Donnell, Kathleen M., 'Embracing Artificial Intelligence in Architecture', *Embracing artificial intelligence in architecture*, (2 March 2018), https://www.aia.org/articles/178511-embracing-artificial-intelligence-in-archit.

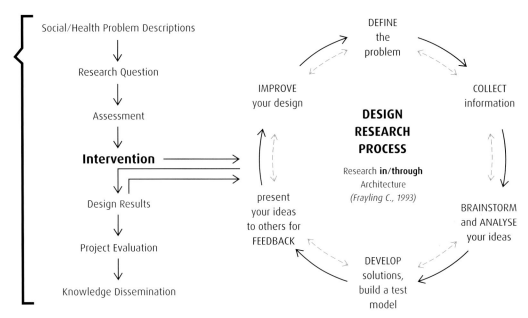

A major aspect is to reach a consensus of terms and terminologies, within health sciences and design disciplines. Architects enjoy an extensive lexicon of creative buzz words such as: pastiche, building envelope, fenestration, Corbusian, stylobate, permaculture, exurbia, blobitecture, and thousands more, describing a parallel universe. It is frequent for greater audiences to find architectural phrases and full sentences, just incomprehensible. There was no better way for me to illustrate this than citing the testimony of Greg Hudspeth, a long-experienced builder dealing with architects:

"as a builder who has been in the industry for over 20 years, ... I have a running list of words and phrases that the architects we work with are using. I spend a portion of each day stripping away the fluff and overly complicated explanations and descriptions for simple ideas. It is the biggest waste of time...[140]"

It seems that developing a communication process across disciplines is fundamental. Transdisciplinarity, as key for Healing Architecture, demands that all team members work together in early planning phases to understand social and health problems, relevant to the project, and formulate questions that seek being answered through design. Working together from the beginning definitely raises the stakes of having excellent results, it avoids information loss along the serial chain of specialists – very typical for conventional planning.

As proposed in the graph above, E-bD research includes architectural designs as experiments that obeying its own nature and laws. It allows non-designers involved in previous steps, to concede objectively whether "the experiment" affects health-related outcomes or not. Certainly an evaluation step most architects for health are unwilling to do especially when working in silos.

Public health as a discipline can also benefit from this kind of research. It is a field with difficulties in conducting studies that include environmental interventions, therefore knowing little about the effectiveness of designs on health. Its traditional approaches usually focus on individual dispositions and socio-economic factors rather than state, condition, and configuration of the physical environments (both natural and built) in which people live. The few studies that prove designed spaces (such as playgrounds) can enable and foster health behaviours (such as physical activity), also demand that more detailed analyses be made[141].

In E-bD research, we can ponder renewing knowledge between public health and urban studies to properly develop concepts until now lacking of scientific grounds e.g. healing gardens, healing landscape, and healing architecture.

Regardless of how E-bD will develop, its importance is critical for standards and policy. Sustaining Healing Architecture principles, scientifically, will be useful to inform competition briefs (as the ones prepared by The Danish Architects Association); and to redefine accreditation mechanisms, such as BREEAM Healthcare; LEED for Healthcare; and Green Star Healthcare (licensed by the Green Building Council of Australia).

In the German context, this kind of systematic research would aid the German Sustainable Building Council (DGNB – Deutsche Gesellschaft für Nachhaltiges Bauen e.V.) in developing its certification profile, called Neubau Krankenhäuser, which integrates Healing Architecture as a concept.

In improving its policy, E-bD research can update quality assurances on hospital designs such as ASPECT (A Staff and Patient Environment Calibration Tool) or the NHS knowledge-based assessments, which support governmental agencies and healthcare providers in generating building guidelines. Some have been initially advanced upon systematic reviews on healthcare design, commissioned in England, Denmark, and Holland between 2000 and 2009[142].

Evidence-based design research model for Architectural Interventions.
"Get your facts first, and then you can distort them as much as you please." – Mark Twain

140 Scott, Rory, *150 Weird Words That Only Architects Use*, 19 October 2015, http://www.archdaily.com/775615/150-weird-words-that-only-architects-use.

141 Lakes, Tobia, and Burkart, Katrin, 'Childhood Overweight in Berlin: Intra-Urban Differences and Underlying Influencing Factors', *International Journal of Health Geographics* 15, no. 1 (22 March 2016), p. 12, https://doi.org/10.1186/s12942-016-0041-0.

142 Lawson, B., and Phiri, M., 'Hospital Design. Room for Improvement.', *The Health Service Journal* 110, no. 5688 (20 January 2000), pp. 24–26; Frandsen, Anne Kathrine, et al., *Helende Arkitektur*, Institut for Arkitektur og Design Skriftserie, Nr. 29 (Aalborg: Institut for Arkitektur og Medieteknologi, 2009); van den Berg, *Health Impacts of Healing Environments; a Review of Evidence for Benefits of Nature, Daylight, Fresh Air, and Quiet in Healthcare Settings*.

Lekshmy Parameswaran | Jeroen Raijmakers

People-centered Innovation in Healthcare

Our lifestyles are increasingly out of balance and we are placing our health at risk with our unhealthy habits. We are ageing as a population and more likely to suffer from chronic diseases as we do so. As a result, our healthcare systems are under increasing pressure to deliver costly and complicated care. Yet, with their limited resources and traditional models, they are already struggling to meet existing demand. In short, the healthcare industry is in crisis and facing paradigm change. However, there are plenty of opportunities for innovation.

Our World in Transformation
We are living in a world that is undergoing immense changes, driven by socio-dynamic forces so rapid that we struggle to keep up, let alone feel any sense of control over our futures. These changes affect everything – from energy to transport, from food to health.

Every few hundred years in Western history there occurs a sharp transformation. Within a few short decades, society rearranges itself; its worldview (paradigm), its basic values, its social and political structures, its arts, its key institutions. Fifty years later there is a new world.
Peter F. Drucker, Post-Capitalist Society

As individuals we are surrounded by increasing convenience and choice, and expect to achieve more quality of life with less effort. 'Quick-fix' solutions fulfil our immediate needs; the short term is a lot easier to grasp when the future seems so unpredictable. Social structures are disintegrating and re-forming, family units are dispersing, yet there are more ways to connect with more people than ever before. Traditional institutions like the state, the Church and the hospital, which once helped us navigate our lives and make personal choices, are no longer equipped to lead us through this complex world; we are left feeling insecure and compelled to find our own new support systems. Technology might liberate us through timesaving solutions, but our bodies are beginning to show signs of the stresses and strains of such a lifestyle. We are becoming less 'productive', which in turn weakens the very economies that support such a way of life. This is not a sustainable situation.

An estimated 13.4 million working days a year in the UK are lost to stress, anxiety and depression, and 12.3 million to back and upper limb problems. 11.5bn Great British Pounds in 2002 was paid out in wages to absent employees and on additional overtime and temporary staff cover.
Paul Roberts, 2004 IHC report
'Absenteeism – Industry's Hidden Disease'

Our Lives at Risk
Amid all this change, do we really know how we feel? We are confused about our own priorities and about what makes sense for us as individuals. Unhealthy lifestyle habits have led to an increase in disease risk factors: over-consumption and a poor diet; reduced physical activity as we perform more desk-bound jobs; increased stress as we deal with change and overwhelming choice; and insufficient sleep and relaxation as we find it harder to unwind and switch off from the buzz.
Even if we do hear and recognise the warning signs of our ill health we may not know what to do about it, and even if we do, we may not be convinced that we can change our own lives, so accustomed have we become instead to adapting ourselves to suit our circumstances. We are an increasingly grey population, and are generally living longer – albeit afflicted with chronic diseases. This is true for both the developed as well as the developing world. The adaptations we are required to make to adjust to new health levels, as we age or develop chronic illnesses, are amplified in light of these socio-cultural changes.

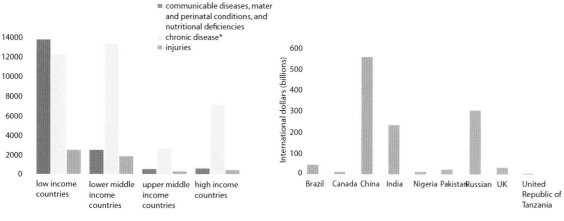

Graphs showing a rising incidence of diseases in both the developed and the developing world: diabetes mellitus, stroke, cardiovascular disease, obesity and cancer

Projected deaths by major cause and World Bank income group, all ages, 2005

Projected foregone national income due to heart disease, stroke and diabetes in selected countries, 2005–2020

Our Healthcare Systems at Breaking Point

These dynamic forces of change batter the healthcare industry from different angles. As the population becomes ever larger, more widely distributed and diversified, and as we suffer the costly consequences of increasingly damaging lifestyles, the result is ever greater pressure on a system already straining to breaking point. Traditional models of healthcare are breaking down and being challenged as they simultaneously face pressure to adapt while confronting several systemic obstacles to change.

Participants (in healthcare) compete to shift costs to one another, accumulate bargaining power, and limit services. This kind of competition does not create value for patients, but erodes quality, fosters inefficiency, creates excess capacity and drives up administrative costs.
　　　　　Michael E. Porter and Elizabeth Olmsted Teisberg, Redefining Health Care

Clinical teams are facing increased demand to perform more efficiently, consistently and safely in delivering improved outcomes. They need to be able to stay up-to-date with clinical advances, and communicate effectively with patients as well as an ever more networked global peer group. Workforce recruitment, retention, burnout and ageing are all factors confronting today's healthcare providers.

The nursing workforce is ageing. Over the next 10 to 15 years (industrialised) countries will experience a large exodus of nurses from their workforce as nurses retire just at a time when demand for nursing and healthcare is on the rise; one of the reasons being the growth in the older population.
　　　　　International Centre for Human Resources in Nursing, An Ageing Nursing Workforce

Hospital management teams face impossible choices to balance cost and quality of care. Healthcare processes tend to be complex, filled with redundancies and characterised by bottlenecks that ultimately affect quality and effectiveness. League tables and new quality-of-care targets pressurise facilities to perform in new ways and with tougher financial and legal constraints. Healthcare institutions are steeped in tradition, organised in silos and clearly-defined professional hierarchies. As care shifts outwards into new areas, delivering optimal care means working across settings and with new care roles. This requires time-consuming organisational change, to create an open and innovative mindset that can foster new modes of professional collaboration. Hospitals were once the sole portals of healthcare, reflecting the identity and lifestyles of the communities they served. Now, however, they must differentiate their services from new competitors such as pharmacies and standalone clinics that occupy the healthcare space between hospital and home. Hospitals strive to formulate distinct healthcare brands in order to generate loyalty among both patients and staff. People bring their expectation for choice with them when they access healthcare services. They may confidently challenge clinical opinion, shop around globally for healthcare solutions and consider alternative approaches. Yet, at the same time, they need support in personal healthcare decision-making because they still experience a certain degree of anxiety in this new, fragmented care landscape. It would be no exaggeration to conclude that, given these factors, the global healthcare industry does indeed face a crisis, a paradigm change.

The Innovation Challenge for Healthcare

Today's healthcare landscape is a challenging arena. Organisations are in search of successful and sustainable innovation strategies to differentiate them from the competition and create viable solutions that offer improved healthcare experiences for patients and care

Hospitals

Nebula lighting

Ambient experience

providers in the short to longer-term. At the same time the financial underpinning needs to be sustainable. Many challenges in healthcare demand a diverse mix of skills, knowledge and competencies which is beyond the capability of most individual businesses. Companies therefore have to think in terms of new models of innovation that include partnerships, acquisitions or strategic alliances to equip themselves for the healthcare challenges ahead.

Design is no longer just about form but is a method of thinking that can let you see around corners. Design Thinking is the new Management Methodology.
Bruce Nussbaum, Founding Editor, Innovation & Design, Business Week

Healthcare Innovation Context
Traditionally the main driver of healthcare design had been in the professional domain, with a focus on the needs of the clinical end-user. However, as new societal needs emerged and reshaped healthcare:

- Healthcare designers, working to improve the human experience in healthcare, intuitively realised that to truly impact patient experience in a hospital they would need to design beyond the boundaries of the medical equipment and consider the design of the environment around the machine. Ambient Experience started as an internal design research project which explored the value of such a design strategy.

- Researchers who had been tracking global socio-cultural changes began to see the emergence of a whole new field of opportunity in personal healthcare. An internal design research project was initiated to explore new value propositions for Philips, one of the key players in the field of healthcare design. It was also decided to invest and participate in a 4-year EU-funded consortium research project called *MyHeart*, whose aim was to develop personal healthcare applications for the prevention and management of cardiovascular disease.

- Interaction and product designers researched methods to take people's experiences as the starting point for innovation in design. For example, Nebula is an interactive projection system designed to enrich the experience of going to bed, sleeping and waking up. The aim was to create an atmosphere in the bedroom that encourages and enhances rest, reflection, conversation, intimacy, imagination and play.

The way in which Philips Design detected and translated the trends taking place in society and culture demonstrates two key strengths that have innovation value; firstly, an innate ability to comprehend socio-cultural change from a human perspective, and secondly to translate this understanding into tangible value propositions.

Key Characteristics of a successful Innovation Mindset
Philips Design has developed a people-focused innovation approach that is driven by qualitative research, and which applies design thinking and skills to identify and respond to innovation opportunities across the full spectrum of healthcare, from consumer to professional. It is a flexible approach that can serve a range of business processes from pre-development and business strategy through to product development and brand communication. It also supports both short- and longer-term innovation horizons.

Focus is on creating the optimal conditions for innovation in healthcare, which has more to do with team dynamics and mindset than standard processes and tools.

Some of the key innovation questions within healthcare include:
- Why do people struggle to change their lifestyles to live healthily, even when they know the facts about healthy living?
- What stops care teams from being able to deliver optimal care and how can obstacles to improved quality of care be overcome?
- Where are the cost bottlenecks in care delivery? Why and how do they form? Can they be eased?
- What will people want from clinical technologies such as on-body bio-sensing or genetic screening? How will this impact professional practice?
- How can healthcare stakeholders make strategic decisions when the forces of change in healthcare are so diverse and the future so unpredictable?

Here we describe the key characteristics of a successful healthcare innovation mindset and introduce the approach and a set of methods and tools used for healthcare innovation research, design and consulting. Project examples will be used as illustration.

Challenge
Consider people and their health in the context of daily life, and seek to understand the real impact on lifestyle over time.

Confront
Untangle the complexity of each healthcare journey without trivialising or over-simplifying the clinical or lifestyle context.

Collaborate
Form multi-disciplinary teams with knowledgeable stakeholders: patients, families, patient organisations, clinicians, insurance companies, technicians, marketing professionals, engineers, designers and market researchers.

Trust
Ensure the flexibility for appropriate disciplines to lead certain phases of the innovation process when this can enable the team as a whole to take the next step.

Act
Watch how the healthcare landscape is changing, learn about the drivers and obstacles, and advise partners about taking managed risks.

Learn
Assess the impact of insights, concepts and prototypes on human experience, clinical outcome and business metrics in order better to define and achieve innovation success.

Understanding a changing Healthcare Landscape
Healthcare trends research in Philips Design has become more interactive and participatory, often combined with insights research and innovation design activities to make better sense of complex trends. Collaboration with experts from relevant fields is essential to be able to gain a quick, in-depth understanding of a healthcare topic. Typical desk research and expert interview tools are often supplemented with multi-disciplinary working sessions with experts and professionals who bring additional relevant trends from business, clinical science and technology. This allows initial insights to be validated and enriched to form a more robust and shared set of hypotheses and assumptions. The intention of trend studies is to provide broad and rich inspiration and research-based directions for innovation teams to consider; this is often a backdrop to in-depth people and context research. The translation of trends findings into interactive innovation tools, such as framework posters or trends cards, allows teams to interact with the content, play with it and trigger critical thinking. The Future Hospital was a project that mapped the future healthcare landscape onto the historical timeline of hospitals. Analysing several key trends in lifestyle and healthcare – and considering a proprietary set of key socio-dynamic forces shaping our world – a framework was established to describe the possible ways in which hospitals could evolve in future. The intention was to offer foresight to guide the evolution of the strategic relationship of Philips with hospital customers, as well as a broader context in which to understand its business in an evolving healthcare landscape. Personal Healthcare Landscapes 2016 was a trends activity in the My Heart project. Combining expert interviews, desk research and a one-day workshop with healthcare policy and industry experts, Philips Design facilitated the shaping of four future healthcare landscapes. The landscapes were used as a fore-sighting tool, to challenge and sharpen

Hospitals

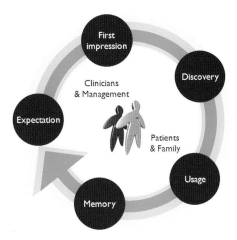

A model to understand people's healthcare experiences in a clinical setting, from expectation through to memory

a number of cardiovascular healthcare propositions and assess their business potential in an uncertain future.

The landscapes capture many of our hopes and fears about our future and will hopefully give policy makers food for thought as they try to make choices for us.
Business Expert quote, My Heart

Socio-cultural scanning techniques were also applied to shed light on the context of cancer in the Ambient Experience Oncology project. This led to a broader understanding of the perceptions of cancer in society and provided a valuable backdrop to in-depth investigations with cancer patients and care teams. A trendscape of selected key trends, validated together with experts, was used to discuss the possible impact and consequences of disruptive trends for the overall business landscape relating to colorectal cancer screening.

Engaging and understanding Multiple Stakeholders
Health can be an emotional and confrontational topic that reveals people's insecurities and leaves them feeling vulnerable. Talking to patients and their loved ones about healthcare experiences requires specific skills and sensitivity. Healthcare is by nature an unpredictable climate and one that can contain challenging circumstances or contexts. Working with care teams whilst they are delivering care requires credibility, flexibility and discretion.
Faced with such challenges, Philips Design has developed a number of appropriate methods and tools to understand people and health.

Research Tools to understand Needs and Values
Research tools have been developed that use the best of both people research and design research techniques. Simple, visual tools are most successful at empowering patients and care providers to share their stories in safe, easy and yet enriching ways. People become the creators of their own narrative. Some patients have described research sessions as comforting and cathartic.
For clinical care teams, generative sessions using visual and interactive tools help to reveal bottlenecks in care delivery that they might feel individually and subjectively but struggle to pin point specifically and collectively. Visual research tools provide a shared language between research and design disciplines, assisting in the more accurate translation of research findings into design directions.

Capturing Experiences over Time
To consider healthcare experiences realistically, research teams need to account for changes in needs, attitudes and behaviour over time. Shadowing and ethnographic observation tools have been effective at capturing key moments in the healthcare journeys of both patients and clinical staff. Imaginaries have also proved to be powerful tools which allow people to describe their personal transformations – physical, emotional and social – as their health changes. Panel session exercises with patients and staff have used the timeline interactively, asking people to project their issues, needs, thoughts and desires directly onto the framework. By sharing the notion of a timeline, researchers and designers find another common vocabulary to increase collaboration and shared understanding.

Research Tools to evaluate Insights and Concepts
Philips Design's multiple-encounter approach is a valuable tool for experience research, and has been used effectively with patients in personal healthcare research activities. It allows researchers to capture the chronological development of people's needs, issues and behaviour. It also fosters dialogue to create expert user groups, building the trust that can lead to deeper, richer insights. As a project progresses, insights, concepts and even prototypes can be tested through this methodology

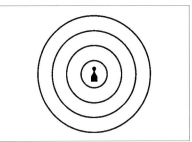
Template

Exercise execution

Exercise completion

Imaginaries allow patients to describe with candor and dignity the sensitive and painful changes they experienced during their cancer treatment (source: Philips Design)

Societies

	Empowered Society	Communal Society	Experience Society	Fear Society	Irresponsible Society
Summary Description	Self-referential, Explorative, Change lifestyles, Fitness & success	Belonging, New networks, Co-creation, Multi-disciplinary	Journeying, Adventure, Sensing, New things	Surviving, Control, Being secure, Outside threats	Egocentric, Avoid ownership, Wasteful, Splurging
Key Societal Drivers	Technological developments, Knowledge economy, Cost pressure on healthcare	Breakdown of safety networks, Open source & innovation, Towards a care-continuum	Rise of the Creative class?, Mental stimulation overload, Reconnection with the body	Speed of change & developments, Emerging economies	Affluenza, From nurture back to nature?

Healthcare matrix

		Empowered society	Communal Society	Experience Society	Fear Society	Irresponsible Society
Healthcare response		Consumer care	Connected care	Pampering care	Reassuring care	Public care
Hospital role		Flexible Supplier	Facilitator	Hospitality	Health Icons	Health planner
Hospital form		Catalogue shop	Osmosis	Stage vibrant	Safe haven	School
Patient view		Customer King	Expert patient	Care Receiver	Victim	Indulgers
Professional view		Service Provider	Expert teams	Steward - Host	Guardian	Herder
Possible directions		Health Avatar	Cultural Care	Experience groups	Total care chains	Health Education
		U 2 Care, Care 2 U	Osmotic Spaces	Health journey	Branded care	Mass health screen & planning
				Fun flow	Genetic fears	Mobile health Regions

The future hospital

to allow for iterative enrichment of new propositions. As the healthcare industry faces new measures of clinical outcome, quality of care and performance-related incentives, there is an increasing shift towards evidence-based approaches in healthcare design. Qualitative and quantitative methods to assess the impact of insights, concepts or solutions are carried out with third-party research agencies to offer objective evaluation according to industry measurements and standards, such as those being developed for clinical experience testing by the Center for Health Design in the USA.

Models to understand People's Experiences

Philips Design has created three main models to support understanding stakeholders in healthcare. They provide valuable starting points in the scoping and analysis phases of a healthcare innovation project. Used in combination, these three models offer businesses a powerful and flexible framework for understanding people's needs in healthcare for any disease, context or situation. The models take into account that healthcare experiences impact lifestyle, that people need support during any type of health transition, and that healthcare journeys often take place across many different locations yet people perceive and recall the experience as a single entity.

Capturing Care Contexts

Understanding people's experiences and needs demands ways to also capture their environmental context and conditions. This includes the pace, rhythm and flow of activities and behaviour as well as specific contextual qualities of multi-sensorial experience: lighting, textures, sounds, layout, objects, style, signage and scents. As care settings expand beyond the hospital to include new areas, so too do research tools need to evolve to effectively capture and communicate the diverse qualities that shape the experiences people have when staying in, working in and visiting these spaces. With the realisation that design could positively impact people's healthcare experiences by also considering the healthcare environment itself, design research techniques were integrated to enrich and deepen the level of contextual research. Research teams were created in which new mixes of disciplines were brought together in order to get a better grasp of the context. Architects and interior designers were able to draw connections between spatial needs and design qualities, whilst interaction designers saw links between activities, spaces and tools. Ethnographers provided yet another layer of contextual information, using observation to identify people's behavioural patterns in relation to healthcare contexts. A context is also characterised by the pace of experience and activities that take place; an A & E department at midnight in a 24-hour city has a different pace from a chemotherapy waiting room at midday or a neo-natal intensive care unit in the early hours of the morning. Philips Design's healthcare innovation teams could sense the differences intuitively and developed new audiovisual ways to capture and communicate these insights to understand the context at a deeper level.

Creating Experience Flows

Different layers of insights – stakeholder-related, contextual, informational, clinical, economic and technological – are gathered during the research stage. Philips Design generates the stakeholder and contextual insights itself, but other layers of insights are collated from the project team which includes clinical scientists, technicians and economists. Synthesis of these multiple layers of information can be complex, and analysis often requires a collaborative, iterative and multi-step approach. Co-analysis with project partners, client team and sometimes healthcare stakeholders can be most effective, bringing experts together to process insights and define strategic conclusions. Philips Design translates such insights in a distinct and unique way compared to other innovation disciplines. Design thinking helps to make sense of these multiple layers of insights

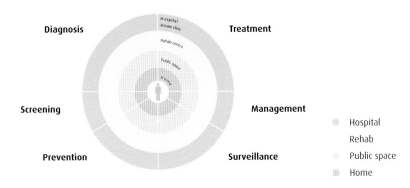

Patient experience zones as a *care landscape*

Experience design considers people, spaces and enables over time

and allows the processing of what can often be complex, inter-dependent findings. Two key principles guide this synthesis and articulation of insights – to be interactive and to be visual. For example, visual representation can reveal further depth of insight into innovation territories by combining multiple perspectives and creating a holistic view the human experience of the situation.

The Experience Flow poster is a healthcare innovation framework and the primary tool used to articulate multifaceted insights. Complex healthcare journeys can be visually represented, discussed, challenged and enriched by multidisciplinary innovation teams. In essence the tool allows the team to walk along the healthcare journey, as seen through the eyes of multiple stakeholders: patients, loved ones and clinical care teams. This framework is successful at making sure that everyone in the innovation team is properly aware with regards to what they know and what they realise they do not know or need to know. As such, it can serve as a stimulus for teams to further share and exchange knowledge, adding to the framework and helping complete the picture of understanding together. Supporting tools can be used to articulate specific layers of insights: mood boards, stakeholder maps, personal profiling, issue cards and trendscapes can all be created and used flexibly depending on the project requirements.

Developing Value through Design Skills

Designing for Healthcare Experiences

Experience design combines several disciplines to create a total system solution which harnesses product, interaction, user interface, environment and multimedia design. Using a proprietary experience design model – which considers healthcare experiences from expectation to memory – the idea is to create the right combination of people, spaces and enablers over time. Enablers can include technology services, services, and contextual elements as sound and light.

Healthcare designers are involved in the research stage of innovation projects, so that during the design phase they can empathise with stakeholder experiences and translate needs into solutions which maintain the context of the original insight. Through synthesis workshops they collaborate with the multi-disciplinary innovation teams to identify areas of opportunity that emerge from the Experience Flow framework where innovation could bring positive change. These areas then become the focal points for Philips Design's in-depth ideation and concept development.

There is a certain degree of magic required during the creative process, so it is important to establish the conditions in which a team feels relaxed, informed, inspired and free to think outside the box. It also helps to provide relevant yet surprising creative triggers as described in the previous section. These are fed into the ideation process at key moments to maintain the creative pace.

The results of the design phase can be on a product, systems or strategic level. As this essay began by explaining how healthcare is facing systemic and paradigm change, healthcare innovation projects increasingly deliver system-level solutions and visions. This means that existing touch-point point solutions are integrated into future visions to enable a feasible development roadmap. Storytelling is the ultimate result of ideation. 2D storyboards, animated Flash demos or 3D renderings are used to communicate initial experience scenarios and proposed new healthcare experiences.

Developing an idea through hands-on prototyping allows designers to get a feel for the experience at firsthand. This is carried out in close collaboration with technicians, engineers and scientists, as well as external technical experts from diverse disciplines such as fashion and textiles, graphic animation and computer generation. The tools, techniques and possibilities offered by these cross-disciplinary collaborations often serve to further stimulate creativity and innovation opportunities.

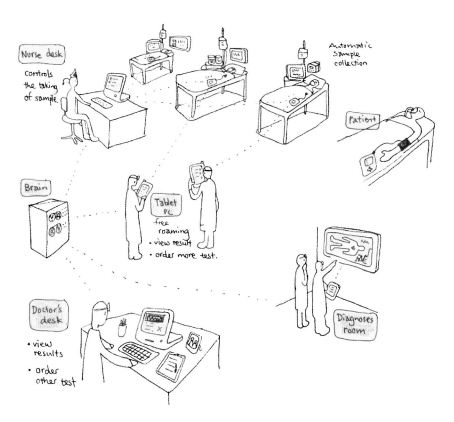

Storyboard scenario from a research project on molecular diagnostics

The human aspect of healthcare experience guides the ideation phase and the translation of insights to design themes, qualities and solutions. However, there is another important factor at work within the healthcare design team; a shared set of professional and brand-related values that safeguard the human-focused integrity of the solutions.

Communicating measurable End-User Value
We assume that applying business values to healthcare comprises fundamental human values, essential clinical values, brand value and – possibly most important of all given the cost constraints on today's healthcare systems – value for money. In this final section we share two key case studies from Philips Design's healthcare innovation portfolio that were developed using the innovation approach described above. They are examples of how Philips Design can contribute to measurable end-user value in healthcare through the creation of tangible design propositions for clinical professionals as well as for patients and their loved ones.

Ambient Experience Paediatric CT Suite
In 2004 Philips Design was asked to improve the paediatric CT scanning experience. Advocate Lutheran General Children's Hospital (Chicago, US) participated in a co-creation process. A typical examination room is cluttered with clinical equipment, including a scanner that could easily appear frightening to a small child. The task of the CT technician is to ensure that as many children as possible complete a successful examination quickly and safely. Children need to lie still and hold their breath during the procedure to ensure optimum image quality. This is difficult for many sick children who are likely to be in pain, restless and upset. Sedation is used but this compromises patient safety and affects throughput time for the procedure as additional recovery time is needed. Applying the concept of human-focused healthcare and a research-driven innovation approach, Philips Design proposed a new CT system design to support children, their parents and the CT technician from patient preparation to completion of the examination. Two powerful insights to improve the child's experience came from research with paediatric specialists: narrative engages children in experiences and provides a natural role for parents in storytelling; and if a child understands what is going to happen and why, in simple terms, then he or she is more likely to be compliant.

The Kitten Scanner is a toy CT scanner. By choosing a toy and placing it into the scanner, children are shown an animated story to help them understand the procedure in an entertaining way. They can see that if the toy is

Hospitals

Reading room 20/20 I-space

Reading room 20/20 We-space

Reading room 20/20 U-and-I-space

The kitten scanner

Next simplicity's celebrating pregnancy experience demonstrator was created through an iterative process of interaction prototyping to determine the optimum way of conveying 'sense and simplicity'

shaken in the scanner the image distorts so they know they must lie still to get a good image. The toy selected by the child is used to trigger personalisation of the examination room with animated projections and lighting effects. The technician can use these effects to guide a child through the procedure. Independent qualitative and quantitative evaluation proved that the solution had significant clinical and experiential impact. Sedation rates for children aged 3–7 years were reduced by 30–40 per cent, which represented a marked improvement in patient safety. Meanwhile efficiency/throughput of treatment planning was improved by 15–20 per cent.

Reading Room 20/20, RSNA 2007

In 2007 Philips Design created an experience demonstrator to communicate a vision of the reading room of the future for the Radiological Society of North America's (RSNA) exhibition. A reading room is a professional space in a hospital's radiology department where radiologists read imaging tests to provide diagnoses. It is typically a quiet, dark and busy environment shared by several radiologists, although it should be accessible to serve all the hospital's clinical needs. Philips Design carried out seven hospital visits to gather stakeholder and contextual insights. We identified a need to support personalisation of workspaces, remote collaborative working, improved mental and physical comfort and the possibility to concentrate on work alone as well as in conference with peers. Insights were translated into three key design strategies: an I-space to create a personalised atmosphere, a U-and-I-space to support collaboration with colleagues, and a We-space to support remote collaboration. A prototype enabled visitors to access a tangible experience of this vision of the future.

Radiologists can personalise the lighting, sound levels and temperature of their workspace. They can use voice control, gestural and touch screen interfaces projected onto the reading room tabletop to view images alone or with colleagues. They can share image files, manage their workflow and consult peers via a video-conference system projected onto the room wall. Intuitive interaction and integrated technologies enhance comfort, reduce stress and improve concentration and quality of work experience. Philips Healthcare evaluated The Ambient Experience Reading Room of the Future quantitatively and qualitatively, with over 200 clinical visitors to the demonstrator at RSNA 2007. They were positive that the concept would improve their overall work experience in terms of a quieter workspace in which to concentrate and improved ways to share clinical data and collaborate remotely with peers.

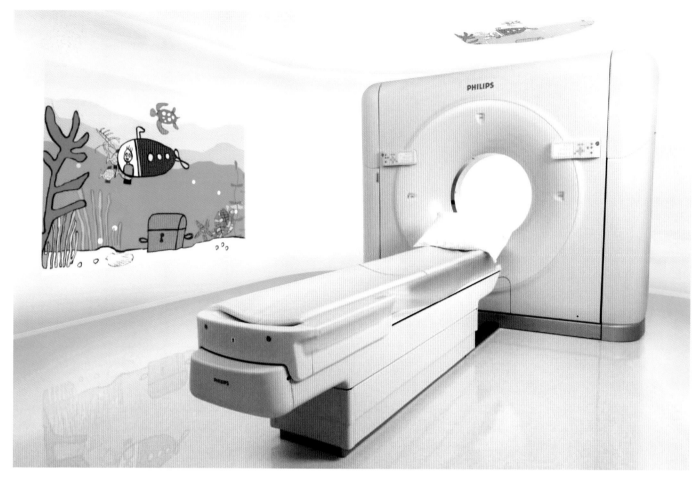

Advocate Lutheran General Children's
Hospital, Chicago/USA
(all sources: Philips Design)

Right page:
Patient receiving a medical scan
(photo: Mark Kostich)

Conclusions

The healthcare industry is in crisis and facing paradigm change. Within this context of paradigm change lie opportunities for innovation. The research-driven innovation approach in healthcare delivers business value by:
· Understanding the changing healthcare landscape to ensure innovations;
· Involving patients, families and clinical staff in the innovation design process to deepen the value of solutions and accurately uncover needs, issues and values;
· Carrying out on-site research which supports the belief, understanding and commitment of multi-disciplinary innovation teams;
· Layering stakeholder, contextual, clinical, economic and technological insights to build up a rich picture of the healthcare innovation landscape;
· Applying design thinking to unlock complexity and design skills to reveal opportunities for innovation;
· Translating insights into tangible and experiential propositions with design qualities that maintain integrity with respect to human needs;
· Promoting a bold and open attitude to innovation that gives room for each discipline to excel and enjoy the freedom to innovate.

The business benefits include:
· Creating a common vision to generate strategic alignment;
· Relevance for a range of business processes and innovation timeframes;
· Leveraging the collective knowledge and skills of a multi-disciplinary team;
· Generating the confidence to make choices within a changing healthcare landscape;
· Engaging healthcare customers in co-creating their future;
· Creating credible research-based consultative selling tools;
· Creating different branding opportunities for hospital customers.

The innovation approach continues to evolve as Philips Design's healthcare teams apply it in new global markets, work with new stakeholders, partners and clients, investigate new areas of healthcare and identify opportunities for new technologies. With the healthcare landscape changing continuously, there is no shortage of innovation challenges nor opportunities to improve people's experience of healthcare.

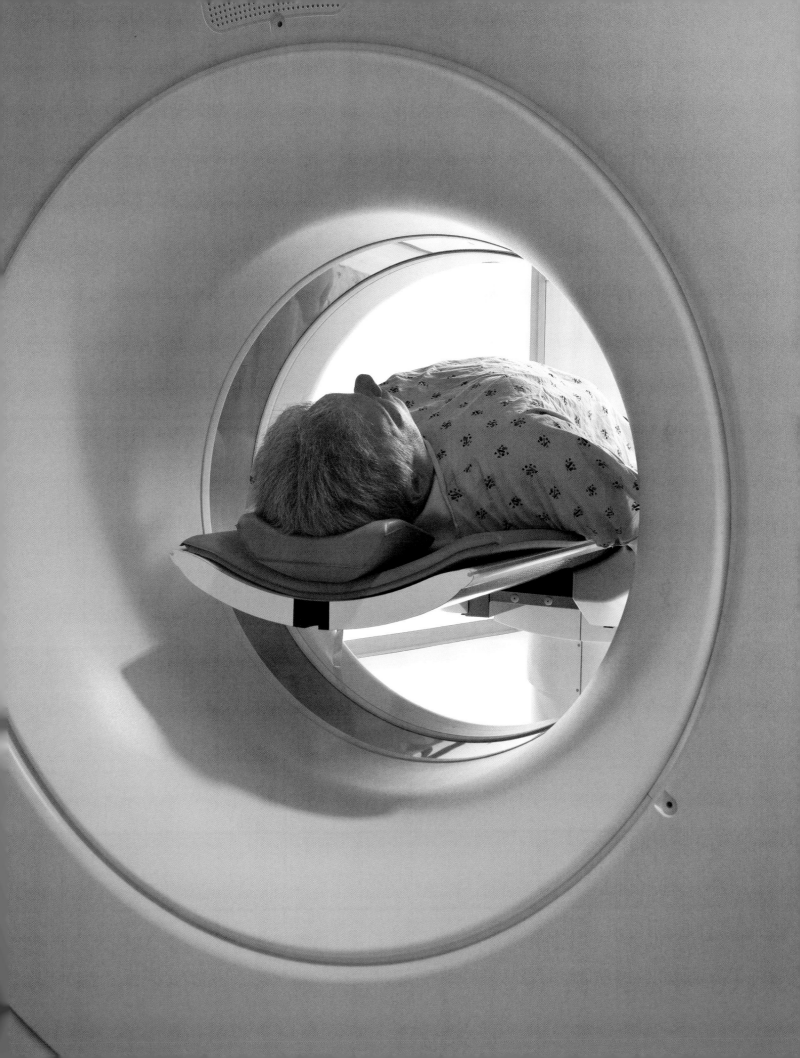

Peter Pawlik | Linus Hofrichter

The Out-Patient Department

Definition and History of Out-Patient Departments

If one searches for the term "ambulatory" or "out-patient area", in the standard text *Subdivision of the Hospital into Functional Areas and Posts,* one won't find anything. The DIN (German Industry Norm) 13080 is far too influenced by the West German past practice of strictly separating ambulatory and in-patient areas. According to Axel Hinrich Murken's textbook of medical terminology[1], the Latin root of this term, *ambulans,* means going around, taking a walk. *Meyer's Encyclopaedic Lexicon*[2] explains "ambulatory" as "wandering, roving about, not tied to a specific place, without a fixed residence". Ambulatory or out-patient treatment is defined as "passageway treatment in the doctor's practice or in a clinic, in which the patient looks for the doctor, as opposed to in-patient treatment". In the first volume of *Meyer's Lexicon,* published by the Biographical Institute in Leipzig in 1924[3], we read the following under the entry "Ambulanz": "Roaming about, the slightly mobile facility in war medical care for first aid, first established in the 15th century by Isabella for the Catholics, then later introduced in other armies. In the Germany army the expression is no longer used. Here the term "Ambulanz" designates the field dressing-stations for the troops, as well as the emergency corps and the field hospital."

Despite this dictionary definition, one can establish that walk-in or out-patient areas were indeed in use at this point in Germany and Austria, and were even newly built then. An example is the AEG-Klinik in Berlin on Hansaplatz. Its architect was the famous government master-builder Heinrich Schmieden, who wrote a series of publications and a book on hospital architecture in 1930."Out of an old private sanatorium located in the Hansaviertel part of town, the works sickness fund of the General Electricity Association (AEG) and its daughter societies had an out-patient hospital with clinic built in 1926 which had 65 beds on two storeys ready for the admission of health plan patients."[4] Thus the hospital with its in-patient treatment was economically combined with an out-patient area. The existing diagnostic and therapeutic facilities such as x-ray diagnosis, x-ray therapy, laboratories and a light-therapy area were available to ambulatory as well as bed-bound patients. Another early example of a walk-in facility is represented by the workers' health plan of the committee of the Vienna Business Association. The building, constructed by architects Fritz Judtmann and Egon Riss in 1926, combines an office building with an out-patient hospital in line with the hygiene regulations of the period. A centrally placed stairwell with direct access to large lounges offers an easy overview to the visitors to the therapeutic department on the ground floor and second floor. The offices and business areas are on the third and fourth storeys. The striking feature of this building is that it has a septic and antiseptic operating room, but no other nursing facilities apart from a recovery room.

Development of the Hospital Out-Patient Department since 1945

Hospital out-patient departments developed differently after 1945 in a divided Germany. In the Federal Republic, walk-in treatment was usually carried out by resident doctors in their own practices. Except for university clinics, walk-in treatment was the exception in hospitals. A necessary change from a walk-in to in-hospital treatment was therefore always bound up with a change from the local physician to a hospital doctor. This change of physicians could be avoided in the Federal Republic when treatment could be carried out in the practices of the referring physician and any requisite in-hospital treatment then followed under contract. Regardless of the legal and insurance issues arising from out-patient treatment in hospitals, groups of rooms were also created in hospitals to allow for the examination and treatment of both bed-bound and ambulatory patients. A proven principle for the successful combination of these areas is the two-corridor solution, with one ambulatory and one bed-bound side of access, as in the St.-Joseph-Krankenhaus

1 Murken, Axel Hinrich: Lehrbuch der medizinischen Terminologie. Stuttgart 2003
2 Bibliographisches Institut: Meyer's Enzyklopädisches Lexikon. Mannheim/Vienna/Zurich 1971
3 Bibliographisches Institut: Meyer's Lexikon, Leipzig 1924
4 Schmieden, Heinrich: Krankenhausbau in neuer Zeit. Kirchhain 1930

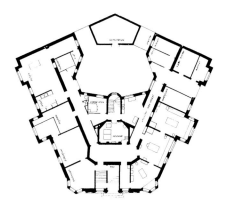

AEG-Klinik Berlin, ground plan,
architect: Heinrich Schmieden (1926)

AEG-Klinik, Berlin, street view,
architect: Heinrich Schmieden (1926)

in Berlin Tempelhof (architects: Planungsring Dr. Jüchser + Pawlik) completed in 1982. In the German Democratic Republic on the other hand, there was a considerably closer link between out-patient departments, the so-called polyclinics. These were mostly specialist medical out-patient facilities in (district) hospitals which generally covered the specialist medical spectrum of the given hospital. They were found in close proximity to the hospitals and could therefore make use of all the hospital's facilities. Repeat examinations of patients were generally avoided because the patient's medical records were available to the hospital in case further treatment was required. The administration was also connected to the hospital. During the mid-1960s the United Health Facilities (VGE) were set up in rural districts. All the health facilities of a district, from the hospital with its polyclinic, the state doctors' and dentists' practices, the provincial walk-in centres, homes and nursing homes for the aged, children's day-care centres and children's clinic facilities were under a central administration. The close linkage of the polyclinic with the hospital remained. The use of the infrastructure of the hospital also continued, especially of the central laboratory and x-ray departments through the polyclinic doctors' practices. Until the formation of the VGE, the specialist departments of the district polyclinic were usually led by the head doctors on the wards there; they held consultations with the medical service under their administration. There was a close intertwining here between out-patient treatment and possible in-patient care under the same medical personnel. Important specialist departments in the polyclinic were gradually filled by specialist physicians on a full-time basis. These departments primarily included surgery, internal medicine, gynaecology and paediatrics; general medicine had already been occupied on a full-time basis since 1949/1950. Those doctors active in the polyclinic worked at least one day a week in the in-patient area as well. In 1982 the VGE was dissolved on orders from the ministry. The hospital together with the polyclinic was transformed into a "District Hospital Functional Unit – District Polyclinic". The functional unit was an autonomous structure with its own administration under the direction of a medical and a finance director. The state doctors' and dentists' practices and all other facilities were under the direct oversight of the district physician under the central organisational leadership of "Ambulatory Public Health". With this new structure, the hospital with its polyclinic attained even greater importance for the medical care of the population. During the entire time-period represented, the so-called referring doctor system continued. Specialist physicians from state specialist practices, such as in the area of ophthalmology, as well as doctors from private establishments (ear, nose and throat specialists) made use of the operating room facilities including the complete infrastructure of the hospital and polyclinic.

The Situation of the Out-Patient Department Today
For years, discussions in the field of public health have been determined by constantly rising costs. Critics of previous health systems with a relatively strict division between the out-patient and bed-bound areas have long pointed to successful international models. Some Scandinavian countries and the Netherlands have successfully practised a flexible hospital system fluctuating between the ambulatory and stationary areas for a long time. Resident doctors already partially cover out-patient needs in the hospital and continue to treat their patients in case of a continued in-patient care, similarly to the earlier practice in the GDR. Another advantage of this system is that high-quality medical equipment, usually already there, can also be made available for out-patient examinations and is therefore put to more efficient use. The high proportion of out-patient departments in hospitals has many causes. These are partially to be found in demands from the social statute book, but also partially in the justifiable concern of hospital administrators that they remain capable of competing in the future market. The reason is the introduction of a new account

Arbeiterkrankenkasse, Vienna, ground plan, second storey, architects: Fritz Judtmann, Egon Riss (1926)

Arbeiterkrankenkasse, Vienna, Reception room on third storey, architects: Fritz Judtmann, Egon Riss (1926)

Arbeiterkrankenkasse, Vienna, street view, architects: Fritz Judtmann, Egon Riss (1926)

settlement system in Germany for hospital services. In 2003, Diagnosis Related Groups (DRG) were introduced as a lump-sum payment system according to section 17 b of hospital law (KHG). The Australian Diagnosis Related Groups provided the models in this case. According to this system, the payment for a large proportion of hospital services is now fixed. The payment remains the same regardless of the length of a patient's stay for the treatment in a given case. Formerly a hospital collected payment, according to the established daily rate, for as long as the patient remained there. Already today it has been established that the duration of patients' stays in hospitals has become significantly shorter since the introduction of the DRG.

Hospital experts predict a reduction of beds of 20–30 per cent over the coming decade. If this development does indeed take place, a proportion of hospitals will not be able to survive economically. It is therefore all the more understandable that hospitals are considering using empty space for other purposes. For example, the German Hospital Society (DKG) has recommended that the approximately 2,200 German hospitals develop service centres and also present a broad spectrum of out-patient services alongside in-patient care. In many places, either out-patient departments have already been integrated in hospitals or built next to the hospital as independent service centres. Cooperation with resident doctors, who receive attractive rental offers in the immediate vicinity of the hospital, is also part of the trend. They offer the possibility of using the expensive medical-technological infrastructure of the hospitals, thereby making it more economical and also generating additional in-patients from this environment.

Hospital Planning in the Age of the DRG
Today's hospitals must be characterised by a perfect patient-flow control system; the optimal organisational forms resulting from this have had architectural consequences as well. The classical terms of "emergency room" and "out-patient" are no longer applicable; today one speaks of a central admissions area (ZAB in German), a central emergency admissions (ZNA) or of an interdisciplinary admissions unit. We are experiencing a renaissance of the admissions station directly connected to the central admissions area. Medical inspectors today determine the processes, also pointing out where financial losses are caused on which places in badly structured hospitals.

It is therefore clear that the DRG has an influence on more economical ground plan layouts. Hospitals with examination and treatment areas distributed over several storeys or resulting in long horizontal distances are therefore to be evaluated critically. The DIN 13080 subdivision of the hospital into functional areas and posts of special responsibility is prone to misunderstandings in this regard, for it still divides the areas into general admissions and emergency care (1.01), clinical doctors' service (1.02), functional diagnosis (1.03) and administration (3.0). Many authorities and institutions therefore have difficulties in evaluating up-to-date plans, especially since the assignment of functions in the classical sense has disappeared. Today, hospitals must no longer be compared in terms of the number of beds, but rather by the number of cases.

A hospital can only survive if the diagnosis of the illness takes place quickly and accurately. That is why a patient process must take place according to definite rules. All patients, with the exception of those for psychiatry, paediatrics and maternity, go through the same process in the hospital. All patients, whether emergency patients, selective patients, able to walk or bedridden, are informed of a definite examination cycle immediately after their arrival. Modern admissions units also complete the medical and administrative admissions simultaneously in order to save time. Any patient who lies in bed on the ward too long or too soon without a clear diagnosis costs the hospital executive money, since ultimately the case must be paid for, not the length of stay. Medical

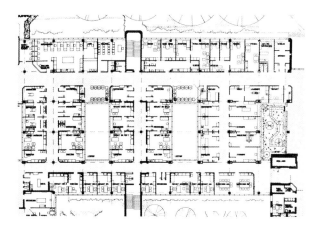

St. Joseph-Krankenhaus, Berlin, Medical administration with out-patient and in-patient corridor, architects: Planungsring Dr. Jüchser + Pawlik (1982)

clarification must take place for each patient before he or she can be assigned to a ward. Today it is more important than ever to decide, beyond a shadow of a doubt, whether in-patient, partially in-patient or walk-in treatment is appropriate. In earlier times, the term "out-patient" was frequently regarded critically by hospital planners and architects because provincial guidelines for hospital financing were strictly differentiated between out-patient and in-patient care. Only since the last health reform has a certain change in thinking taken place which is now influencing planning. The formerly strict separation was loosened in favour of an interlocking of both types of care. The essential and purposeful element in the development of modern out-patient departments is above all interdisciplinary work. The constantly increasing specialisation of doctors in past years has caused the patient as a whole entity to be overlooked. Often a patient would lie on a certain ward for days without a clear diagnosis because departmental thinking dominated interdisciplinary work for the good of the patient. In order to understand the new concepts, technical modes of behaviour in the course of the process must be clarified together with building structures. Such processes are elementary for the planning. The subject can be vividly illustrated by describing various specific patient cases. In connection with the competition preparations for the hospital in Munich in the part of town called Harlaching, the TeamPlan bureau from Tübingen described the various courses of the patients similarly to the following account:[5]

The Course of a Patient with a Diagnosis

The patient, capable of walking, enters through the main entrance where he immediately recognises an information or direction area. Bedridden patients reach the hospital through the driveway for the bedridden. The first station is important and decisive for the patient's orientation, regardless of whether he or she comes with or without a doctor's referral. He or she must be received by a competent medical post. The central admissions area and the interdisciplinary emergency department are ideally suited for this. During regular working hours and with a doctor's orders, the patient's first stop is the main office of the central admissions area. The patient's data are here entered into the computer and thereby administratively admitted. In the examination rooms near the central admissions area, the patient is given a thorough specialist medical examination; further examinations can be carried out if necessary. For the continuing diagnosis, the areas of functional diagnosis, endoscopy and x-rays should be closely connected to the admissions area. If patients come with a doctor's referral outside working hours, they are guided directly to the interdisciplinary emergency department and given a specialised medical examination. Further diagnostic measures are carried out in the adjoining functional areas if necessary (e. g. x-ray).

The administrative admission and the processing of the patient's paperwork takes place, in this case, in the interdisciplinary emergency department. According to the diagnosis and type of patient (ambulatory, partially stationary, pre-stationary, stationary) the patient is either discharged to go home, attended to again or admitted on an in-patient basis. If an in-patient stay is probable, the patient is usually temporarily transferred to the admissions care department for final clarification. Even if a short-term observation is necessary and the patient can go home again immediately afterwards, staying in the admissions care department has a purpose. The installation of such a unit can replace an Immediate Care Unit (IMC) in a small hospital; the purposeful completion with day-clinic functions, such as chemo-out-patients or out-patient surgery, can guarantee an easing of the workload in such a unit. Personnel devote themselves to such purposeful synergies. If a hospital stay is necessary, the patient is transferred to the medical centre that corresponds to their diagnosis.

---

[5] More information on the competition under www.teamplan.de

Hospitals

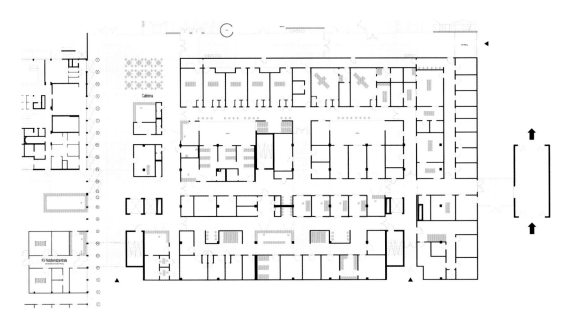

Interdisciplinary central admissions unit in Krankenhaus St. Marienwörth, Bad Kreuznach,
architects: sander hofrichter architekten (2005), ground plan

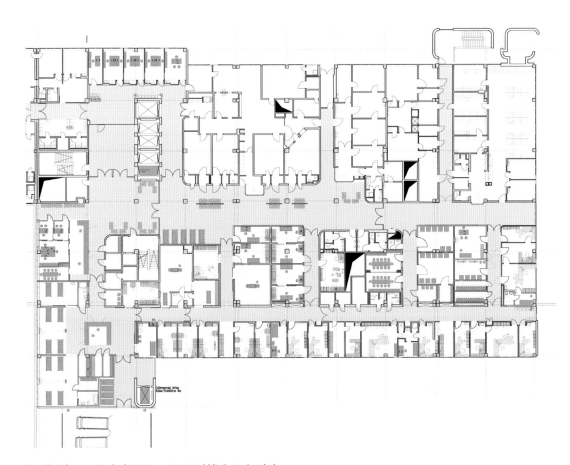

Interdisciplinary central admissions unit in Stadtklinik Frankenthal,
architects: sander hofrichter architekten (2005), floor plan, second storey

Altnagelvin Hospital in Londonderry, design: HLM Architects with Hall Black Douglas Architects (photo: Christopher Hill)

Harborview Medical Center in Seattle, emergency entrance (photo: Candice Cusack)

The Course of a Patient without a Diagnosis

The following section describes the normal course of a patient, of "normal" appearance and capable of walking, with unclear symptoms and diagnosis. The first priority with this patient is to establish the diagnosis as rapidly as possible. This is especially important when bearing the DRG in mind. During regular working hours and with a doctor's referral, the patient must go to the main office of the central admissions area, where his data are immediately fed into the computer. The patient is given a specialised medical examination in the examination rooms of the central admissions area; further examinations are carried out if necessary. The result of the examinations is digitally communicated to the examining physician where possible. The various examination and treatment areas (gynaecology, urology, internal medicine) are available during working hours for necessary consultations or discussions with family members. A patient without a doctor's referral turns to the interdisciplinary emergency department either during or outside working hours. The specialised medical examination takes place in the examination and treatment rooms of the interdisciplinary emergency department. The ensuing procedures take place during regular working hours, as for patients with a doctor's referral.

The patient is administratively admitted, and his paperwork processed, in the central admissions area and in the interdisciplinary emergency department. If the diagnosis remains unclear despite completion of a specialised medical examination, clarification remains the first priority. According to the situation, the further diagnostic process can be ambulatory, partially stationary or stationary. If only a short-term observation period for the patient is required, and if the transfer to the ward is unclear, then the patient can remain in admitting care for up to 24 hours for clarification. Afterwards the patient is either discharged or transferred to a ward.

The Course of an Emergency Patient

The course of an emergency patient has not essentially changed during the age of DRG. The patient usually enters the hospital via the driveway for the bedridden. A medical emergency case with or without a doctor's referral is brought to the interdisciplinary emergency department either in or outside regular working hours. Here is where first aid, acute diagnosis and stabilisation take place. The administrative admissions and processing of the patient's paperwork are done by the staff of the interdisciplinary emergency admitting department.

The Course of a Patient to Be Treated in the Out-Patient Department

For the sake of completeness, we shall now describe the course of a patient to be treated primarily as an outpatient. His first stop is the main office of the central admissions area; here the patient is examined in the examination rooms during office hours. If a health centre is attached, these examinations can also take place there. All further examinations are carried out in the examination and treatment areas of the hospital (e.g. x-ray or functional diagnosis). These results can then be immediately accessible in case of a later hospital stay.

This double utilisation of rooms and equipment for ambulatory, partially stationary and stationary patients is highly economical. In all these processes, the most important thing is that the patient be diagnosed competently and quickly; he or she must then receive the treatment that will help to get well as quickly as possible. The basic rule of thumb is "walk-in treatments before hospitalisation," since this not only saves money but is also usually in the patient's best interests. An in-patient admission is frequently unavoidable, but many illnesses which formerly required long stays in hospital require only a partially in-patient or even a simple out-patient treatment nowadays.

Wolf Dirk Rauh

The Operating Theatre

From the Anatomy Theatre to the Operating Theatre
A form of architecture specifically intended for carrying out operations can be traced back to the 16th century. Padua is considered the birthplace of modern medical research, for it was there in 1594 that the first room, a part of the local university, was used for planned operations on the human body. The anatomy theatre was conceived in Padua, initially constructed with Venetian support and in this manner developed as a prototype for the operating room for one of the highest medical arts – invasive surgery – a prototype which was to remain valid for centuries. Without the dissection of human bodies at the universities of the Italian Renaissance, the immense expansion of scientific knowledge would be unthinkable. Since opening up the body was against the laws of the Church, those interested in medicine at that time found themselves involved in illegal activities; the bodies of executed criminals were transported directly from the sites of execution by underground waterways to the anatomy theatre.

The anatomy theatre at Padua became the architectural model for medical research centres throughout Europe. Between 1610 (Leiden, the Netherlands) and 1875 (Greifswald, Vorpommern, Germany), numerous anatomy theatres inspired by Padua were created where medicine was studied. A famous example, in which the central positioning of the operating room is unmistakable, is found in the Sommerlazarett of the Berlin Charité. The unparalleled importance of the operating room for training and observation is easily deduced from the ascending rows of seats, the arrangement allows each observer unrestricted view of the operating process. In order to enable the observation of operations at first-hand, two-storey operating rooms were built until the mid-1960s. Two examples of this construction method, equipped with observation galleries, are the Columbia Presbyterian Hospital in New York City (1956) and the Barraquer Dental Clinic in Barcelona (1971). With the technical development in the area of communications, the typology of the anatomy theatre also changed. At the so-called operating theatre, surgical operations were relayed around the world with modern cameras positioned in adjacent rooms or over the internet; the connection to the original Paduan model was only retained by the English designation of "theatre".

Air Procurement and Air Treatment
Since the mid-nineteenth century, hospital design, and especially that of the operating area, has been determined by the ventilation of the buildings, due to the recognised danger of infection. Pavilion hospitals were built all over the world as a direct consequence of these insights; all their individual buildings and patient wards were equipped with good through-ventilation. Satisfactory room ventilation was the first priority; increasing mechanisation and technological advances made possible warming, cooling, increased water vapour, drying and other air-treatment including filtering.

Today, a disturbance-free, vertical stream of filtered air directed from the ceiling, free of suspended particles, is standard practice. The field of direction is designed to be as wide as possible, so that the patient, operating team and instrument tables are caught in a germ-free or nearly germ-free displacement stream lacking in turbulence. The issue of whether preparation of the instruments should be carried out in the operating room or in a separate room (and then only under separate air streams directly from the ceiling) is controversial in many cases the answer depends on the intended frequency of operation or on related factors.

Daylight and Artificial Light
Over the past hundred years, the design development of the operating room, the presence of daylight played

Anatomy theatre, Padua, 1594

Anatomy theatre, Leiden, 1610

Anatomy theatre, Greifswald, 1875

a dominant role alongside the ventilation. At the beginning of the twentieth century, bright, sunlight-filled operating rooms were built, frequently with overhead lights as in painters' or sculptors' studios. The Krankenhaus Detmold demonstrated in exemplary fashion the importance of good lighting in the operating room with glare-free northly light. The operating rooms became the brightest in the hospital, and could often be clearly identified from the outside thanks to their wide glass surfaces, and not only in Detmold. With the introduction of high-quality artificial lighting systems, the rooms shifted increasingly back towards the inside of the clinic building in the ground plans. Architectonic and technical developments, however, did not always run in parallel. During the 1970s grotesque ideas also arose in the area of artificial lighting for operating rooms. One example was that of building 70 light projectors into the ceiling and walls of an egg-shaped operating room. The operating table was to be moved so as to change the illuminated area on the patient's body. Today's standards include hanging gas discharge-lamps, halogen lights or light diodes which primarily result in elimination of shadows, stark contrasts and colour variability.

Surface Expansion and Deep Building Ground Plans

The development of high-quality mechanical ventilating systems and artificial lighting systems, as well as the move away from natural lighting had far-reaching consequences for the design of surgical departments. Pre- and post-operative usable surfaces and pedestrian walkways influenced the typical 1970s ground plans of operative departments. Special rooms assigned to or placed in front of the operating rooms were built for every function: the administration of anaesthesia, recovery room, washing and changing room, rubbish storage, instrument preparation room and sterile goods storage. There has been an enormous increase in the surface area of the examination and treatment areas during the past hundred years, especially in surgical departments. At the beginning of the 20th century, the surgical department at the University College London had two operating halls consisting of 25 different rooms with a total area of about 400 square meters.

Atmosphere and Upgrading of the Work Place

During the 1980s and 1990s, transparency and sunlight were rediscovered in the architecture of surgical areas; the trend towards unlit rooms located within hospital buildings, which could be observed until then, waned in many places. For some years architects no longer designed operating rooms without windows, as for example in the Hautklinik Hornheide in Münster. On the one hand, the priority is natural sunlight and a possible view, and on the other, the general spatial and atmospheric qualities of the work place. The reason for this trend is obvious: while the average stay of the clinic patient is limited to about four days and may well continue to get shorter, the average "stay" of the staff is up to four decades. Atmosphere is therefore especially important to the staff; indeed, it is a genuine need. Due to the high standards of hygiene, building and medical technology demanded, there is a real danger that room groups of a low aesthetic standard will be built, especially in the area of surgery. For this reason, it is important to use means of construction such as natural materials, colours and light intensively in the work place of staff with particularly heavy responsibilities and exposure to stress, e. g. in the surgical, maternity and other treatment rooms. Glazing and/or transparent openings between operating rooms can contribute to an improvement in the spatial quality, making an overview possible and thus leading to an increase in efficiency. In certain cases, several operating areas or so-called operating cabinets can be arranged in a large space, thus creating completely different spatial configurations.

Hospitals

Hospital, Detmold, around 1905

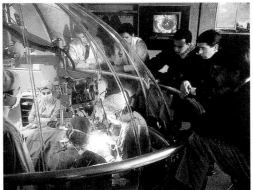
Barraquer Dental Clinic, Barcelona, 1971

Front Zones and Sluices

In the development of departments of surgery, their degree of separation from the remainder of the hospital plays an important role. Numerous ground floor variants and spatial plans indicate a very strict sealing-off of the surgical area through complex air and spatial sluice systems which are not justified by efficiency oriented, hygiene requirements. For several years there has been a change in thinking here as well; the American Institute of Architects, in its 2001 guidelines, called only for the avoidance of pedestrian through-traffic. In Germany there was a new planning recommendation as well: in the year 2000 the Robert Koch Institute in Berlin published guidelines in which the subdivision of surgical departments into zones rather than defined rooms is recommended. For some areas, the demands on the purity of airspace were reduced; as a consequence, these areas can now be much more flexibly planned, also saving considerable expense. An exchange zone for staff and equipment belongs in every surgical department. This zone contains changing rooms for the staff and/or public cloakrooms and possibly safety deposit boxes, patient storage rooms as well as procurement and disposal rooms. In the patient sluices, the person being operated on is transferred from their bed onto the transportable operating table, with or without mechanised help, since the bed must not be brought into the operating area for hygiene reasons. Some clinics are considering whether the patient bed is a suitable means of transport for the future and how patients can be transported into the operating room by other, more dignified means.

Reception and Recovery Rooms

In every discussion about new surgical departments, the question of the operating room front zone for reception and recovery rooms is of importance. Separate rooms for recovery are in general no longer required thanks to improved methods of anaesthesia; however, special reception rooms remain a component of many spatial plans. The necessity of separate reception rooms is justified by the intensification of operating schedules and the optimisation of work loads. The spectrum of possible reception rooms or areas ranges from a central, open holding area in front of the operating rooms to single rooms assigned to each operating room as reception areas. An example of an open holding area is the Herzzentrum in Karlsburg/Vorpommern; it is placed in front of the operating rooms, without a barrier or sluice, and washing areas for the cardiac surgeons are located next to the reception places. An open holding area is possible in this special case, since the patients there have already been anaesthetised. The degree of openness and necessary separation are to be judged according to the kind and number of medical disciplines operating in the surgical department and the type of operations carried out in connection with these. This aspect must be reconsidered with each project. In the new building of the interdisciplinary surgical department for the clinics of the City of Cologne in Holweide, for example, each operating room has been furnished with a separate reception room placed in front, in order to avoid undesirable eye contact between patients as well as noise disturbance from the patients and anaesthesiologists. At the same time, an especially efficient density of utilisation of the operating rooms, a disturbance-free operating routine and an economical process of operation should be facilitated.

Anaesthesiologists maintain that the administration of the anaesthetic no longer determines the critical time allocation, but rather the preparation of the instruments. In view of the additional time-consuming placement and covering of the patient in the operating room, the sense of a reception room is often questioned. In some new clinic buildings, therefore, the conventional front zone placed before the operating room is disconnected

Columbia Presbyterian Hospital, New York, 1956

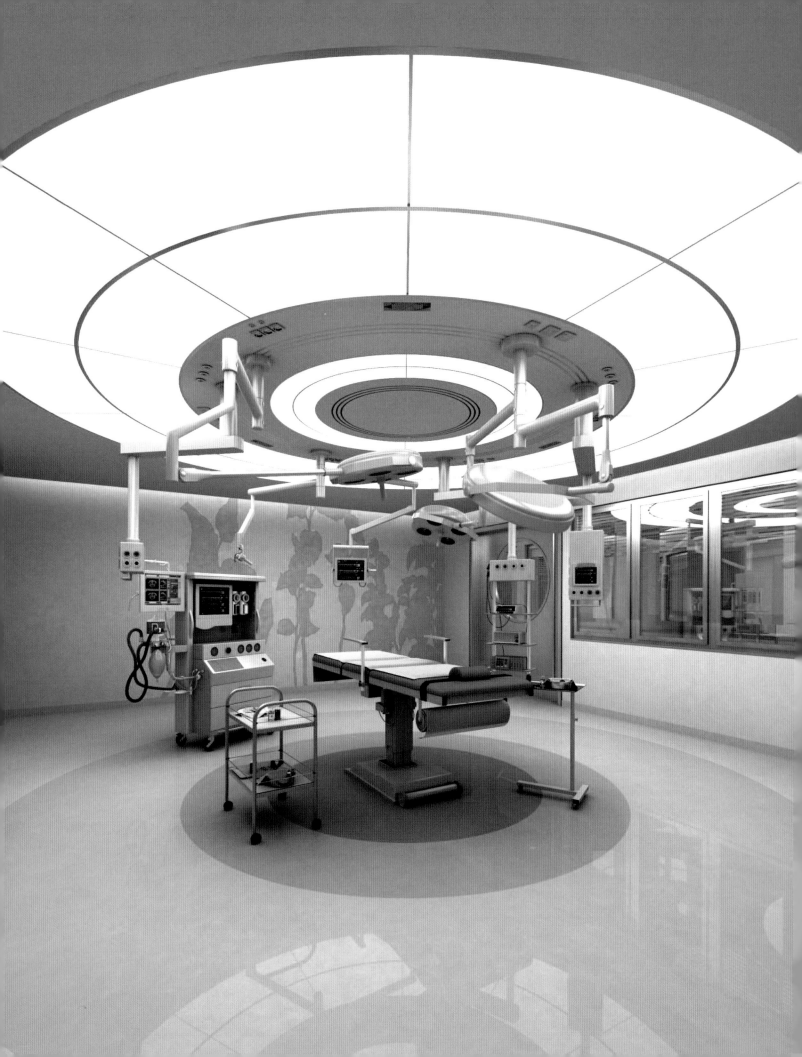

Hospitals

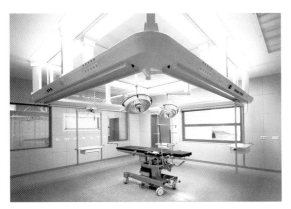

Operating room in the Klinik Holweide, Cologne
architects: Rauh · Damm · Stiller · Partner

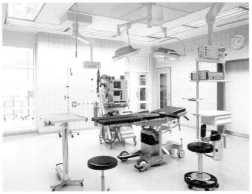

Operating room in St. Josefs-Hospital, Bochum-Linden,
architects: Rauh · Damm · Stiller · Partner

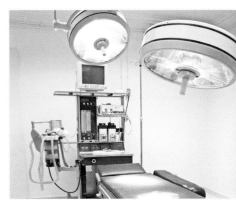

Operating room with monitoring equipment
(photo: Paul Vinten)

from the operating room in order to flexibly use washing and reception areas for several operating rooms. In 2002 Franz Labryga added a new wing, for pre- and post-operative utilisation, onto a remodelled storey of the existing examination and treatment building for the new construction of a surgical department in the Klinikum Saarbrücken. This provides space for ten operating rooms, all with ample sunlight. The patients are placed under anaesthesia at central posts or directly in the operating rooms; a preparation room for instruments and a workroom are placed between two operating rooms.

Operating Ambulatory Patients

A goal of health policy is the inclusion of out-patients surgery in the hospital's spectrum of services. Due to differing organisational structures, divided and also spatially separated surgical departments are sometimes built or planned for stationary and ambulatory processes. Conventional, simple room arrangements for resident surgeons and orthopaedists are often built. In other clinics, differentiated changing, preparation and recovery rooms are installed for out-patients, in front of or next to the central surgical department. Some clinics attempt solutions in which ambulatory patients are admitted to general care wards, similarly to bed-bound patients, where they are cared for before and after the operation.

The Intensive Care Room

Hospital architects have tried hard, in collaboration with physicians and hygiene specialists, to think of ways to reduce costs in the building of expensive surgical areas. Alongside reduction in space, these ideas centre round simple ground plan systems and combined room utilisation, in order to attain operationally efficient and hygienically sound work routines. With a reduction in the strong constructional dividing lines between the surgical and other functional areas, and with the application of knowledge of clinical hygiene based on evidence, today's often isolated surgical department, the so-called jewel in the crown of hospital architecture, can in many cases be integrated into a succession of technological intensive treatment rooms. After all, the spatial and technical requirements of radiology diagnostic and treatment rooms, for IT and PET tomography, nuclear medicine, endoscopes and heart catheter measuring areas are almost identical to those of operating rooms. New standards in quality for continually evolving minimally invasive surgery and micro-therapy will lead to a new definition of the operating room as an intensive treatment room in the future; complex endoscopy treatments will take place there as well, alongside conventional surgery.

Left page:
Operating room of a medical centre with monitoring equipment,
design: Meuser Architekten, 2011

Franz Labryga

The Area of Nursing and Care

Origin of the Term and Present-Day Meaning

The etymology of the German verb *pflegen* ("to nurse, to care for") is unclear. It originally meant "to be responsible for something, to engage oneself for something". Later the meaning "to take care of, to care for, take pains for" evolved out of this; it is closely related to the noun *Plifcht* ("duty"). Some linguists are of the opinion that the verb could be a borrowing from the Latin *plicare, according* to which theory, nursing and care would mean "to fold up, entangle, involve, concern and occupy oneself with something", from which the concepts of "common, binding, caring, responsible, busy and everyday doing and action" may have evolved. Other sources maintain that the term *Pflege* turns up in 11th-century Old German: according to these, *Phlega* meant an "observant or solicitous (physical or mental) development and an occupation with the aim of health and well-being"[1.] The basic senses of these early forms of the terms *pflegen* and *Pflege* still apply today, despite transformations over time, or have resurfaced in our consciousness after having been temporarily repressed. Following the unfortunate development of "hospitalisation syndrome", the effects of which on patients have been described by Rohde as "loss of psychosocial roots, relative depersonalisation and infantilisation"[2], in the words of a memorandum issued by the Robert Bosch Foundation nursing and care today constitute "an interaction between a caring person and one who is in need of care". The task is "to perceive the human being in his entire existence" and "to support the maintenance and recovery of self-care competence"[3]. From birth until death, the person in need of care and nursing is to be carefully accompanied, informed, advised, encouraged and comforted. Today these tasks are, like most personal services, bound up with ever greater economic challenges.

Gaining Insights through Patient Surveys and Questionnaires

In order to properly care for and nurse patients, medical staff must know their wishes and needs as far as possible. Despite some (surely justified) reservations about the methods of conducting patient surveys (which patients? At what time in their "career?"), this approach is gaining in importance. Patient surveys or questionnaires are carried out by institutes, researchers, hospital directors and PR consultancies, and are increasingly becoming a permanent fixture in quality management. The subjects covered in such surveys are the areas of medical and nursing care, the infrastructure, how the procedures run (particularly reception and discharge), tidiness and cleanliness, but also the functional and building components. This last area encompasses questions about the number of beds per room, the possibility of admitting relatives, the set-up of the rooms and their sanitary arrangements, the sitting rooms on the ward, ancillary services such as prayer room, cafeteria, kiosk and hairdresser, patients' garden, technical equipment, orientation in the house and atmosphere enhancing measures (ranging from lighting and colour scheme to art in the hospital). Unfortunately, the results of patient surveys conducted in hospitals have only been published – if at all – in rudimentary form. Well-known institutions, too, such as the Warentest Foundation[4] have difficulties passing these individual results on to interested parties in detailed form through an extensive "patient survey concerning hospital care". With regard to this difficulty, the questions relating to patient satisfaction developed by the Picker Institutes in Boston and London[5] and the publication on this subject by Applebaum, Straker and Geron[6] are lucky exceptions. One can only hope that the recommendations of legislators that reports on quality

1 Grimm, Jacob; Grimm, Wilhelm: Deutsches Wörterbuch. Vol. 13. Leipzig 1889
2 Rohde, Johann Jürgen: Der Patient im sozialen System des Krankenhauses. In: Ritter-Röhr, Dorothea [Ed.]: Der Arzt, sein Patient und die Gesellschaft. Frankfurt am Main 1975, p. 167–210
3 Robert Bosch Stiftung [Ed.]: Pflegewissenschaft.Grundlegung für Lehre, Forschung und Praxis.Denkschrift. Stuttgart 1996
4 Stiftung Warentest: Patientenbefragung zur Krankenhausversorgung. Berlin 1995; Stiftung Warentest: Eingeliefert – ausgeliefert? test (1995), number 11, p. 1200–1206
5 Gesundheitsdirektion des Kantons Zurich. Projekt LORAS. Zurich 1999
6 Applebaum, Robert K.; Straker, Jane K.; Geron, Scott M.: Patientenzufriedenheit. Benennen, bestimmen, beurteilen. Bern 2004

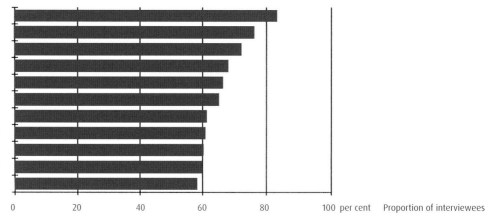

Most frequent constructional-functional shortcomings declared in surveys
(source: Research Project Workplace Hospital)

Nursing standards developed so far:
• Decubitus prevention
• Discharge management
• Pain management
• Prevention of falling
• Furthering of continence

7 Arbeitsplatz Krankenhaus. Gesundheitsförderliche Maßnahmen für das Personal (Assistance: Franz Labryga). Berlin 1993
8 Applebaum, Robert K.; Straker, Jane K.; Geron, Scott M.: Patientenzufriedenheit. Benennen, bestimmen, beurteilen. Bern 2004
9 Schiemann, Doris; Moers, Martin: Entwicklung und Anwendung nationaler Expertenstandards in der Pflege. In: Dieffenbach; Harms; Heßling-Hohl; Müller; Rosenthal; Schmidt; Thiele [Ed.]: Management Handbuch Pflege. Heidelberg 2005
10 Dahlgaard, Knut; Stratmeyer, Peter: Patientenorientiertes Management der Versorgungsprozesse im Krankenhaus. Pflege & Gesellschaft, 10 (2005), number 3, p. 142–147

of care are publishing on the internet starting in 2005 will make a permanent contribution to the recognition of strengths and weaknesses in the operational-building structures as well as in the implementation of patients' wishes.

Gaining Insights from Staff Surveys
Compliance with patients' wishes is a fundamental service – above all for nursing staff and doctors; all other hospital staff play secondary roles in this respect (management, technicians, porters, cleaners, personnel which frequently function as "information carriers"). In contrast to the patients, whose presence in the hospital is in general counted in days (and will surely be further reduced in the future), the staff often remain in this place over many months and years, indeed sometimes for their entire professional lives. A large treasure-house of experience is collected during this period of time – it would be foolish not to make use of it. Within the framework of the project "Working-Place Hospital" conducted by the Research Union for Public Health at the Technical University in Berlin,[7] over one thousand questionnaires anonymously filled-in by hospital staff were evaluated. The results, considered representative according to basic demographics principles, are to be taken seriously and referred to when future measures are implemented.

Of the numerous aspects, only those will be mentioned here which have to do with planning and construction. In their response to the question of constructional-functional shortcomings, over 80 per cent of the personnel indicated long distances between departments as a shortcoming, followed by workrooms that were too cramped and poor room planning. When asked about constructional-technical provisions, staff were least satisfied with air-conditioning, ventilation and soundproofing. Nearly two-thirds of those questioned rated the importance of a liveable arrangement of the workplace high or very high, over 90 per cent of the personnel were not involved in the constructional planning and almost 85 per cent said that they would very much like to have a say in any future planning of their working areas. What architect can, in future, ignore such unequivocal statements by staff in relation to definite constructional demands? What managing director can afford to ignore their staff's express desire for collaboration in the creation their own workplace? Employees in the area of nursing and care expressed their opinions concerning the colour of the rooms in a complementary survey. Those who desire more colour in the hospital are forced to stop and think again when confronted with the fact that most of the personnel voted for white in workrooms, corridors, patient rooms and sanitary areas, and over 50 per cent in the sanitary rooms. In the first three groups of rooms, yellow, green and blue colours followed; in all four room groups, the red tones were considered the least desirable.

Qualification of Nursing Care
For several years, especially since the introduction of the new remuneration system of the DRGs (Diagnosis Related Groups = a summary of diagnoses and procedures in clinically homogeneous treatment case groups which are also homogeneous in terms of expenditure)[8], the qualification of nursing care activities with patients is one of the nursing staff's most urgent tasks. For years in several European countries, intensive efforts to raise the quality of care have been made. As a result, in 1999 the conference of health ministers of the German provinces agreed to instigate "the development of a unified quality strategy in the area of health". This was the incentive for the development of national evidence based care standards. They were developed by experts who paid special attention to transparency.[9] With the help of these standards, the quality of medical service provision is being defined and evaluated. At present, there are five care standards: prevention of bedsores, pain management, discharge management, prevention

Hospitals

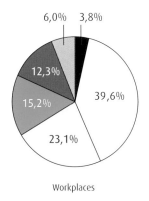
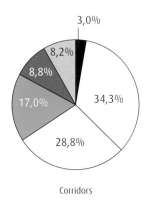
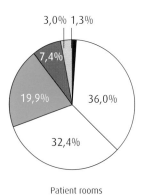
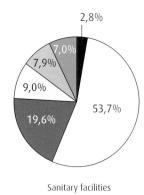

Colour preferences for the main areas expressed by nurses and medical staff
(source: Research Project: Working-Place Hospital)

of falling and encouragement of continence – of all which are among the worst problems in nursing and care. The last two include provision of patient and sanitary rooms.

Organisation of Improved Care

Hitherto it was common for patient care on the ward to see a great deal of conflict generated by misunderstandings, unnecessary double shifts, unnecessary waiting and lack of information. The cooperation between doctors and nurses depends on minimal in friction. "Cooperative process management" can serve as a model for the new regulations governing the working relationship between doctors and nurses. According to this model, the principles of work distribution between the two professional groups are to be to work collaboratively to optimise patient care.[10] Better working conditions and more efficient practices result in advantages for the patients. Alongside this important process of efficient cooperation between doctors and nurses – in the past there was often a shared office for the doctor on duty and the nurses – the usual ward procedures should be analysed in detail and improved as much as possible, before plans are finalised for any new construction or renovation project.

Creation of Functional and Beautiful Rooms

The importance of architecture in the maintenance and recovery of health has been confirmed in many ways over the past 15 years e.g. in the results of investigations carried out by the WHO model project "Health in the Hospital,"[11] contributions such as "Design as Therapy,"[12] "Health-Encouraging Architecture,"[13] "Architecture – a Healing Factor"[14] and numerous reports and discussions at congresses. Their message naturally applies to all buildings concerned with public health; however, it especially applies to the nursing ward because that is where patients spend the majority of their time during their stay. The ethos of a hospital is most clearly revealed on the ward. There, one can sense the rank to which the patient is assigned in the hospital hierarchy. The area, rooms and equipment placed at his or her disposal, the quality of the finish and the materials used – all these reveal what the patient is worth to the hospital, i.e. its owners and operators. In some places, customary may of course be exceeded – in one hospital, the architects installed quality of the finish marble sinks with gold-plated taps, an example of "social magnificence" that, in this case, was not much more expensive than usual. Most patients perceive an acknowledgement of their importance in such signs. What this is about, then, is the basic attitude of those involved in the planning and construction process. In decisions pertaining to functional construction, there is general agreement that things should work properly. The requirements for this are a detailed knowledge of the procedures that will undertaken there and experience gained through extensive practice. For those not directly involved in the hospital life, participatory observation is a considerable help. One architect, an example worthy of emulation, was asked to plan a building for the handicapped. He sat in a wheelchair for a week first in order to "experience" how these patients felt.

Patient Rooms

Although the term "area" is widely used in hospitals the designation "room," meaning the area for the patients, has been universally with its homely connotations of "bedroom" and "living room". We associate warm materials with a room (German: *Zimmer*), after the original meaning of the word which derives from the Middle High German *zimber*, meaning wood for building.[15] Patient rooms developed out of patient dormitories and over time the large rooms became ever smaller, with ever fewer patients. Today there are official guidelines for standards for usual hospital room size suggesting practical solutions for specific areas. Such standards have been developed in the Federal Republic of Germany within the framework of a research assigned by the Federal Ministry of Labour.[16] The Viennese Union of Hospitals then

11 Ludwig Boltzmann-Institut für Medizin- und Gesundheitssoziologie (Pelikan, Jürgen M.; Lobnig, Hubert; Nowak, Peter): "Wie ein Gesundheitsförderndes Krankenhaus" entwickelt werden kann. Vienna 1995
12 Monz, Antje: Design als Therapie – Raumgestaltung in Krankenhäusern, Kliniken, Sanatorien. Leinfelden-Echterdingen 2001
13 Rauh, Wolf Dirk: Gesundheitsfördernde Architektur. Krankenhaus Umschau (2002)
14 Rauh, Wolf Dirk: Architektur – ein Heilfaktor. Deutsche Bauzeitung (2001)
15 Bibliographisches Institut: Duden. Etymologie, vol. 7. Mannheim 1963
16 Labryga, Franz et al.: Untersuchungen zur Einführung von Standards für die Bauprogrammplanung Allgemeiner Krankenhäuser. Forschungsberichte 23 and 75 (3 vol.) des Bundesministeriums für Arbeit und Sozialordnung. Bonn 1979 and 1982

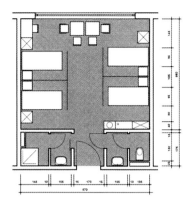
Four-bedded rooms
(source: Planning manual)

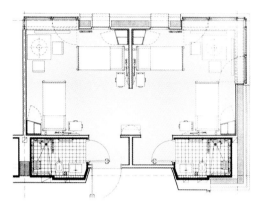
Double rooms convertible to four-bedded rooms,
Städtisches Krankenhaus Forchheim,
architects: Rappmannsberger, Rehle und Partner

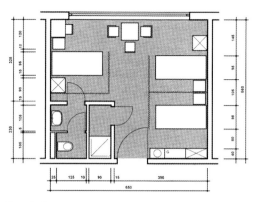
Three-bed rooms
(source: Planning manual)

adopted their recommendations on standards, producing a planning manual for hospitals and nursing homes.[17] The following sections are examples of these standards as they have been developed by an Austrian-German team under this author's scientific leadership. They illustrate basic principles, intended to establish the optimal areas required for properly functioning procedures which have been included in the spatial planning programme. Bearing in mind the intentions of the spatial programme, the architect is confronted with the task of creatively designing patient rooms within this given area.

Four-Bedded Rooms
In newer hospitals one hardly ever finds a room for general care with more than four beds. With regard to nursing care, this size has the advantage of efficiency: one nurse can take care of four patients in one visit. Each patient has its own "corner". The possibility of a variety of modes communication is given. The four-bedded room requires only 12 square metres per bed within a total usable area of 48 square metres. So, in terms of investment and operating costs, it is the most economical alternative. The arrangement of curtains to preserve each patient's privacy is worth mentioning as these have been in wide use in the Anglo-Saxon countries for a long time. Many patients find the noise disturbing when numerous people are together in the same room. The floor plan for the Städtisches Krankenhaus Forchheim offers a flexible solution with two combinable two-bed rooms; this can considerably reduce or even eliminate the disadvantages of the usual four-bedded room without the need to forgo the nursing advantages.[18] The patients themselves can decide whether or not they want to make use of the opportunity to interact. The solution with two sanitary rooms requires a corresponding additional outlay.

Three-Bedded Rooms
The room has both advantages and disadvantages in terms of group dynamics. On the one hand, the number of patients is ideal for playing skat; on the other there is the potential for conflict because one patient could easily become excluded. When the beds are arranged three in a row, as was nearly always the case during past decades and is, fortunately, hardly ever planned any more, the patient in the middle in particular is at quite a disadvantage. He or she has no corner of his own and his private sphere is subject to considerable disturbance. The two-bedded row in a three-bedded room offers help here; this arrangement can fit into other patient rooms of various sizes. The standard version with 35 square metres of usable space, is just as economical as the standard four-bedded room. However, the usable space set out in the guidelines for four- and three-bedded rooms somewhat exceeds that in the other customary areas. Alongside larger room sizes, the arrangement of the shower is especially beneficial, since it reflects the desire for a high level of individual personal hygiene.

Double Rooms
According to surveys most patients prefer a double room. Apparently, for many people, the presence of a "neighbour" provides sufficient opportunity for communication, but especially a desired measure of security. Whether it happens by chance, or whether the nursing personnel succeed in bringing together two like-minded people, then favourable conditions will already be in place for them to spend a harmonious and health-promoting time together. The double room is therefore the most important basic cell of the hospital; for architects and designers a thorough familiarisation with its arrangement is worthwhile because the room has the highest multiplication

17 Labryga, Franz; Mejstrik, Wilhelm; Staudinger, Charlotte et. al.: Planungshandbuch für Krankenhäuser und Pflegeheime. (5 vol.) Wiener Krankenanstaltenverbund [Ed.]. Vienna 1997–2004
18 Lensch, Henning: How to increase flexibility in hospital design. Lecture. Groningen 2005

Hospitals

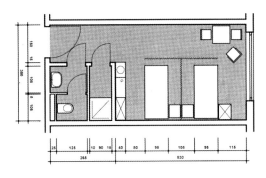

Double room
(source: Planning manual)

Double room, Klinikum Dessau,
architects: Eling + Novotny Mähner

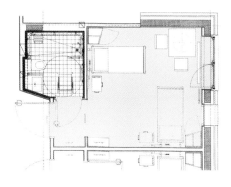

"Handicapped friendly" double room, Städtisches
Krankenhaus Forchheim,
architects: Rappmannsberger, Rehle und Partner

factor. The standard example of a double room reveals a usable area of 29 square metres including the sanitary rooms. This asset, somewhat above the usual standards, results largely from the special spatial arrangement of the shower here. Further special features of this standard layout are the patient wardrobe right next to the bed, a separate cupboard for visitors and a small wash-basin with cabinets above and below. This small wash-basin is the result of numerous discussions and will surely spark further ones. If hand hygiene is really to be taken seriously, then the easy access to this part of the patient room is surely beneficial. Also worth mentioning is the place for a refrigerator next to the window.

Hospital architects have continually made efforts to upgrade the classic double-bedded room by accentuating certain aspects of the spatial requirements. For the new construction of the Klinikum Dessau in 1992, architects working within a limited budget developed a double room which started a new trend in spatial creation in the city of the Bauhaus.[19] In this room each patient has his or her own partitioned off space, each with its own window; the fight for a place in the sun is thus a thing of the past. But it is not only access to light, but also the view out onto the green areas which every patient can now enjoy – a new quality in a double room. The arrangement of the beds required a room axis of 5.20 metres and led to a sectionalised façade. With a very limited usable space of 24.2 square metres, one must remember that, according to the official regulations showers are not planned for each room and therefore not included as an area. The solution for the three-bedded room in this project, following the same strategy, is note-worthy, involving once again a window for each patient. In the Städtisches Krankenhaus Forchheim there is a "disabled-friendly" room for two patients.[20] The subtle linguistic difference between "disabled-friendly" and "appropriate for the handicapped" hints that the stricter requirements of the latter were not entirely met, but that the handicapped do fare better here than in the usual double room. This possibility is attained by a greater area of usable space (about 1.5 square metres more) than in the standard double room. The arrangement of the beds is interesting: the patients' heads are metres apart. Each patient has its own bed area (bed, patient wardrobe and night-table) and its own window. At 25.5 square metres including the "disabled-friendly" toilet facilities, the usable area is very small. If financial constraints had been less severe, the architect would have surely enlarged the distances from the beds to the walls, and also found a better solution for the closing of the toilet door.

A spatially very unconventional double room was built in 1998 in Agatharied.[21] The patients' own areas within the room can provide an uninterrupted view of the landscape or a more secluded position with more comfort. It is an advantage that both patients retain their view of the entrance door. A spatially multi-formed double room was also built in the Kreiskrankenhaus Jugenheim.[22] Here, the bed areas are rearranged opposite each other, resulting in two room corners next to the bay window. The built-in patient cupboards are another advantage, as is the possibility to separate off the bed areas by a curtain. Both patients have an unobstructed view of the window wall and door, even if the inclined position of the room to the floor only allows for this solution in alternate rooms; in the remaining rooms, the view of the door is limited. The architects for the Klinikum Süd in Nuremberg have implemented solutions that are a far cry from the previous norms for double rooms; as early as 1994 they designed they built double rooms without the beds next to each other.[23] The rooms allow a variety of bed positions. With the help of movable patient cupboards, the room can be divided up, so that the individual patient's needs for

Double room, Klinikum Dessau, sketch,
architects: Eling + Novotny Mähner

19 Eling + Novotny Mähner: Städtisches Klinikum Dessau-Alten. (Assistance: Franz Labryga). Anröchte 1992

20 Lensch, Henning: How to increase flexibility in hospital design. Lecture. Groningen 2005

21 Nickl & Partner: Kreiskrankenhaus Agatharied. Munich 1998

22 Junghans + Formhals: Kreiskrankenhaus Jungenheim. 1994

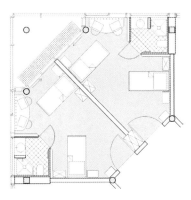
Double room, Krankenhaus Agatharied, architects: Nickl & Partner

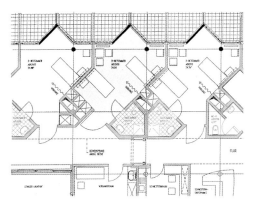
Double room, Krankenhaus Jugenheim, architects: Junghans + Formhals

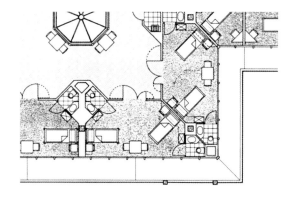
Double room with individual bathrooms, Klinikum Nürnberg-Süd, architects: Beeg, Geiselbrecht und Lemke

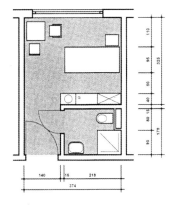
Single room
(source: Planning manual)

23 Beeg, Geiselbrecht und Lemke: Klinikum Nürnberg-Süd. Munich 1994
24 Wischer, Robert; Rau, Hille: Ein- oder Mehrbettzimmer im Akutkrankenhaus – Analyse ihrer Tauglichkeit. Stuttgart 1984
25 Labryga, Franz; Mejstrik, Wilhelm; Staudinger, Charlotte et. al.: Planungshandbuch für Krankenhäuser und Pflegeheime. 5 Vol. Wiener Krankenanstaltenverbund [Ed.]. Vienna 1997–2004
26 Grober, Julius: Die Einrichtung der Krankenstation. Berlin 1912
27 Diskussion nach einem Vortrag im Berliner Krankenhaus-Seminar (Baumgarten, J. and Labryga, F.) Berlin 1998
28 Schuster und Pechthold: Klinikum Nürnberg-Nord, Neubau West. Munich 2003

communication and privacy can be met. Each patient has the same view out of the window (which is as wide as the room) and to the large dividing doors.

Single Rooms

The smallest room in the hospital is the single room. In Germany today a maximum of 15 per cent of all beds are in single rooms on the wards. Where there are 35 beds, that would be five single rooms; but usually there are only four and many wards have only two. However, since the need for single rooms is increasing, their numbers are bound to raise. With further reductions in duration of stay, there is a rise in the percentage of the seriously ill who are best cared for in single rooms. Patients in danger of infection and of infecting others must be isolated for a certain period of time, and our present straitened economic circumstances increasingly force patients to work from their hospital bed. Patients also increasingly express the desire to have family members with them – not only for many hours a day, but also overnight. Isolation is requested more frequently in the case of illness; many people want to be alone and are prepared to accept the cost. Finally, single rooms are required for the dying, because calm palliative care without the burden of fellow patients can only be guaranteed in such a space. These needs add up to about 25 per cent of the bed areas, and in some other areas surely up to 50 per cent. A recent research suggests that it would be best to do away with shared rooms altogether. But even this study mentions the advantages of multi-occupancy rooms: no loneliness in a crisis and the ready opportunity for social interaction. We must surely agree with the final recommendation that the constructional requirements should be met so that "the patient can choose between a single and a double room"[24]. According to the planning manual, the standard for a single room is a usable area of 18 square metres – large enough to accommodate a relative sleeping there if need be.[25]

Bathrooms

The standards of accompanying sanitary arrangements have risen in tandem with the development of patient rooms. Then the well known hospital architect Julius Grober wrote in 1911: "We can get by with two toilets and one urinal for 20 to 25 men and with three toilets for as many women."[26] Today's norm is a bathroom accessible from the patient's room, with a wash-basin, WC and shower. The discussion in Germany of a few years ago, which was prompted by cost, about the installation of showers has (almost) come to an end. The need for additional space can be kept to a minimum through careful planning; the increased cleaning costs should be seen as a contribution to the attainment of hotel-like standards. In the meantime it has been frequently decided to give each patient its own wash-basin for psycho-hygienic reasons. Provision of two separate bathrooms of the double room is possibly a first for Germany in this context. In the Klinikum Süd in Nuremberg, each patient has his or her own wash-basin and WC. At the time, the clinic's director of nursing had good reasons for installing double WCs and influenced this decision.[27] Now one might assume that this solution in 1994 was a unique flight of fancy that came to an abrupt end after investment and operating costs had been calculated – but such was not the case. In 2004 another clinic was built in the north of Nuremberg in which the double rooms were also equipped with two individual bathrooms. The architects had been instructed to provide them following a patient survey, solving the construction problem with the help of prefabricated bathroom units.[28]

Patient Room Groups

Simply multiplying the same kind of patient rooms does not necessarily lead to optimal design solutions. The formation of patient room groups of different sizes, or the incorporation of subsidiary rooms, often results in proper function and appropriate room design. In a study project,

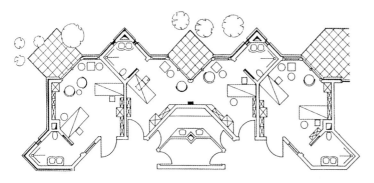
Patient room group with six beds,
architects: C. Kromschröder und H. Rauh (TU Berlin)

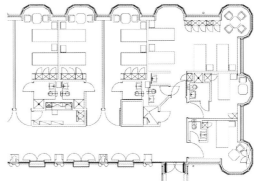
Patient room group, Behring-Krankenhaus in Berlin-Zehlendorf,
architects: Planungsring Dr. Pawlik

four patient rooms with a total of six beds were combined, forming a lively room sequence.[29] The ratio between the beds in the double and single rooms was 4:2, i.e. two double rooms and two single rooms. The number of single rooms corresponds to the requirements of the future. The combination of a four-bedded room with two doubles and two singles, each with specially assigned sitting areas and a nurse's workroom, also led to favourable spatial conditions with short distances in the Berhling-Krankenhaus Berlin-Zehlendorf.[30] Unfortunately there is only one shower for the ten beds. A lounge area much enjoyed by the patients is to be found in a patient room group at the Auguste-Viktoria-Krankenhaus in Berlin.[31] The patients occupying the two adjoining double rooms can meet here, and also sit together with their visitors by prior arrangement. Curtains shield them from prying eyes. The arrangement of the sitting room makes possible the location of a nurses' office near the patients. An even more generous solution for a sitting room was realised in care units for patients requiring a low degree of care at the Zentralklinik Bad Berka.[32] The sitting room for four patients is located between the (spatially economical) two-bedded rooms, offering for the taking of meals and for a refrigerator much appreciated by the patients. A theoretical approach to the flexible use of patient rooms was introduced by Meyer in 1974 in his dissertation.[33] Several possibilities for interaction are available to the patients, depending on their inclination and mood, thanks to adjustable wall elements that can be easily moved and are not too costly. In this model each patient has his own bathroom.

Ward Nursing Stations and Offices
It has proved advisable to divide rooms belonging to a nursing unit into main rooms, smaller side rooms, adjoining rooms and staff rooms.[34]

Main Rooms
Alongside the patients' rooms, the other rooms used by the patients count as main rooms. These are the sitting room, examination and treatment room and patient bath. Sitting rooms are areas in which the patients spend time, talk to each other, read, play, make things and watch television. These rooms are equipped with comfortable chairs and/or sofa and cupboards. Lighting, plants and pictures play an important role in creating a comfortable atmosphere. Besides a larger room, there should be smaller sitting areas on the ward for intimate conversations as well as a place where drinks and fresh fruit are available. Altogether the sitting room should comprise a usable area of about 1 square metre per patient. In some nursing units, sitting rooms are also used for taking meals; as in a hotel, a buffet is set out there every morning and evening for those patients who are able to get there unaided. The social aspect of this should not be underestimated; prerequisites are appropriate tables and seating possibilities. In some hospitals there is a small conversation room, where patients can speak with the doctor, nurse or visitors without being disturbed. 8 square metres of usable space are sufficient for such a room. In the examination and treatment room, the duty doctor takes care of the patients from one or two nursing wards. The room is equipped according to the specific branch of medicine. It requires a usable area of 16 square metres. Where patients are no longer separated according to their specific illness, the room must be larger, since interdisciplinary equipment requires a usable space of about 20 square metres. A patient bath usually suffices for the patients of two wards. It should be equipped with an adjustable tub (to go as high as possible), a hoist, a shower, a WC also usable by the disabled and a wash-basin. Its usable area should not be less than 18 square metres.

29 Kromschröder, C.; Rauh, H.: Studienprojekt im Institut für Krankenhausbau der Technischen Universität Berlin (Assistance: Franz Labryga). Berlin 1985
30 Pawlik, Peter: Behring-Krankenhaus Berlin-Zehlendorf. Berlin
31 Baumann, Thomas: Auguste-Viktoria-Krankenhaus Berlin (Assistance: Franz Labryga). Berlin 1988
32 Wilhelm, Wolfgang: Zentralklinik Bad Berka. Bettenhaus Bad Neustadt/Saale 1995
33 Mayer, Walter: Analyse, Entwicklung und Bewertung von Alternativen für den Normalpflegebereich des Allgemeinen Krankenhauses. Dissertation Technische Universität Berlin, Berlin/Nuremberg 1972
34 DIN 13080: Gliederung des Krankenhauses in Funktionsbereiche und Funktionsstellen: 2003

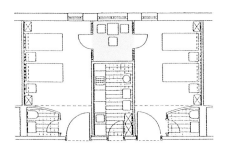
Patient room with adjoining sitting area and nurses' station, Auguste-Viktoria-Krankenhaus, Berlin, architect: Thomas Baumann

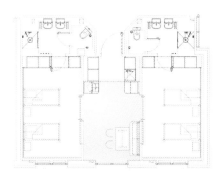
Communal area between two patient rooms, Zentralklinik Bad Berka, architect: Wolfgang Wilhelm

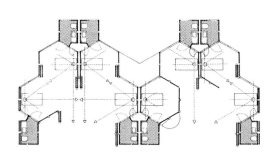
Patient room group ideal for flexible use, design: Walter Mayer

Smaller Side Rooms

The required usable areas of the smaller side rooms depend on the number of patients being cared for. The following values given for orientation are based on a ward size of 32 beds. The areas for smaller side rooms are often calculated too sparingly and thus hinder efficiency. This is especially true of storage rooms and closets. These should not be too large either, however, to avoid establishing a hoarding mentality for supplies. A nursing unit which is optimally furnished with smaller side rooms can be present on floors where there are no materials or equipment lying around. The following smaller side rooms are required: a clean working room (12 square metres), four unclean working rooms as close as possible to the patient rooms with WC (8 square metres each), a kitchen for the preparation of drinks and light meals (10 square metres), a storage room for clean goods (14 square metres), a tool and equipment room (16 square metres), a disposal room for unclean goods (10 square metres) and a cleaning cupboard (6 square metres). Each ward requires a preparation room for the disinfection and cleaning of the bedsteads, so as to avoid the usually costly transport to a central bed preparation area. The room should have places for clean and unclean beds and provide areas for preparation (36 square metres). A room should be reserved on each ward for the deceased which should be large enough for relatives who wish to say farewell to their loved ones (8 square metres).

Staff Rooms

This room group includes the service and sitting rooms, as well as clothes-changing and washrooms for all those involved in patient care. The main nursing station, located as centrally as possible and allowing for plenty of daylight, must have enough room for an information desk as well as for the extensive paperwork. The information desk is designed to be completely open, without glass dividers, to allow communication. The extra expense for sprinkler installations is accepted by an increasing number of hospitals. A location next to the clean nurses' work room and a direct access to the personnel sitting room is desirable. The usable space for the main nurses' station should be about 30 square metres.

Adjacent to this, a service room for the ward director with a usable space of 16 square metres and a quiet location has proven its worth. This is where work-plans can be decided and discussions in small groups can take place. The expansion of this room into the personnel sitting room allows for discussions during shift changes without the usual spatial drawbacks. The doctor's service room should be located as close as possible to the examination and treatment room, and also near the main nursing station, in order to facilitate cooperation (12 square metres of usable space). The staff sitting room mentioned above should not be too small. Alongside ample seating possibilities at the table, a counter for the preparation of meals and drinks is necessary, as well as a refrigerator and closable drawers for important utensils. The staff changing rooms with lockers, wash-basins, showers and WCs are either centrally placed in the basement or decentralised near the workplace. Staff questionnaires have revealed a clear preference for decentralised facilities. A usable value for orientation would be 1.5 square metres per employee, which assumes that each staff member has his/her own subdivided wardrobe.

THE WARD

Functional Requirements

The ward, made up of patient rooms and service rooms, is an organisational unit in most hospitals which is multiplied in the same way according to the hospital size. The simplest form is vertical stacking, as in the Auguste-Viktoria-Krankenhaus in Berlin. However, two, three, four or more wards are frequently arranged on one level. In this case, certain rooms which are less often used can be shared: examination rooms, multipurpose rooms (e.g. with gym equipment), patient baths, sitting rooms and

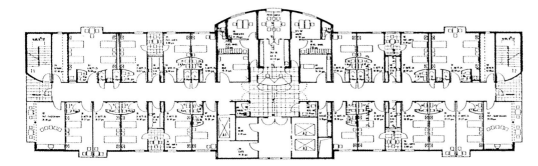

Ward accommodating 34 patients, Auguste Viktoria Hospital in Berlin, architect: Thomas Baumann

visitors' WCs, supply and storage rooms as well as rooms for the deceased. If the wards are not strictly organised by medical specialism, a certain flexibility in the wards has many advantages. Patients on a full ward could be transferred to the neighbouring ward; this principle encourages a better balance amongst the wards in terms of bed occupancy. The arrangement of at least two wards per level is in any case favourable.[35] There are, of course, also solutions with three, four and more wards. The functional basic principle behind this is that each ward must remain free of pedestrian through-traffic.

Hygiene Requirements
With the exception of special units, nursing units are not areas requiring special protection from infection, nor does infection emanate from them. The Federal Health Board's on hygiene requirements in the building of nursing units[36] impose a distance between beds of at least 1 metre. Facilities for the cleaning and disinfection of bedsteads and urine bottles should be near the patient rooms. It has also been established that waste cupboards (for unclean goods) are hygienically unsound. It would be desirable to give each patient its own wash-basin. Centralised patient changing facilities have also proved unsound, according to the guideline. In changing rooms, double wardrobes for street-clothes and hospital-clothes are required.

Operational and Constructional Milieu Formation
The feeling of being at home in the hospital is produced in patients by various factors: their ability to organise their daily routine, their sense of physical comfort and social safety, the knowledge that they are looked after by competent staff who can take care of them in case of discomfort and offer them appetising meals as well as their ability to make full use of the room, all within a liveable designed environment.
In a hospital in which all these desires are fulfilled upon arrival the patient is introduced to a doctor and a member of the nursing staff. The patient can always turn to these people if he or she has any questions or problems. The patient receives information about the other people who are taking care of him, the facilities and procedures in the hospital.
When assigned the "apartment-room," the patient's wishes in respect of room size are met as closely as possible, and staff will try to find appropriate roommates. The patient is woken in the morning at the pre-agreed time. Meals are taken at the times which are customary at home, and the patient's eating habits are respected: drinks and light snacks are always available. The patient is looked after by expert doctors and is informed punctually and in detail about all procedures of examination and treatment. The consultant's ward round is one of the highpoints of the day in the clinic, because the encounter with the doctor in charge creates security, trust and optimism. The technical media available today for information, communication and entertainment are at the patient's disposal. Visits are allowed throughout the day; relatives can also stay the night in hospital. The patient is cared for socially and spiritually, and also has opportunities for quiet times and for worship. Entertainment is offered in the form of small in-house concerts, readings and fun with clowns. Those who have regained their health are discharged in a friendly manner, while the chronically ill are sent home with customised care packages.
The dying are accompanied; relatives and friends can come and visit them one more time. The hospital directors treat all employees well, ensuring a good working atmosphere and in-service training. With the help of health guidelines, the personnel feel healthy and well. Staff are goal-oriented and quality conscious; furthermore, they do not go on fear of losing their jobs.
The architect can support the development of this still unfortunately rare, operationally "healthy" hospital environment through a number of constructional measures. The most important elements are:

35 Eling + Novotny Mähner: Städtisches Klinikum Dessau-Alten. (Assistance: Franz Labryga). Anröchte 1992
36 Kommission für Krankenhaushygiene und Infektionsprävention: Anforderungen der Hygiene an die funktionelle und bauliche Gestaltung von Pflegeeinheiten. Bundesgesundheitsblatt 24 (1981), number 13, p. 212–214

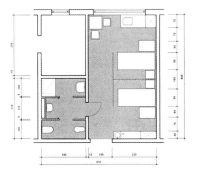

Double room in maternity nursing and care

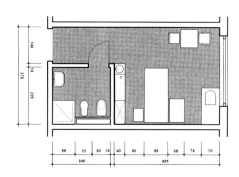

Single room in maternity and newborn nursing and care

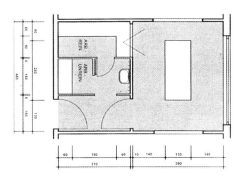

Single room in intensive treatment with anteroom
(source: Planning manual)

- the selection of an environmentally sound apartment-like location in a natural, beautiful landscape
- the arrangement of a well-balanced room-plan, providing appropriate living space for patients and personnel; rooms for confidential conversations, rooms
  for the reception of relatives, rooms for quiet periods and for worship, rooms for the dying, functionally appropriate workrooms and sitting rooms, sufficient rooms, according to size and number, for storage and deposit, and service rooms (patient library, games rooms, cafeteria, kiosk, coiffeur)
- the design of clear floorplans that permit easy orientation and simple walking systems
- the subdivision of the buildings with consideration for the human scale
- daylight for all rooms
- colour design which takes into account the physiological state of each user, whereby the ceilings (many of which are in patients' view) require special consideration
- the use of warm, tactile materials and surfaces
- attention to many small details which make a stay in hospital easier: architectural interventions serving for better orientation, name-plates on patient room doors with a place where a much loved picture-postcard can be stuck, the possibility of erecting screens between the beds for privacy, a place for visitors' coats, mirrors, sufficient room for personal objects (small drawer next to the bed), pin-up wall for a patient's pictures and postcards, own television, telephone, clock, sufficient electrical outlets, possibility of regulating daylight and artificial light, letter boxes with stamp machines, transport wagons for patients' baggage etc.
- inclusion of live plants and works of art, exhibitions or a sculpture garden.

Environmental design has a particularly high value for a hospital for patients generally have reduced resistance to illness. The personnel are often under great stress with some, in danger of burn-out. A good design will surely have a healthful influence here.[37]

Functional Posts in the Area of Nursing
The DIN 13080, developed for planning and building tasks, deals with the differentiated representation of functional posts, which have varying characteristics due to their different specialisations and requirements. In what follows only some of the particular characteristics of functional posts can be shown.

General Care
The constructional units discussed so far (patient rooms, ward operational rooms and wards) are components of the functional aspect of general care, a DIN-approved designation which can replace the otherwise generally used term "normal care". The general wards offer constructional features for "normal" patients of internal medicine, surgery, gynaecology, maxillo-surgery, ophthalmology, ear-nose-throat medicine, dermatology, neurology, orthopaedics, and urology.

Maternity and Neonatal Care
In just a few hospitals has it been recognised that a patient room with a woman in a maternity bed and a newborn baby is, strictly speaking, a double room and therefore requires more usable space than a normal single room. The same is true of two women in maternity beds and two newborn babies. The standard example of a double room for women in maternity beds and neonatal care with both continual and intermittent occupancy requires a usable space of 33 square metres, additional 4 square metres as opposed to a conventional double room.[38] For the single room (24 square metres) the difference is 6 square metres, since here, additionally, a space to allow for the presence of a relative was planned. Further aspects of maternity and neonatal care please refer to the chapter *Obstetrics*.

37 Labryga, Franz: Das Krankenhaus – ein Zuhause für Patienten und Personal. In: Krankenhaus im Friedrichshain 125 Jahre. Berlin 1999
38 Labryga, Franz; Mejstrik, Wilhelm; Staudinger, Charlotte et. al.: Planungshandbuch für Krankenhäuser und Pflegeheime. 5 Vol. Wiener Krankenanstaltenverbund [Ed.]. Vienna 1997–2004

Hospitals

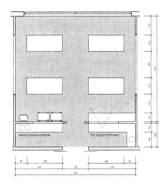
Four-bedded room in intensive care with nursing station

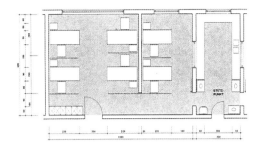
Six-bedded room for dialysis with nursing station

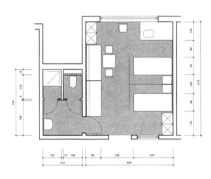
Double room for infants and children with two accompanying persons
(source: Planning manual)

Intensive Medicine
As the functional post with the most beds, intensive medicine used to be classified under the area of nursing and care. Due to the heavier workload involved in this type of nursing, the designation "intensive medicine" was preferred to "intensive nursing and care". Intensive medicine belongs to the specialised parts of the hospital where hygiene requirements have a particular significance in terms of construction and function. The Federal Board of Health's Hygiene Requirements for the functional and constructional design of units of intensive medicine (intensive therapy)[39], provide differentiated planning instructions. These units must be near to and on the same level as most of the patients. As to size of the units, a minimum of 6 and no more than 16 beds are required. If the size of the hospital allows, the units are to be separately equipped for the operating areas and for paediatrics, transplants, burns and toxin units. According to the degree of danger of infection, five patient groups are differentiated with different spatial requirements: there are three groups of intensive treatment patients (A1, A2 and A3) and two groups of intensive observation patients (B1 and B2).

Experts in hygiene could prove that hospital infections in intensive care units occur three to four times more often than for the overall hospital. For this reason, the guideline imposes single rooms above all, since multiple room occupancy encourage the spread of germs. The minimum measures set out in the DIN (see margin) at first met with much criticism, but have since been widely accepted and put into practice in new intensive care units. As far as this goes, one must take note of a commentary on the guideline published in 1998 by the Robert Koch Institute, in which these aspects are once again confirmed, also with regard to apparatus-related development for intensive treatment patients of groups A1 and A2 (these are patients in particular danger of infection, e. g. patients after transplantations and long-term respiration patients).[40] On the other hand, with patients in groups A3, B1 and B2 (these are intensive treatment patients who are at no particular risk of infection and patients under observation in intensive care), it is possible to depart from the guidelines. The right-hand illustration on the opposite page shows a standard single room with a front zone consisting of a clean and an unclean workroom with a conduit function. The usable space consists of 32 square metres altogether. One must consider that the space here is also determined by a special overlong bed. The room almost fulfils the requirements of the guideline for patients in groups A1 and A2. The standard four-bedded room, with a front zone of a workroom and a nursing station, also has overlong special beds, but reduced individual space which is acceptable for patients in groups A3, B1 and B2 with 74 square metres of usable space altogether.

An intensive unit requires additional rooms, in comparison to a ward for general care:
- Anterooms in some single rooms with facilities for hand disinfection, for changing protective clothing, facilities for unclean work and for disposal (at least 10 square metres)
- a larger treatment room, in which, for example, small-scale surgical procedures can be undertaken
- Area with safe-deposit box for personal property
- Place or room for a laboratory
- Staff changing rooms, if possible with sanitary rooms which are to be designed as conduits in the area of access depending on the type of intensive medicine
- Patient delivery area to transfer patients into a special bed for intensive care
- Areas in patient rooms for the admission of relatives who also stay the night
- Changing rooms with lockers for use as a wardrobe for visitors
- Sufficiently large provision channels and disposal channels.

The functional area of Nursing and Care is subdivided into eleven functional posts according to DIN 13080 [34]
1. General Nursing and Care
2. Maternity and Neonatal Care
3. Intensive Medicine
4. Dialysis
5. Nursing and Care–Infant and Children
6. Nursing and Care–Infections
7. Nursing and Care–Mentally Ill
8. Nursing and Care – Nuclear Medicine
9. Nursing and Care–Admissions
10. Nursing and Care–Geriatrics
11. Day Clinic

39 Kommission für Krankenhaushygiene und Infektionsprävention: Anforderungen der Hygiene an die funktionelle und bauliche Gestaltung von Einheiten der Intensivmedizin (Intensivtherapie). Bundesgesundheitsblatt 38 (1995), number 4
40 Guideline: Anforderungen der Hygiene an die funktionelle und bauliche Gestaltung von Einheiten der Intensivmedizin. Bundesgesundheitsblatt 41 (1998), number 6, p. 272

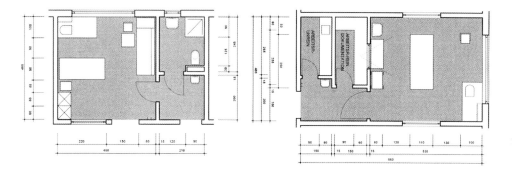

Single room for children's and youth nursing and care (with accompanying person) and channel

Single room for children's intensive medicine with two nursing stations

Six-bedded room on the neonatal ward
(source: Planning manual)

Minimum distances for patient rooms
- from the foot of the bed to the wall: in multibed rooms 2.20 metres, in single rooms 1.60 metres
- from the head of the bed to the wall 0.80 metres
- from the outer edge of the bed to the wall 1.50 metres, between the beds 2.25 metres

41 Schmieg, Heinzpeter: Intensivpflege im Krankenhaus – Überlegungen und Vorschläge zu neuen Formen baulicher Ausprägung. Dissertation. Stuttgart 1984
42 Labryga, Franz; Mejstrik, Wilhelm; Staudinger, Charlotte et. al.: Planungshandbuch für Krankenhäuser und Pflegeheime. 5 Vol. Wiener Krankenanstaltenverbund [Ed.]. Vienna 1997–2004
43 Kommission für Krankenhaushygiene und Infektionsprävention: Anforderungen der Hygiene an die funktionelle und bauliche Gestaltung von Dialyseeinheiten. Bundesgesundheitsblatt 35 (1992), number 12
44 Erste Europäische "Kind im Krankenhaus"-Konferenz: Charta für Kinder im Krankenhaus. Der Kinderarzt 21 (1990), number 12

Intensive units belong to the especially sensitive room groups, whose medical and communications technology, logistics, function and constructional design are undergoing rapid development. The quality of intensive medicine depend on competent and continuous medical and nursing care. Since in general it is linked with a high work load, every technical and constructional option should be used to make the treatment and caring processes easier. Architecture can also contribute to this through a properly functioning, transparent and friendly design. Schmieg has convincingly proven in his dissertation that the room has an influence on the human being as an individual therapeutic factor, especially in intensive care units. [41]

Dialysis

The dialysis unit treats, patients with chronic kidney insufficiency, terminal infectious kidney failure ("yellow" dialysis) and acute kidney failure. Dialysis is necessary when the kidneys are temporarily or permanently unable to process the waste products and foreign substances in the body, and to preserve the balance of electrolytes, acid-base and hydratation. The "artificial kidney" replaces the vitally necessary process of the kidneys. For chronic dialysis, six-bedded rooms are suitable with appropriate space between the dialysis couches which must be accessible from both sides. To preserve privacy, the dialysis areas should be separated. The standard example of a six-room unit[42] stipulates a usable space of 70 square metres. For the infectious "yellow" dialysis, two- and three-bedded rooms should be available. A spatial separation of infected from uninfected patients is recommended, under the guideline of the Federal Health Board.[43] Fourteen dialysis couches have proven an optimal number in terms of the size of a dialysis unit. The need for smaller side rooms is greater than in a general care unit: moreover, a room is needed for water purification, a room for small operations, sufficient storage space and an office for dialysis preparation.

Neonatal and Infant Nursing Care

For premature babies, the newborn, infants, children and teenagers the presence of a relative for as long as possible is essential for recovery. According to the Charta for Children in Hospital, children have the right to have their parents or another relative with them at all times.[44] a double room with a nappy-changing table and two couches for accompanying persons requires a usable space of 36 square metres. The room shown on page 59 (right) with a sanitary room on the side and the window to the corridor offers children confined to their beds a view of those playing outside. The same is true of the 30 square-metre single room for patients in danger of infection to whom a channel has been prescribed. The door of the patient room, leading directly to the corridor, only serves to transport beds. For several children's care units, additional playing, schooling and sitting rooms are required, if necessary also a speech therapy room and an office for teachers, psychologists and social workers. The unit for children's intensive medicine also belongs to the functional post for the care of infants and children where at most 10 beds in four-bedded, double and single rooms are included. The single rooms have a usable area of 40 square metres and are equipped with a nappy-changing table and bath unit, a clean and unclean workroom and a channel (p. 60 centre). The functional post for the newborn and infant care also includes units for neonatology, in which a size of 16 beds has been found to be optimal. This hospital area should be close both physically and in terms of atmosphere to the labour and delivery wards and those for maternity and neonatal care. Eight-bedded, six-bedded and single rooms are suitable as patient rooms in terms of optimal operational efficiency. An anteroom belongs in the six-bedded room with storage cabinets and working places for paperwork, as well as working areas with nappy-changing and bathing units. 61 square metres of usable space are required. Enough space should be available to screen off neonatology single-bedded rooms with one couch for a relative:

Hospitals

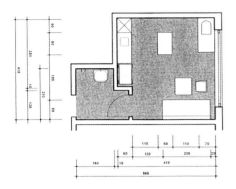
Single room in the neonatal ward (with an accompanying person) and channel

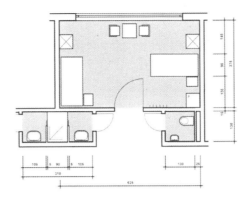
Double room for the mentally ill

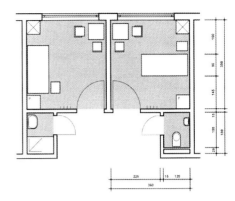
Two single rooms for the mentally ill (source: Planning manual)

20 square metres of usable space are required. For the relative, a sitting room with a kitchenette is required in all children's units.[45]

Infectious Diseases Unit

The guideline published in 1979 by the Hygiene Commission of the Federal Board of Health for the functional and constructional design of infectious diseases units is presently being revised by the Commission for Hospital Hygiene and Infection Prevention of the Robert Koch Institute and should be newly released in 2006. Despite the effective medication with which infection illnesses can now be treated in general wards, special units for the care of infectious diseases continue to be maintained. In some functional areas of nursing, one or two isolation rooms with a front zone and channel are provided. In the Federal Republic of Germany there are also some units with filtration systems and outlets for rare, seriously contagious infectious diseases such as Ebola virus, Lassa fever and haemorragic fever. The extent to which further special units for infectious diseases are installed in individual countries is dependent on the decision of the respective health ministries. It is to be hoped, with ever closer relations to regions of infection, that no new infectious diseases will be appearing here which would require more specially equipped care units.

Care of the Mentally Ill

There are different types wards for the mentally ill (psychiatric acute illnesses and chronic illnesses, gerontopsychiatric illnesses, serious mental handicap, alcoholism and drug addiction, as well as day and night care for psychiatric patients) and are operated as so-called open wards; an exception is the nursing ward for forensic mental illness. Moreover, there is a series of not fully stationary functioning equipment. In comparison with the patient rooms of the other functional posts, a standard double room is shown here with a usable area of 27 square metres and a patient room unit with two single rooms with a usable space of 32 square metres. In all the rooms, the beds can also be placed with the long sides against the walls, in order to create a more liveable atmosphere. For the mentally ill it is generally the case that they need sufficient "territory" around them so that conflicts with other patients can be avoided. There is also a need for greater differentiation of room availability for activities and seclusion.

Nursing – Nuclear Medicine

In this functional post, national and international radiation regulations are in force due to the direct exposure to sources of radiation and the ionising radiation emanating from them. Accordingly, the functional post must be built in a connected, but clearly sealed-off area without through-traffic. For the patients, double and single rooms are suitable, containing a work area for the personnel that must be screened off from the patients by protective architectural interventions. A sanitary room with wash-basin, WC and shower must be furnished for each patient. The double room should be so planned that the localised dosage emanating from the other patients does not exceed a certain level. The standard double room requires 43 square metres of usable space; the single room requires 20 square metres. As for additional rooms, the ward needs a dosage room for the administration of nuclides as well as an examination room and measuring room. A decontamination facility requiring a great deal of space is necessary, guaranteeing that the waste water for the administration zone does not exceed legally permitted levels for the daily average concentration of radioactivity for all liquids within the designated area for nuclear medicine.[46]

Nursing Admissions

In some hospitals, patients are examined at a central location upon admission and observed for there for hours or several days. In terms of organisation there are night-time admissions and long-term admissions in

45 Labryga, Franz; Mejstrik, Wilhelm; Staudinger, Charlotte et. al.: Planungshandbuch für Krankenhäuser und Pflegeheime. 5 Vol. Wiener Krankenanstaltenverbund [Ed.]. Vienna 1997–2004

46 ibid.

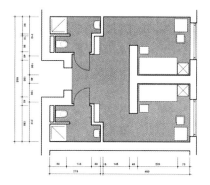
Double room in nuclear medicine

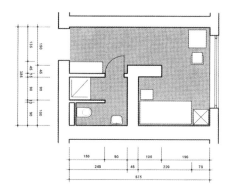
Single room in nuclear medicine

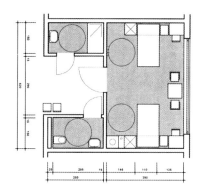
Double room for geriatric
(source: Planning manual)

which the following patients are admitted: emergency and accident patients, alcoholics, drug addicts and suicidal patients up until transfer or discharge, undiagnosed patients and ambulatory patients who require short-term care and observation following diagnosis or therapeutic intervention. The patient rooms and ward workrooms correspond to those in general care; an additional sanitary room is required for the cleaning of very dirty patients. The decision to instal a ward for admissions nursing depends on the size of the hospital and therefore on the possibility of economically operating the ward.

Geriatric Nursing
The present demographic development of our society leads to an ever higher proportion of elderly patients in our hospitals who must be cared for on special nursing wards due to their geriatric illnesses. Double rooms are primarily suitable for their stay here. They offer a high degree of individuality and desirable communication. Some double rooms should be available to accommodate patients in wheelchairs, i.e. the surfaces must allow turning circles with a diameter of 1.50 metres in front of the beds and in the sanitary rooms. The standard example for double rooms requires 32 square metres of usable space, for single rooms 19 square metres. As for ward size, 24 beds are recommended, since this kind of nursing unit offers the best conditions for efficient care. Compared to a general ward, the sitting room should also be suitable for the taking of meals and usable wheelchair patients. Moreover, a therapy room for physio- and ergotherapy should be provided.[47]

Day Clinic
Nursing wards increasingly contain units for bed-bound patients that only care for patients during the day. These wards are usually on the ground floor and should be easily accessible. Since most of the patients are able to walk, the patient rooms only play a subsidiary role. Patients in a day clinic require spaces to lie down for rest periods and a sufficient sitting rooms for various activities. According to the area of medical specialisation, corresponding treatment rooms are required.

Rehabilitation
Rehabilitation units no longer belong to the field of nursing in hospitals. Daniel Gutmann deals with this subject in his contribution entitled "Rehabilitation" (vol. 2). Similarly to the patient rooms in the main hospital, three-bedded, double and single rooms are also introduced for rehabilitation patients at the Vogtareuth[48] treatment centre and a patient room group for early rehabilitation with a connecting sanitary room is provided at the Allgemeines Krankenhaus St. Georg in Hamburg[49].

Developments and Tendencies

A View over the Borders
German public health and the care of patients in particular are of a high, internationally recognised standard. But since optimal performance has not been reached in every area, it is worth looking out for innovative solutions which in future might also be applied in German hospitals. This contribution contains only some examples from the area of construction. In parts of the USA the patient room has developed from a room with three places to a room with five places.[50] The fifth place, the "family support zone," allows for the admission of family members – a possibility that will surely also gain in importance in our hospitals as well in future. In Toronto, Canada the patient rooms on general care wards are equipped with large windows extending to the corridor, so that patients can take part in the life of the ward, but also screen themselves off by means of curtains. These windows give older patients in particular a strong feeling of security, because the nursing staff come by often and check up on things. In Japan there are predominantly four-bedded rooms and single rooms on the

47 ibid.
48 Beeg, Geiselbrecht, Lemke: Behandlungszentrum Vogtareuth (1989–1996)
49 Voermanek, Katrin: Stefan Ludes Architekten. Bauten und Projekte 1995–2003. Hamburg 2003, p. 105–117
50 Patient Room Zones. The International Hospital Federation. Vol. 29, number 1

Hospitals

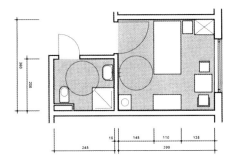

Single room for geriatric illnesses
(source: Planning manual)

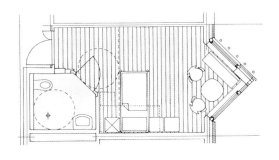

Single room for rehabilitation, Behandlungszentrum
Vogtareuth, architects.: Beeg, Geiselbrecht und Lemke

Double and three-bedded room for rehabilitation,
Behandlungszentrum Vogtareuth, architects: Beeg,
Geiselbrecht und Lemke

nursing wards in two different size-types with relatively small dimensions.[51] In Switzerland, in the Bezirksspital Schwarzenburg, a nursing unit was developed in 1988 which already incorporates many of the advantages discussed here: in the patient room group there is a subdivided four-bedded room with a high degree of individuality for each patient, also double rooms and on the same floor-plan single rooms of the same quality; each patient has its own window. Nurses' working rooms and storage areas are directly integrated with the patient rooms groups, so that the walking distances for the staff remain short. By today's standards, the only shortcoming is the lack of showers assigned directly to the patient rooms.[52]

Research

In line with developments in the USA and some European countries, teaching posts and institutes for specialised areas of nursing and care, care management, and research into care have been established at German universities and technical colleges over the past few years.[53] Alongside teaching, scientific investigation of many previously unknown areas could be developed. At the front of care research are questions with a direct application to practical nursing. But in the year 2005 the subject "Room Design in Bed-Bound Care" was raised which provided unequivocal results for the most frequently named wishes of the nursing personnel: more consistent separation of the side-rooms from the ward service room, more seating possibilities, more storage possibilities and large working areas, brighter and friendlier wall surfaces, larger rooms as well as better lighting.[54] a close cooperation between all those directly or indirectly involved in the caring process must still be encouraged: nursing staff, doctors, management technicians and architects they all that to talk and work together. Many questions are still to be answered; developments already integrated into German medical practice and other countries must be examined, and innovative directions must be pointed out

to improve nursing and care for the good of the patients. The furtherance of this aim with public means must be increased; in so doing, the authorities should surely bear in mind the great public economic significance of the health market.

Future Constructional Measures

The present structural debates will inevitably lead to constructional modifications in hospitals and, with them, of areas of nursing and care. Especially the reduction in length of stay caused by the DRG will further limit the number of hospital beds. This means reduced (because mostly uneconomical) or empty nursing units; in the worst case entire hospitals may have to close. In all cases it must be considered, before it is too late, how the empty building can best be put to use. A slight reduction in the number of beds can sometimes result in an improvement in spatial conditions. For example, unfavourable three-bed rooms could be occupied by only two patients. Other uses must be found for empty wards in general, for example an expansion of the care spectrum in the area of rehabilitation or by using the empty areas for the increased spatial requirements of the areas of examination and treatment. If Germany's hospitals are to develop into integrated service centres, or better, still into health centres, changed constructional structures are required alongside the legal regulations. Hospitals that are unable to compete because they have reacted too late will be used for other purposes or torn down.

Care in the Year 2030

The Viennese model project "Health and Hospital" under the aegis of the World Health Organisation (WHO) pointed out the following causes of the "sick" system known as a hospital: antiquated organisational structures, the growing number of elderly people, the erosion of social networks, the increase in chronic illnesses, greater personal demands, the speed of medical-technological development, the lack of integration of care and the

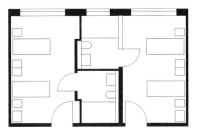

Patient room group for early rehabilitation, with adjoining sanitary room in between Allgemeines Krankenhaus St. Georg, Hamburg, architects: Stefan Ludes Architekten

51 Katta Public General Hospital, Journal of Jina. The Japan Institute of Healthcare Architecture (2003), number 138, p. 17
52 Hofrichter, Linus: Hospital rooms: Where they've been, where they're going. Lecture. Groningen 2005
53 Robert Bosch Stiftung: Pflegewissenschaft. Grundlegung für Lehre, Forschung und Praxis. Denkschrift. Stuttgart 1996; Höhmann, Ulrike; Krampe, Eva-Maria; Kohan, Dinah: Von der "weiblichen" Pflege zur "männlichen" Wissenschaft? Forschungsprojekt des Hessischen Instituts für Pflegeforschung. www.hessip.de/Veröffentlichungen (December 2005)
54 Conzelmann, Johannes; Dacaj, Salih; Duemmel, Sarah et. al.: Raumgestaltung in der stationären Pflege. Zwischen Chaos und Verantwortung. www.pflegeforschung.net (December 2005)

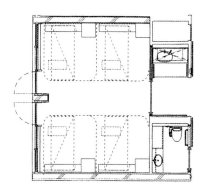
Four-bedded room, Public General Hospital, Katta, Japan

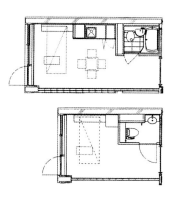
Single room, Public General Hospital, Katta, Japan type a and type b

Patient room group with double rooms in Spital Schwarzenburg, Switzerland, architects.: Atelier 5

The hospital system has become ill from the fact "that it has no philosophers in its midst. That means it thinks too little about basic things, such as: what is happiness, actually? The modern interpretation of happiness is that pain must be eliminated, that illnesses must be eliminated. But illness and pain cannot be eliminated, they are fundamental parts of life, they always were and will remain so forever. That does not mean that illnesses should not be fought, it only means that there's no point in fighting illnesses with the aim of totally eliminating them someday, or to take on this goal as a health system or individual, the idea that one must be free of all negative factors. That is too demanding; it asks too much of the individual human being, the doctor and the hospital".[58]

constant increase in expenses. In the wake of this, the project has come up with numerous contributions to the development of a health nurturing hospital.

Meanwhile there is an international network of health-fostering hospitals to which some German hospitals belong.[55] The Budapest declaration of the network contains numerous aims which also apply to the area of nursing and care, such as: to develop supportive, humane and stimulating living conditions, especially for long-term patients and the chronically sick; to provide more information, communication and educational and training programmes for patients and their relatives, also improving their quality; to create healthy working conditions for all workers and – especially relevant to this contribution – to further healing processes by means of interior and exterior design.[56]

For a number of years Berlin physician Markus Müschenich has concerned himself with the thinking about hospital development extending even further into the future. One cannot but agree with his hypothesis that hospitals, in contrast to other large enterprises, do not carry on future research, although these ideas are elements of a survival strategy upon which hospitals must orientate themselves before it is too late. Müschenich and Richter stated at the German Doctors' Forum in 2005 that: "Whoever does not regard himself as a victim, but as a creator of change, will begin to develop strategies – so as not to end up a loser."[57] Müschenich's hospital of the future, the Concept-Hospital (a linguistic borrowing from the ConceptCar), must look beyond the boundaries of its own specialised departments. The future will be entirely webbed: the patient comes with his digitalised illness data to a friendly reception lady who hands over a clinic mobile phone to him, over which his examination appointments are communicated to him. With the help of wireless technology, doctors, nursing staff and patients can remain in contact during all procedures and call up any and all medical and management-technical data at all times. The patients live as in hotel apartments, to which relatives are also admitted; rich patients go to a luxurious suite for post-operative treatment. If sufficient public and private means are available, further ideas and desires concerning operational and constructional design of nursing units can be given substance: better information for patients and relatives concerning their state of health, the discharge strategy and after care programmes for the home; availability of secretarial services; the use of nursing robots to take over time-consuming and physically taxing nursing tasks; more single rooms with sleeping provision for relatives; a broad selection of differently designed patient rooms (making possible a "change of scene" for long-term patients); more facilities for personal hygiene (private wash-basins and WCs with handles on opposite sides); more attention to (perhaps even alterable, according to mood) colour; more facilities for fitness and well-being (e.g. massage stools). Even more important are the measures for certification and upgrading of nursing professions, but also for others active in the area of nursing, with whom verbal exchange without a language barrier should be possible. Personal attention to patients is of the utmost importance; it cannot be replaced by any of the measures previously mentioned. The essential thing is being able to talk about illness, pain, death – and life.

55 Ludwig Boltzmann-Institut für Medizin- und Gesundheitssoziologie (Pelikan, Jürgen M.; Lobnig, Hubert; Nowak, Peter): Wie ein "Gesundheitsförderndes Krankenhaus" entwickelt werden kann. Vienna 1995
56 Budapest Deklaration des Internationalen Netzwerks Gesundheitsfördernder Krankenhäuser
57 Müschenich, Markus; Richter, Dirk: Was nach der Zukunft kommt – von ConceptHospital zu ConceptHealth. Berlin Medical, p. 20–21
58 Schmid, Wilhelm. In: Brink, Nana: Erste Hilfe. Die Zukunft der Krankenhäuser. Deutschlandradio Kultur, 12 February 2005

Hartmut Nickel

Obstetrics

In ancient times the art of midwifery was highly an esteemed by society, as evidenced alone by its Latin designation, *nobilitas obstetricum* (*nobilitas*: nobleness, aristocracy; *obstetrix*: midwife). In contrast to the medical profession which was also open to slaves, midwifery could only be practised by fee women. In ancient Greece, the midwife was known as *maia* (Greek for mother). The Germanic peoples were familiar with the *hev(i)anna* (Old High German *hevan: heben* – raise; *ana*: *Crolimutter, Ahnin* – grandmother, ancestor), who presented the newborn child on a shield to the father. In medieval times, too, midwifery was the sole preserve of women. In some cities the midwives, or *weise Frauen* – wise women, enjoyed certain privileges and were paid for out of the public purse. However, because of their knowledge, passed down from one generation to the next, they were increasingly subject to accusations of witchcraft, and many of them died at the stake. During the Renaissance there was a return to the art of *nobilitas obstetricum*, but it did not last long. The profession of midwife soon came to be regarded as on a par with the occupation of barber surgeon (the *niederer Chirurg* – minor surgeon), who was not highly during the period. Midwives joined associations of surgeons. And so, over time, midwifery developed into medical obstetrics. The medical historian and doctor Axel Hinrich Murken, in his work *Vom Armenhospital zum Großklinikum*[1] (From Hospitals for the Poor to Large Clinics), includes a description of the development of child birth or maternity clinics. According to his research, the first academic maternity clinics were established during the second half of the 18th century. Around 1800 most maternity clinics were set up in specially adapted residential houses and served as places of training for young doctors or midwives. During the Biedermeier period these institutions gradually experienced increasing difficulties as a result of rampant childbed fever (puerperal fever); due to high mortality rates some of them even had to be closed down. The cause of childbed fever remained unexplained as late as the 1880s and it was only with the discovery of its bacterial origins that the death rates began to decrease. In 1847 at the Allgemeines Krankenhaus Wien (General Hospital, Vienna) the Hungarian doctor Ignaz Philipp Semmelweis (1818–1865) demonstrated how puerperal fever was spread by the hands of doctors and midwives. The maternity department of the Allgemeines Krankenhaus in Vienna adopted his proposals advising that the hands of medical staff should be disinfected with chlorinated lime after every gynaecological procedure and the measure achieved great, if unexpected, success. Following the introduction of the new hygiene regulations, the proportion of deaths among women in childbirth fell from 18 per cent to under 3 per cent. The misunderstandings in the medicine of the period were, however, to delay the general acceptance of his findings, because other doctors assumed that different causes were responsible for the spread of puerperal fever, poor ventilation, for example. For this reason, new maternity hospitals were established in a series of larger German cities during the middle of the 18th century, the design of which was intended to counteract the spread of childbed fever. A period of time elapsed before the lessons learned by Semmelweis became generally accepted.[2]

During the period of industrialisation, the rate of childbirths increased. Up until the beginning of the 20th century, the majority of births took place in the home; in rural areas this remained the case until the middle of the 20th century. As general access to hospital care increased and with the specialisation of medical disciplines, births were relocated more and more to the hospitals. Today in Germany, home births, at less than two per cent, play a very minor role; in the Netherlands the rate is around 30 per cent. It is not known what factors are responsible for the popularity in Germany of hospital births; it may be assumed that with the falling birth rate there is a parallel increase in the demand for safe births on the part of parents-to-be.

Ignaz Philipp Semmelweis
(1818–1865)

1 Murken, Axel Hinrich: Vom Armenhospital zum Großklinikum. Die Geschichte des Krankenhauses vom 18. Jahrhundert bis zur Gegenwart. Cologne 1988
2 Sillo-Seidl, Georg: Die Wahrheit über Semmelweis. Das Wirken des großen Arzt-Forschers und sein tragischer Tod im Licht neu entdeckter Dokumente. Hoya 1984

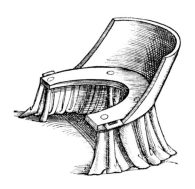

Birthing chair, Middle Ages, woodcarving (around 1500)

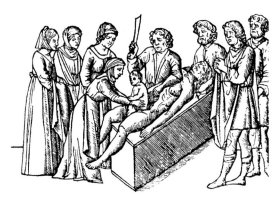

Birth of Gaius Julius Caesar, born by caesarean (sectio caesarea), woodcarving (around 1506)

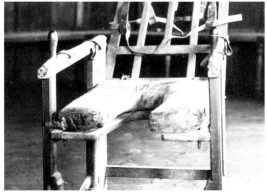

Birthing chair, City Hall of Marktbreit, Bavaria, 16th century

Modern obstetrics in Europe is of a high standard. This is underlined by the extremely low rate infant mortality, 4.16 per 1,000 live births (the worldwide average is over 50) as well as the very low rate of maternal deaths in childbirth, running at 6 per 1,000 births. Within this context it must be taken into consideration that the rising age of pregnant women carries with it associated higher risks for childbirth. These high standards have been achieved through diverse means; in 1987 the scientific advisory committee to the German Medical Council (Wissenschaftlicher Beirat der Bundesarztekammer) published its recommendations on prenatal medicine, which are now a fundamental element of the discipline. Borrowing from the guidelines of the European Study Group of Prenatal Diagnosis, these recommendations cover all diagnostic measures that enable the detection and, where possible, elimination of physical, functional, chromosomal and molecular abnormalities prior to birth. The establishment of these measures alongside the aim of reducing perinatal morbidity and death rates, places the emphasis on the identification of the risk of individual specific genetically inherited diseases.

With the introduction of ultrasound diagnostic techniques, the embryo (and the foetus) has been removed from its prenatal anonymity and, prior to the actual birth, is figuratively made into a documentable individual that, if illness is detected, is transformed into a patient. The human, ethical and legal problems of prenatal diagnostics make significant demands not only on the parents, but also, on the medical team. Specific responsibilities are placed on the experts together with, commonly the planning and organisation, at any early stage, of the joint efforts of gynaecologists, human geneticists, neonatologists and, where required, specialists in other selected areas. As part of their regular examinations in Germany, pregnant women have the opportunity to have three ultrasound screenings that become part of in their pregnancy records (Mutterpass). The first screening takes place between the 9th and 12th week of pregnancy.

From the size of the embryo the doctor estimated date of birth; and, moreover, it can be determined if twins etc. are to be expected. From the second screening, between the 19th and 22nd week of pregnancy, it can be seen if all the organs are in place and if the foetus is developing normally; the position of the placenta can also be determined. In the third screening, between the 29th and 32nd weeks of pregnancy, a check is made on the growth of the unborn child, the functioning of its internal organs and of the placenta, the amount of the amniotic fluid is also determined. This procedure enables high-risk pregnancies to be detected at any early stage, with the birth then scheduled, depending on level of risk, for a specialist clinic with integrated acute neonatal facilities.

The basic and further training of medical staff and midwives is now of a very standard, not least because of the introduction of quality guidelines. Amniocentesis (examination of the amniotic fluid) has shown itself to be another highly effective diagnostic aid alongside ultrasound scanning. Supplemented by the chorionic villus biopsy, it is possible to bring forward the prenatal diagnosis from the second to the first trimester. Alongside the medical care provided by gynaecologists and midwives, there are clinical obstetric facilities available nationwide within a tiered care system close to the expectant mother's home. Depending on the progress of the pregnancy and the any known risk, an informed choice can be made between a maternity unit in a local small hospital, with its own dedicated caesarean facility or perhaps a perinatal centre in a specialist or a university hospital. Recently there has been a discernible trend towards home births and out-patient clinic births; in larger cities birth homes provide an alternative to hospitals.

On 20th September 2005, the Federal Joint Committee (Gemeinsamer Bundesausschuss) in accordance with section 91 paragraph 7 Social Security Code V agreed on measures to ensure quality of care for both premature and newborns legislation, which came into effect on 1st January 2006 and led to changes in obstetric medicine. These measures the quality of the organisation, procedures and results of the care of premature and newborn babies, to guarantee nationwide care, provide a differentiated classification of premature and newborns according to the risk profile and, on this basis, to optimise neonatal care together with the reduction of infant mortality and infant disabilities. The admission of pregnant women who do not correspond to the defined criteria of the neonatal care concept (see column in margin) will only be possible in individual cases where adequate grounds are established. The transport of newborn babies should only be undertaken in emergencies. In principle, the goal should be the antepartum transport of at-risk children for whom a postnatal medical intervention is anticipated. Hospitals may provide care for newborn babies within the four prescribed levels, if they have provided the required evidence in the checklist. The evidence of the fulfilment of the prerequisites for neonatal care as a LEVEL 1 perinatal centre, as a LEVEL 2 perinatal centre or as a perinatal priority centre is to be provided on-site vis-à-vis the health insurers as part of their annual hospital rate negotiations.

Procedure of the Clinical Birth
The normal pregnancy lasts 280 days, i.e. 40 weeks on average, and is calculated from the first day of the last menstrual cycle. The relevant examinations allow the date of birth to be forecast relatively accurately but it is not possible to give an exact date; the number of premature births is on the rise. The birth announces itself by the onset of labour pains. Each birth proceeds according to its own specific rhythm, and as a human biological function cannot be pre-planned and managed down to the very last detail. Statements about what is normal result in average values in a statistical sense, together with deviations from the norm. A normal birth, the so-called spontaneous birth, begins with the dilation of the cervix, in which regular contractions open the mouth of the uterus. The cervical contractions come at intervals of between ten to thirty minutes at the outset and increase to around every forty seconds at ever-shorter intervals; the amniotic sac will usually rupture during this phase, i.e. the waters break. Once the cervix is completely dilated, the second stage of the birth commences, the expulsion stage. During this stage the mother bears down and the baby is expelled through the birth canal. The birth process may have to be assisted with the use of a vacuum extractor (ventouse) or forceps. In most cases an epidural (spinal anaesthesia) or acupuncture will be offered for the alleviation of labour pains. Following the birth the umbilical cord will be cut; the newborn baby will be cleaned, dried and placed at the mother's breast. The final phase begins immediately after the birth and ends with the expulsion of the placenta; it can last from 15 minutes to an hour or more. Immediately after the birth (within the first minute) the physical condition of the baby is assessed by means of five simple examinations, the so-called agpar test. These are repeated after five or ten minutes. As soon as possible, ideally before the expulsion of the placenta, blood is taken from umbilical artery to be tested for pH value, CO_2 levels and the base excess. Determination of the acid-base balance from the umbilical arteries and umbilical veins enables the precise evaluation of the intranatal gas exchange. Some 10 minutes after the birth and, where possible, in coordination with any measures required for care of the mother, the subsequent care and initial examination of the newborn baby (UI) is undertaken by the obstetrician. The remainder of the umbilical cord is cut short and clamped, and the baby's weight, length and head circumference measurements are taken for the first time.

In the first half hour after the birth the newborn baby is placed at the mother's breast to feed; following a vaginal birth with operative procedures or following a caesarean section done once the condition of the mother permits it. The first two hours after the birth are spent by the mother (or both parents) and child in the delivery room,

The neonatal care concept

- LEVEL 1 perinatal centre for the care of high-risk patients
- LEVEL 2 perinatal centre for almost 100 per cent area coverage of intermediary care of high-risk patients
- Perinatal priority unit for total area coverage of care provision of newborn babies requiring postnatal therapy prescribed, implemented by high quality paediatric medicine in hospitals with maternity and paediatric clinics
- Maternity clinic with no specialist paediatric clinic where only no-risk full-term pregnancies are catered for

in order to ensure that both are under the constant supervision of the on-duty midwife and/or the obstetrician during this period. The baby must be kept warm. Where no precise clinical assessment can be made of the baby, the delivery room must provide the means to measure oxygen saturation of the baby, even when in its mother's arms.

Care of the Newborn Baby in the Post-Natal Ward
The Society for Neonatology and the German Society for Gynaecology and Obstetrics recommend what is known as "rooming in" for the mother and child, whereby the child remains constantly with its mother. In the event that the mother does not want this, or is simply not well enough the possibility should exist for the child to be cared for outside the room, for rooming in is a service that is offered and is not obligatory. Mother's breast is the natural method of nourishing the baby and breastfeeding alone should be promoted during the stay in the post-natal ward. If no complications arise, mother and child can be allowed to return home after three to five days.

Higher-Risk Deliveries
Deliveries with a higher risk are classified as those involving multiple births, or where there are pathological processes occurring during the pregnancy, as well as the delivery of premature births. For medical reasons, higher-risk deliveries are often performed by means of a caesarean section. The ratio of caesarean sections to vaginal deliveries in obstetrics has risen steeply over past years: in 1990 the proportion was around 14 per cent; by 2004 it had risen to 27 per cent. This means that today, one of every four deliveries is undertaken using this operative procedure. Where a pregnancy has been diagnosed as higher risk, as a rule the delivery should be performed in a medical environment designed for the purpose. State of the art equipment and coordination with other diagnostic and therapeutic specialities are available in perinatal centres which are gynaecological hospitals with an integrated children's clinic and specialist paediatricians trained in neonatology they are, moreover, staffed by personnel trained in the field who have developed the best working routines. The monitoring of premature babies and sick newborn babies is carried out in paediatric intensive care units, where in recent years, the treatment of children has involved increased parental involvement. The isolation of babies for reasons of hygiene where parents were only allowed eye contact with their children is now a thing of the past. Mother and father are a fundamental part of the therapy team because contact with parents, wherever it is possible, forms the foundation of the baby's positive development both physically and psychologically. The transfer of babies from maternity clinics to paediatric clinics – neonatal transport – should be avoided wherever possible in order to minimise the transportation risks to the child. Antepartum transport of at-risk pregnant mothers to an intensive care unit is preferable to the transportation of the neonate. Babies are termed premature when, at the time of birth, the 37th week of pregnancy has not been completed. It is possible today, in a specialist baby unit, to maintain the life of a premature baby corresponding to a birth weight of 500 grams with a good prognosis of survival, from the 24th week of pregnancy.

The Perinatal Centre
A perinatal unit exists where a gynaecology and a paediatric clinic are integrated within a hospital. These suitably equipped centres are responsible for the care of premature babies and sick newborn babies. The spectrum of perinatal care facility encompasses prenatal assistance, obstetric medical services together with the postnatal care of babies and their families. The larger perinatal centres also have paediatric surgical facilities for carrying out surgical procedures on babies whenever required. Perinatal centres are equipped with areas for the provision of intensive paediatric care with respiratory

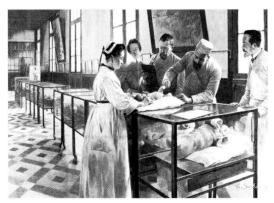
Babies in incubator, Paris (1897)
(photo: bpk bildarchiv preußischer kulturbesitz)

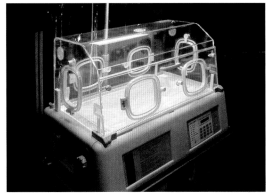
Contemporary incubator (2006)
(photo: L. Hesse Nachf. GmbH)

equipment, as well as rooms in which newborn and premature babies can be cared for in incubators. The incubators provide a sleeping environment with its own individual microclimatic atmosphere to protect the child, and enable the condition of the baby to be monitored at all times. In Germany, thanks to high medical standards, the very advanced state of prenatal diagnostics and the regional availability of perinatal centres, prematurely born babies have a high rate of survival (90 per cent) and a good prognosis for quality of life thereafter. In the larger clinics, such as the Rudolf Virchow Klinikum in Berlin, which sees some 3,500 deliveries per year and is the largest maternity clinic in Germany, the delivery area and the intensive care unit are housed directly together under the same roof; the general paediatric department and the German Coronary Centre (Herzzentrum) are located in the immediate vicinity.

The Neonatal Centre
The neonatal centre is the part of the perinatal centre for which the paediatricians are responsible. Premature babies and sick newborn babies are cared for and treated here. In the intensive care units, with their highly specialised equipment, premature babies can be artificially respirated, given medical attention and cared for. Because of the high value accorded to direct parental contact, the so-called Känguruhen method is actively promoted within the intensive care rooms (literally kanga-rooing, whereby "ruhen" implies being at rest). This entails the parents laying the child on their stomachs whilst they themselves lie back in reclining chairs and this is permissible even where children are still subject to partial monitoring. Alongside the premature babies, sick newborn babies or babies with deformities are also treated here. As a rule, neonatal care units are separated from general hospital floors for reasons of access control and the maintenance of hygiene requirements. Within these units visitors and staff (as well as supplies and goods for disposal) either lock themselves in or are hermetically locked in to keep the area airtight. The hospitalisation of infectious newborn babies or babies at high risk of infection is carried out in ancillary areas with specialised facilities for the disinfection of hands and for changing overalls in an airtight environment.

The Maternity Department in the Hospital
Within a tiered healthcare system, we differentiate between hospitals that provide primary care, regular care and intensive care. Furthermore, there are other specialist hospitals with specific priority areas such as gynaecology and paediatrics. Within this tiered system three models have evolved for the differentiated assessment of the structural requirements of the delivery department.

First Model: Primary Care Hospital (150–200 beds, approximately 500 deliveries per year)
For marketing reasons, the delivery departments of these hospitals, for marketing reasons, offer a variety of delivery methods – from the birthing stool to the Roma wheel and water births. In the field of obstetrics there has long been intense competition between hospitals to attract expectant mothers. Guided tours of the maternity department are now a standard part of the service offered; the service is augmented by, among other things, pregnancy physiotherapy, swimming sessions for pregnant mothers and back-strengthening classes. The primary care hospital is generally equipped with a maternity facility with two delivery rooms, an examination room with ultrasound and CTG, a room for labour pain management and a resuscitation unit. Alone because of the number of specialised beds in the primary care hospital, postnatal mothers are generally accommodated in a mixed gynaecological and obstetrics ward with an attached baby care facility. The hospital's delivery room should be located close to the surgical department due to the need for speed in the event that a caesarean

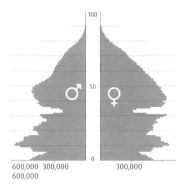

Population Federal Republic of Germany (FRG) and German Democratic Republic (GDR) in 1950: 68m inhabitants

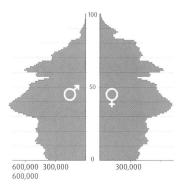

Population in Germany
in 2000: 82,5m inhabitants

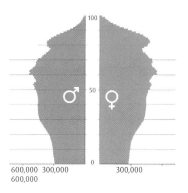

Population in Germany
in 2050: 70m inhabitants
(source: Statistisches Bundesamt)

section is required. It is crucial that there is an easy and short access route from the main hospital entrance, as well as an appropriate level of rest rooms and sanitary facilities for companions. In the majority of cases with this size of hospital, we are talking here about maternity clinics without an attached paediatric clinic or a paediatric clinic that does not correspond to the characteristics of a perinatal clinic within the definition of the agreement of the Joint Federal Committee (Gemeinsamer Bundesausschuss). These facilities should only undertake deliveries of women who have completed the 36th week of pregnancy (36 + 0) where no complications have been diagnosed, which is the case for 96 per cent of all births. All other pregnancies, where it is expected that the child will require treatment, are to be referred to a perinatal centre during the antenatal stage, i. e. prior to the delivery, the exact timing being dependent on the level of risk ascertained. This measure will, as a rule, enable separation of mother and child to be avoided in the event treatment being necessary. The transportation of newborn babies should only be carried out an emergency.

Second Model: General Hospitals (250–500 beds, approximately 1,000 deliveries per year)

Perinatal priority units are found in hospitals equipped with a maternity clinic with an adjacent paediatric clinic. These facilities should have the capability to adequately deal with unexpected neonatal emergencies. Where complications persist, the patient should be transferred to a longer stay specialist hospital. As a general rule, a paediatric clinic offering care should only treat babies born between the 32nd and 36th week of pregnancy. The agreement of the Joint Federal Committee, sets out the following requirements for perinatal priority units:
- The doctor responsible for the newborn baby must have the requisite qualification in paediatrics together with at least three years experience in neonatology
- Facilities for assisting the respiration of premature and newborn babies
- Access to diagnostic procedures such as radiology, ultrasound, echocardiography and EEG
- Round-the-clock on-duty paediatrician
- The perinatal priority unit must adhere to the criteria for referral to a higher grade of care as part of its internal quality management.

The maternity centre of the general hospital has from three to four delivery rooms and the majority of these hospitals are equipped with a dedicated surgical facility for carrying out caesarean sections. Alongside the primary rooms there is also a midwifery service area, a resuscitation unit, examination rooms, a bath if required (where not already located in the delivery room), a waiting area for relatives and suitable ancillary rooms as required. Specially designed and equipped wards are available for the care of women giving birth and for newborn babies.

Women in postnatal care should be accommodated in suitable environment for nursing the baby while recovering, which means there should be a maximum of two beds to each room and, alongside the usual sanitary arrangements (possibly with bidet), there should be baby-changing facilities and, if required, facilities for bathing the infant. Apart from this option, there are areas equipped for the centralised care of newborn babies supervised by maternity and paediatric staff nurses. Nursing rooms and a communal area for mothers to eat breakfast together are already a standard feature of many hospitals. The medical care service is usually complemented by a paediatric department.

Third Model: Priority/Maximum Care (500–1,000 beds, over 1,000 deliveries per year)

The perinatal centres of priority/maximum care hospitals are classified either as LEVEL 1 or LEVEL 2. The medical supervision of the neonatal intensive care unit must be in the hands of full-time neonatologist while medical supervision of the obstetrics unit must be assigned to a full-time obstetrician. The delivery area, surgical area and neonatal intensive care unit must be interconnected or at least located in the same building or in buildings that are linked to each other, to avoid having to use vehicular transport for transfers to the NICU (neonatal intensive care unit). A interconnected arrangement is now essential when planning new building projects. The centre must be equipped with at least four neonatal intensive care areas. Continuous medical and ancillary care must be ensured by shift work arrangements with the permanent presence of a doctor in the intensive care area; a larger proportion of staff (at least 30 per cent) should have completed further training in the area of paediatric intensive care. The areas where services should be available and the special quality assurance processes are listed in the column on the left-hand side of this page.

In Germany the requirements that exceed those asked of a LEVEL 2 perinatal centre which must be met by a LEVEL 1 perinatal centre are listed below:
- In addition to the medical supervisor, his/her deputy must also have completed the requisite standard of training in a neonatal specialism; the deputy of the obstetrics areas must likewise prove that he/she has completed the requisite training with a qualification in special obstetrics and perinatal medicine
- In addition to the medical and ancillary care of LEVEL 2, there should be general access to a specialist in neonatal care
- At least six neonatal intensive care unit places must be available
- At least 40 per cent of the care staff must have completed training in the area of paediatric intensive care
- Centre must be accredited for further medical training specialising in neonatology and specialist obstetrics and perinatal medicine
- An emergency doctor for newborn babies must be available to respond to unforeseen circumstances as a rule he or she should not be on call for other clinics dealing with higher risk births.

The following services that exceed those required of LEVEL 2 should be available:
- Paediatric surgical and cardiological conciliar service
- Human genetics and laboratory
- Implementation of special quality assurance procedures.

In hospitals specialised in intensive care the aesthetics and atmosphere must sometimes take second place to the medically required structures and procedures, first, due to size of the department due to the necessarily high frequency of related operational procedures. In accordance with graded requirements these clinics are responsible for undertaking the delivery of higher risk pregnancies and the care of premature births and sick newborn babies within the neonatal centre. The maternity facilities in these institutions are equipped with four to six delivery rooms, isolated examination rooms with ultrasound and CTG, rooms for labour care, a neonatal examination unit with a labour area and, where the department is not in the immediate vicinity of a surgery facility, its own isolated surgical area for caesarean sections, together with a midwifery area and a rest area.

Due to the size of the facility, there must also be an appropriately sized waiting area with washroom facilities for relatives and a storage room for beds and for supplies and disposal transport requirements. Annexed to the facility are wards suitable for the care of postnatal mothers and newborn babies; the wards for the mothers should be fitted out with only one- or two-bedded rooms to facilitate undisturbed rest and recovery while nursing their babies. The rooms should have a baby-changing area and, possibly, additional baby bathtubs for bathing and washing infants under the guidance of the care staff.

Maternity Facilities

In accordance with the hierarchical structure described above, the medical service provided by clinical delivery also has implications for the construction and organisation of delivery facilities, which can likewise be

Admission criteria for LEVEL 2 perinatal centres

- Prenatal transfer of premature babies with a birth weight from 1,250 to 1,499 grams or premature births between the 30th and 32nd week of pregnancy (29 + 0 to 32 + 0)
- Twins from the 29th to 33rd week of pregnancy (29 + 1 to 33 + 0)
- Serious illness that is linked to the pregnancy
- Insulin-dependent diabetic metabolic disorders with risk for the foetus

Admission criteria for LEVEL 1 perinatal centres

- Prenatal transfer of premature babies with a birth weight of less than 1,250 grams and/or premature births prior to the completion of the 29th week of pregnancy (<29 + 0)
- Multiple births (with more than two children up to the 33rd week of pregnancy (<33 + 0) and with still more children regardless of week of pregnancy)
- All prenatal illnesses requiring postnatal emergency treatment

Kreiskrankenhaus Gifhorn,
spa room

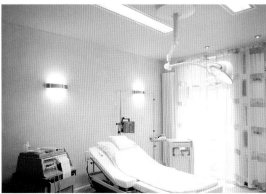
Kreiskrankenhaus Gifhorn, delivery room,
architects: Schweitzer + Partner (2004)

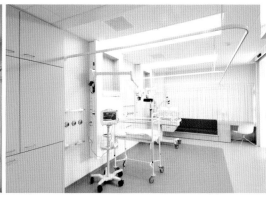
Emma Children's Hospital in Amsterdam,
architects: OD 205 architectuur bv (photo: Mike Bink)

represented within a hierarchical structure. Whereas the maternity departments of the primary care and general hospitals compete with each other in terms of the facilities available and special services for carrying out the delivery, the service provision of the priority clinics and the specialist hospitals is characterised more by the specific medical service on offer. Delivery wards are separated from the generally accessible areas of the hospital and access is controlled; hygiene measures such as overalls, masks and over-shoes are much less likely to be seen. Expectant mothers and their companions can leave the maternity area prior to the onset of the second stage of labour and they may move about freely within the hospital and its vicinity and return at any time to the maternity unit.

Delivery Rooms
Nowadays the delivery room of the first and second models is scarcely recognisable as an area of the hospital; its ambience has rather more in common with a modern hotel. Colours and materials, lighting and the absence of medical apparatus are intended to dispel the tension felt by expectant mothers going into labour. All necessary medical examination instruments are kept in wall closets, the atmosphere is warm and friendly, the delivery beds are spacious as are the delivery baths for expectant mothers and the latter can accommodate companions, if required, where these people are making a positive contribution to the birth experience. Furthermore, there is a work area with a warm lamp and an examination light and there is also the means by which to weigh, measure and to clean the newborn baby.

Care during Labour
Expectant mothers with contractions are accommodated in suitable rooms – two-bedded rooms with sanitary facilities, similar to those for general care. There is the facility to employ the CTG and to induce the birth.

Midwifery Area
All administrative activities are carried out in the midwifery service area. As a rule, it should therefore be illuminated naturally. Depending on the size of the department, from one to three computer workplaces will be required.

Examination Room
Together with an examination chair, the examination room should be equipped with an examination couch for ultrasounds, an examination light in the form of a ceiling or wall-mounted light, a wash stand and a closet for instruments.

Resuscitation Area
The resuscitation area should contain an examination facility equipped with both a warm lamp; as well as medical gases, respiratory equipment and a facility to carry out a blood gas analysis, a transportation incubator for the possible transfer of the baby to a paediatric clinic and a washing facility.

Operation Facilities for Caesarean Sections
The operation room should be equipped with a portable and a stationary operating table, a ceiling supply unit for anaesthesia and surgery, an operating light, a ceiling panel with laminar flow, a wash room for doctors and anaesthetists as well as a preparation room, where required.

Psychological Structural Aspects
Over the past 25 years there has been an unmistakable trend in the development of obstetrics to give birth under the secure medical care of a hospital's facility but, at the same time, within a more homely atmosphere; in a manner of speaking, the traditional home birth has been transferred into the safe environment of the hospital. Relatives, in particular the father-to-be, can be present in most clinics not just at the actual birth but

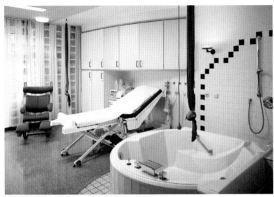

Städtisches Krankenhaus Salzgitter-Lebenstedt, delivery room with birthing pool and birthing chair, architects: Schweitzer + Partner (2002)

Krankenhaus Kirchrode, birthing pool, architects: Schweitzer + Partner (2004)

also in the operating theatre during caesarean sections. The structural arrangement of delivery departments is becoming ever more important and is planned to harmonise with the general development towards family friendliness. The atmosphere should be homelike and the colour choice should exercise a calming influence on the expectant mother (it must not, however, interfere with the colour perception of the medical staff who must be able to detect the signs of jaundice, where present). The technical medical equipment of the delivery room should be readily accessible, but should otherwise not be visible. Natural illumination of the delivery room is to be preferred; if required it should be possible to darken the room. Also, it must be possible, by means of artificial light, to generate different room atmospheres, with illumination that is dimmable, indirect, or precisely directed for medical procedures. The working light that is needed should be arranged so that it does not disturb the patients. Sound absorbent surfaces should be included to improve the comfort of the delivery room. In creating this special atmosphere, so necessary for all, the use of textiles and natural, warm materials such as wood play a positive role.

Care of Postnatal Mothers and Newborn Babies
With postnatal mothers and newborn babies, we are not dealing with patients in the normal sense of the word. The birth of a child in western society, where the birth rate is dropping, is an important event for the family concerned; this is one of the reasons why the mother and child environment must be arranged in a pleasant and comfortable fashion. A relatively high rate of visitors is to be expected in the first days after the delivery. Therefore consideration should be paid to ensuring a room area for mothers that is large enough not just for permanent nursing of the baby with beds for mothers, infant beds and baby-changing facilities but which can also adequately welcome visitors.

Arrangement of Perinatal Departments
Perinatal centres should, as a rule, be arranged according to design principles specifically to the maternity facility and the care of postnatal mothers. In the choice of colours and materials, a degree of harmony must be achieved technical medical equipment. It is crucial to have short access routes between the delivery area, the paediatric clinic and, when required, the paediatric surgical facility, such that these facilities are organised within a coordinated arrangement.

Future Developments
In Germany, a society with an increasing trend for its birth rate (2004: 712,000) to fall far below its death rate (2004: 821,000), attention must be paid to ensuring that, together with the other sociopolitical and family-friendly services, all the medical facilities are in place to guarantee that newborn babies are brought into the world in a healthy and viable condition. For the 90 per cent of all births that are assessed to be non-risk, alongside the delivery services provided by maternity clinics, there are the options of out-patient birth, birthing within a birth house under the supervision of a midwife and home birth.

When compared to international standards, the quality of obstetric medicine in Germany is among the best available; it is the task of obstetrics to maintain these standards and to steer medical developments accordingly. But this is not something that will affect the trajectory of German population statistics with either an increase or reduction in the birth rate. In future, the choice to have children will, in the future, continue to be an aspect of our private lives and it is one that will be made on the basis of developments related to other social parameters. The most that can be achieved in good maternity facilities is the attainment of high medical standards.

Hospitals

St. Marienwörth Hospital
Sander Hofrichter

Johannes Wesling Clinic
TMK Architekten Ingenieure

Helios Clinic
TMK Architekten Ingenieure

City Clinic
Architekten RDS Partner

New Clinic
RRP Architekten + Ingenieure

Johanniter-Krankenhaus
Planungsring Dr. Pawlik

Emil-von-Behring-Clinic
Planungsring Dr. Pawlik

Bundeswehrkrankenhaus
Heinle, Wischer und Partner

Städtisches Klinikum
Heinle, Wischer und Partner

Kreiskrankenhaus
Berg Architekten

Müritzklinik
Schindler Architekten

Städtisches Klinikum
Dr. Ribbert Saalmann Dehmel

Medical Service Centre
Isin Architekten Generalplaner

Katholisches Krankenhaus
TMK Architekten + Ingenieure

Diakonie-Klinikum
Arcass Freie Architekten

University Medical Centre
Arcass Freie Architekten

Robert-Bosch-Krankenhaus
Arcass Freie Architekten

Architect
sander hofrichter architekten Partnerschaft

Operated by
Franziskanerbrüder vom Heiligen Kreuz e. V.

Situated in
Mühlenstraße 39, 55543 Bad Kreuznach

Planning | Construction
2005–2009

Beds
279

Gross ground area
13,798 square metres

Gross volume
49,550 cubic metres

Usable area
5,546 square metres

Total cost
26.5m euros

1 Registration
2 Main entrance
3 Elevation north-west

St. Marienwörth Hospital Bad Kreuznach

The existing St. Marienwörth hospital in Bad Kreuznach has been given a four-storey extension, an old people's home, sheltered accommodation, medical practices and a multi-storey car park so that a versatile health centre has now evolved in its vicinity. The departments in the existing hospital have been largely converted or fully renovated. The decision in favour of the major extension made it possible to carry out all the works while the hospital was in operation while also reorganising access and supply routes within the complex. The new building is similarly proportioned to the existing ward block; the new link, a glass passage between the old and the new buildings, defines the new main entrance. All examination and treatment areas required by the patient before, during and after their hospital stay are grouped neatly together on the ground floor. The central patient admission point is located right next to the new main entrance and medical and administrative patient admissions take place at this point. The first and second floors house two large 30-bed nursing wards with all the comfort of a hotel. Here, warm harmonious colours and adequate natural light create a welcoming ambience. On the third floor an ultra-modern operations department has been created with four theatres and separate preparation and recovery areas for in- and out-patients while a directly adjoining central sterilisation unit ensures high medical and hygiene standards. For emergencies, a large helicopter landing pad is situated on the roof of the new building. A modern hospital depends on the networking of in- and out-patient care. For this reason the new building houses a health centre with five specialist practices so that patients can use the hospital functions on the same level. The result is a hospital with minimum transit distances where the patient is the focus. Optimum patient care is also considered in the parking concept: the new building has an underground car park and a spacious multi-storey car park in the immediate vicinity.

Photos: Markus Bachmann, Stuttgart

St. Marienwörth Hospital Bad Kreuznach

2 3

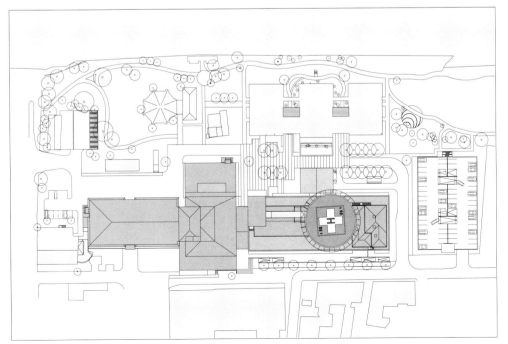

Site plan, scale 1:2,000

sander hofrichter architekten Partnerschaft

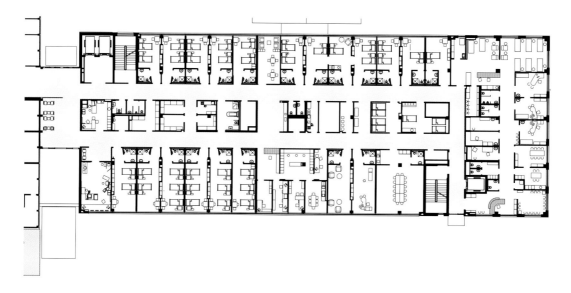

Second floor

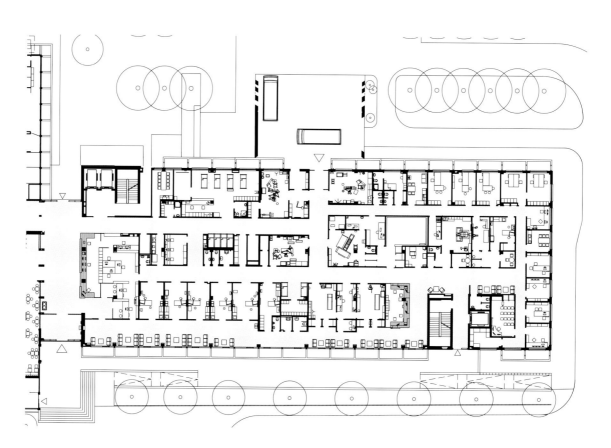

Ground floor

St. Marienwörth Hospital Bad Kreuznach

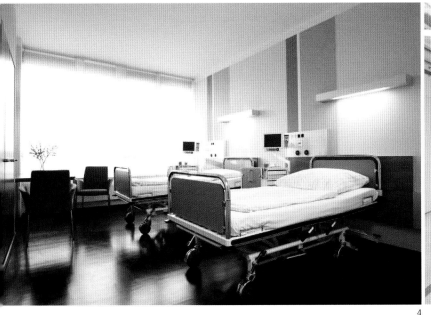

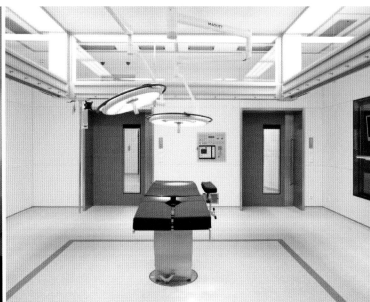

4 Patient room
5 Operating theatre

Plans, scale 1:625

Longitudinal section

Cross-section

sander hofrichter architekten Partnerschaft

Architect
TMK Architekten Ingenieure

Operated by
Mühlenkreiskliniken

Situated in
Hans-Nolte-Straße 1, 32429 Minden

Planning | Construction
2003 | 2004–2007

Beds
867

Gross ground area
98,000 square metres

Gross volume
400,000 cubic metres

Usable area
46,000 square metres

Total cost
230m euros

1 Main entrance
2 Southeast view

✚ ● Johannes Wesling Clinic Minden

The Johannes Wesling Clinic is one of the largest and most modern clinic buildings in Germany. Built in the space of 3 years, the merging of two hospitals on the edge of Minden at the foot of the Wiehengebirge mountains has resulted in a maximum-provision clinic facility with 867 beds and 18 operating theatres. It is distinguished by spaciousness and transparency and along with state-of-the-art technical equipment offers the atmosphere of an art gallery. The basic aim of the design was to create a "clinic in nature" and to this end the striking arrangement of the individual building cubes along two glazed main thoroughfares interlocks with the surrounding landscape. Designer planting between the building modules also works together with the adjacent water features to create a harmonious park. It is an arrangement that makes the building complex appear less solid and overpowering and creates optimum indoor conditions thanks to good use of daylight and natural ventilation via room-high windows. Patients have a view into the surrounding landscape from practically everywhere in the clinic. A particular feeling of well-being at the Minden clinic is created by the reduction of the height of the building to a maximum three storeys: the building and the rooms between are thus designed to human scale. A clear and functionally considered layout means good way-finding and optimum working practices plus ideal distribution of the individual wards and services. Environmentally compatible and sustainable building plays a central role particularly in major building projects so that Minden clinic is based therefore on environmentally-aware planning and resource-efficient building methods. The use of biomass heating and photovoltaic systems enables significant energy savings to be achieved. A geothermal heat exchanger heats the outdoor air in winter and cools it in the summer with no additional energy input. In this way all areas and rooms are easily ventilated and air-conditioned in the summer. This reduces the heat requirement, thus reducing CO_2 emissions.

Photos: Jochen Stüber

Johannes Wesling Clinic Minden

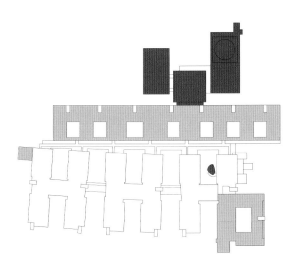

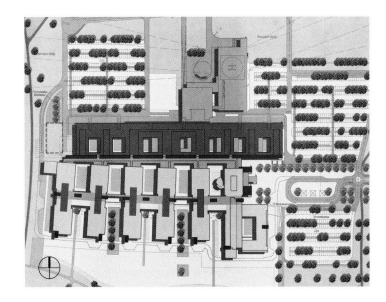

Site plan

TMK Architekten Ingenieure

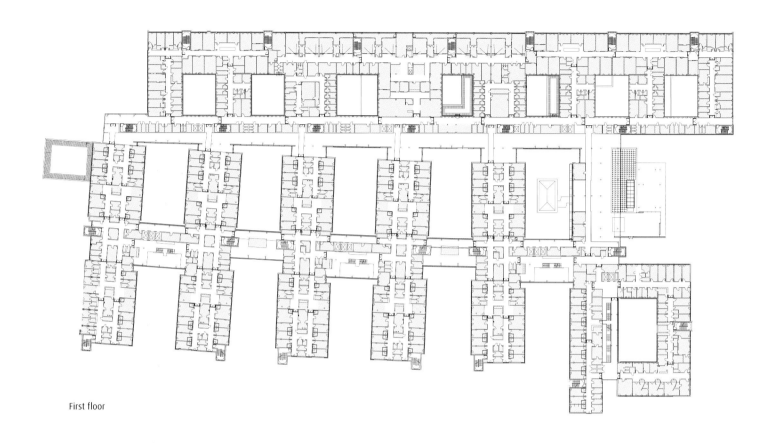

First floor

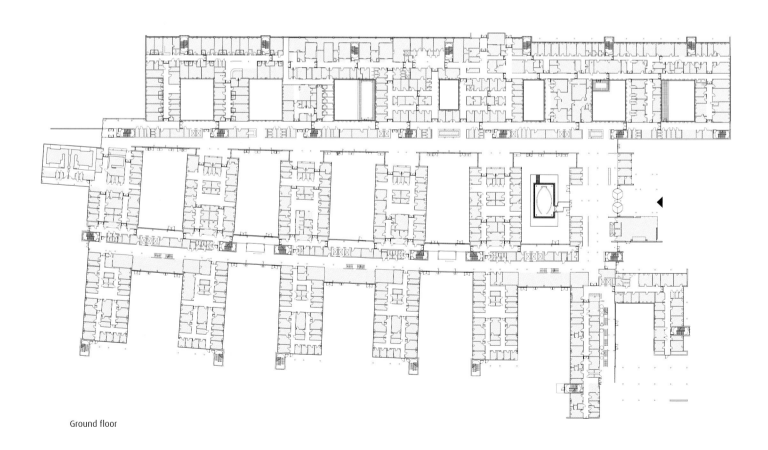

Ground floor

Johannes Wesling Clinic Minden

3 Entrance hall with orientation desks
4 Nursing station, general nursing
5 Façade, detail
6 Patient room
7 Operating theatre
8 Intensive care unit
9 Main hall
10 Café/restaurant
11 Inner courtyard with restaurant terrace

TMK Architekten Ingenieure

Photos: Jochen Stüber

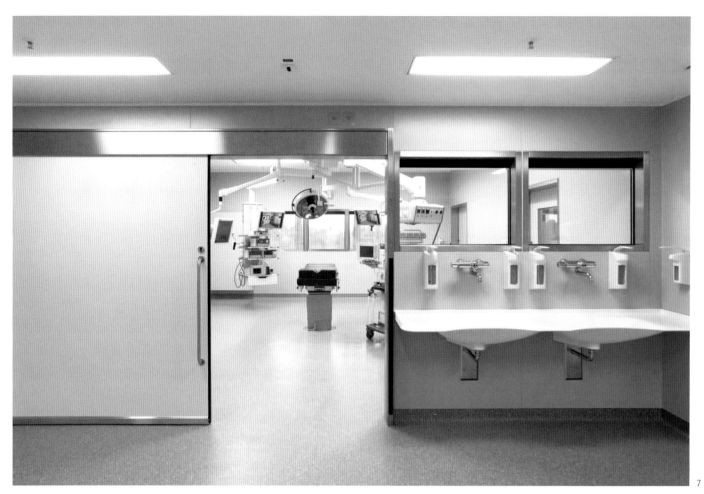

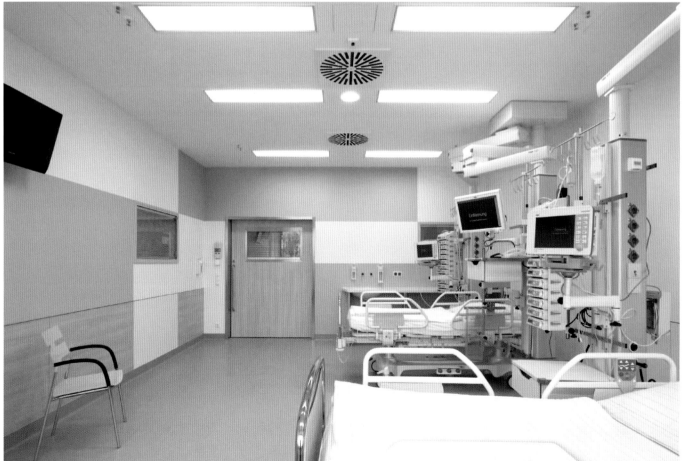

Johannes Wesling Clinic Minden

TMK Architekten Ingenieure

Johannes Wesling Clinic Minden

TMK Architekten Ingenieure

Architect
TMK Architekten Ingenieure

Operated by
Helios Kliniken GmbH

Situated in
Schwanebecker Chaussee 50, 13125 Berlin-Buch

Planning | Construction
2002 | 2004–2009

Beds
1,000

Gross ground area
102,000 square metres

Gross volume
438,000 cubic metres

Usable area
52,000 square metres

Total cost
200m euros

1 Aerial view
2 Elevation main entrance

✛ ● Helios Clinic Berlin-Buch

The new maximum-provision clinic in the Berlin-Buch district marks the end of the historic hospital dating from the 1920s and 1930s, of the Berlin architect and city planning officer Ludwig Hoffman. Taking the site's building and route axes as its reference, the result is a functionally compact and nevertheless transparent building complex with great internal flexibility. The aim was to bring together 23 specialist clinics, institutes, departments and polyclinics which had previously been spread out over five different sites in over 100 buildings. The basis of the design was clear signposting of the medical specialist disciplines, which are displayed on the building clusters. The design was based on basic urban patterns and way-finding systems. The clinic was interpreted as a small town and structured according to the principles of streets (main streets), squares (atria) and houses (blocks). Therefore six ward blocks and a central operating complex with 18 operating theatres are grouped around six four-storey glass roofed atria along an interlinking main thoroughfare. At the start of the main street, patients and visitors are received in a two-storey reception foyer with adjacent restaurant/café and a central chapel. All ward blocks are spatially flexible. Multifunctional spatial units can be connected together as required and used as out-patient areas. Despite the structural unity, friendly colours and light natural materials and forms ensure optimal way-finding. A different colour scheme dominates each atrium. In addition, the atria serve as climate buffers and form the central component of the efficient, sustainable energy concept. A geothermal heat exchanger preconditions the air flowing into the building via the ventilation system while the used air is fed outwards via the atria. A compact design also ensures a good energy balance.

Photos: Linus Lintner

Helios Clinic Berlin-Buch

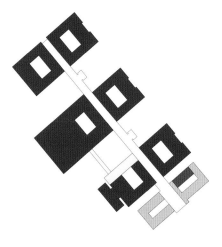

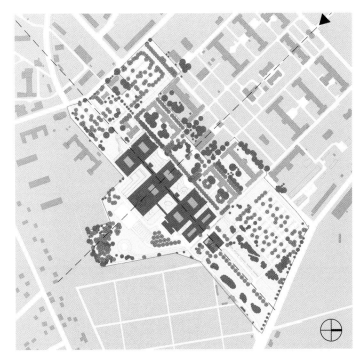

Site plan

TMK Architekten Ingenieure

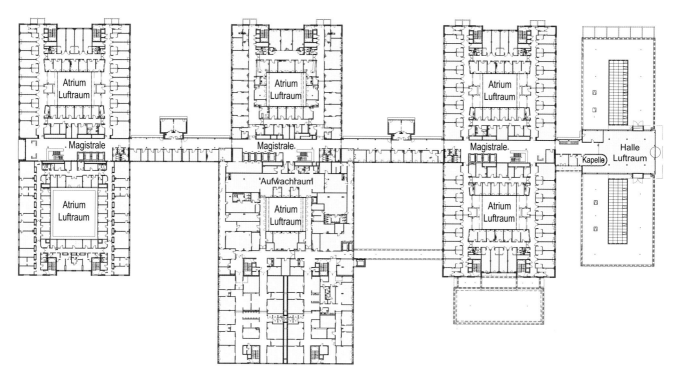

First floor

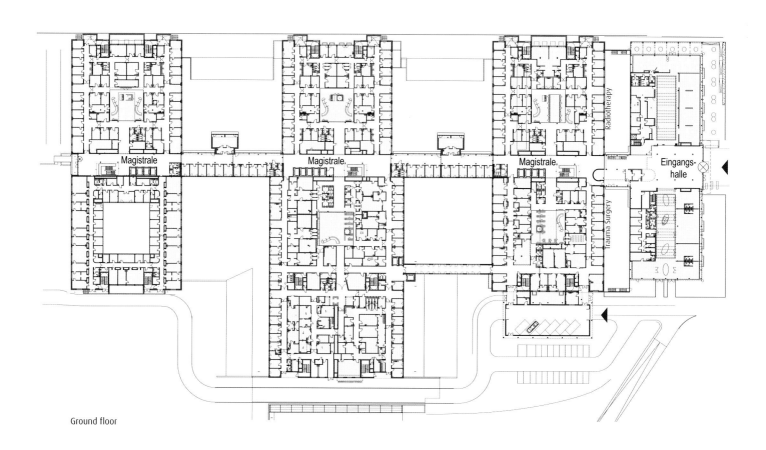

Ground floor

Helios Clinic Berlin-Buch

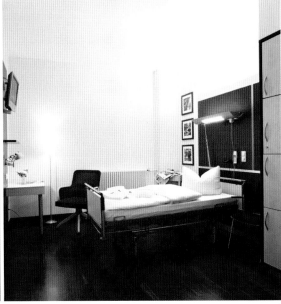

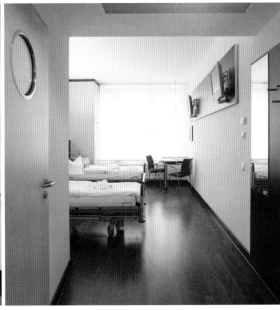

3 Waiting area
4 Patient room
5 View into two patient rooms
6 Patient room
7 Atrium
8 Atrium
9 Entrance hall
10 Main entrance at night
11 View into restaurant
12 Restaurant

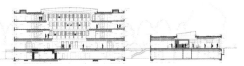

Longitudinal sections

Photos: Linus Lintner

Helios Clinic Berlin-Buch

TMK Architekten Ingenieure

Architect
Architekten BDA RDS Partner
with Koller Heitmann Schütz

Operated by
Klinikum der Stadt Wolfsburg

Situated in
Sauerbruchstraße 7, 38440 Wolfsburg

Planning | Construction
2000 | 2003–2005

Beds
256

Gross ground area
17,600 square metres

Gross volume
67,600 cubic metres

Usable area
8,800 square metres

Total cost
32.4m euros

1 Elevation west
2 Elevation south-east

✚ ∞ City Clinic Wolfsburg

The forty- to fifty-year-old nursing areas of the clinic needed replacing with contemporary, comfortable patient rooms and nursing equipment. The new "G Block" gives the clinic at Wolfsburg eight new, economical general nursing wards with 256 beds in total. As well as reorganising the nursing areas, the unsatisfactory entrance facilities in the clinic have been upgraded thanks to the G Block. A new main thoroughfare in front of the clinic has optimised wayfinding for both patients and visitors as well as separating visitors and out-patients traffic on the one hand from in-patients traffic on the other. The new-builds have been erected without substantially affecting the ongoing operation of the clinic and without interfering with the original building fabric. There are 60 parking spaces in the semi-basement. The south-facing sloping plot also permits level-ground access to a seminar, library and conference area. The main entrance to the clinic is on the ground floor. From the reception area there is an uninterrupted view over the main thoroughfare, the clinic forum, the neighbouring lift groups, the administration waiting area and the cafeteria. Rooms have been built on the east side of the G Block for spiritual resources and social services, and to house the patient media library and commercial service areas. The south wing houses the cafeteria with a large multi-use service area and open-air terraces. The main thoroughfare connects the ground floor and on the first floor the key vertical traffic elements in the old and new buildings by the shortest distance. The four upper storeys each contain two wards of 32 beds which are used for intermediate care and general nursing.

Photos: Rainer Mader

City Clinic Wolfsburg

2

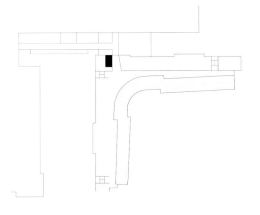

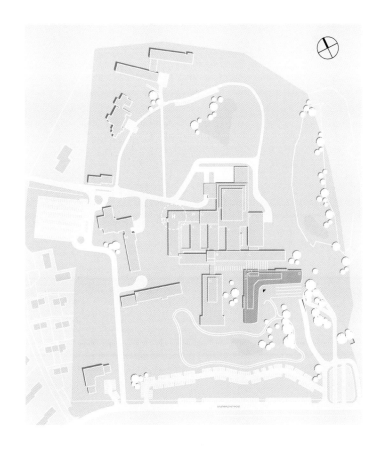

Architekten BDA RDS Partner | Koller Heitmann Schütz

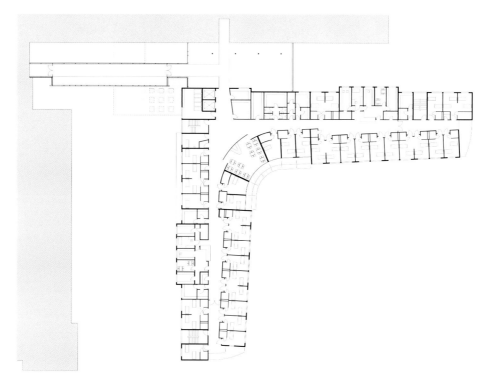

First floor

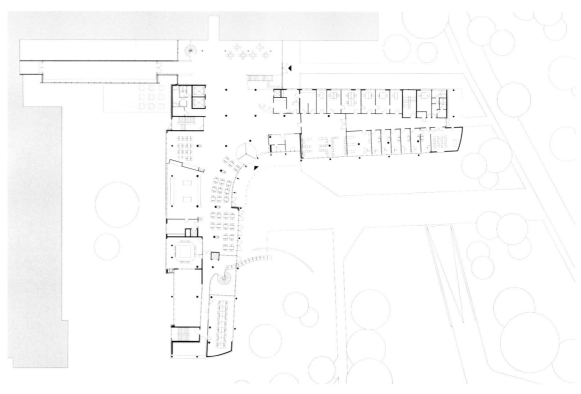

Ground floor

City Clinic Wolfsburg

3
4

3 New link building
4 Nursing service point
5 Main entrance
6 Main hall

Plans, scales 1:1,000 and 1:300

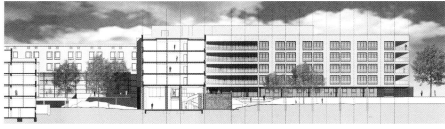

Elevation south

Elevation west

Perspective

Photos: Rainer Mader

Architekten BDA RDS Partner | Koller Heitmann Schütz

City Clinic Wolfsburg

Architekten BDA RDS Partner | Koller Heitmann Schütz

Architect
RRP Architekten + Ingenieure GbR

Operated by
Klinikum Forchheim

Situated in
Krankenhausstraße 10, 91301 Forchheim

Planning | Construction
2000 | 2002–2006

Beds
225

Gross ground area
21,000 square metres

Gross volume
82,000 cubic metres

Usable area | Access ways
10,500 square metres | 7,680 square metres

Total cost
57.3m euros

1 Façade
2 Façade, detail
3 Roof gardens

— ● New Clinic Forchheim

The 5.5 hectares area situated in the flood plain of the river Wiesent furnishes the hospital building to the east with a multi-storey car park, driver access and helicopter landing pad in the central area and a development area for a health centre to the west of the plot. The interlocking of natural and built areas was the design and planning goal. Rear-ventilated aluminium cladding and aluminium/glass façades are the defining materials of the building shell. All roof surfaces are intensively or extensively planted and so make a valuable additional contribution to the building climate and the overall energy balance. The building is designed around the central access hall with elements pushing into the landscape of the neighbouring nature reserve. The nursing areas to the east curve gently out of the basic direction of the building while the examination and treatment areas to the west are based on a strict grid pattern. The building services have been selected based on the guidelines dictated by the site-specific circumstances and the functional requirements taking equal account of energy-saving and environmental sustainability. The building services thereby underline the architectural approach and make a decisive contribution to its implementation with a variety of functional and system-appropriate solutions. The incorporation of the plentiful groundwater resources on the site led to a technical concept based on the heating and cooling potential of the building via concrete core activation. Heat is generated by three low-temperature boilers and eight miniature combined heat and power plants (CHPs). Cooling is carried out using the approximately 10°C spring water available on the site. CO_2 emissions will be around 65,000 kilograms less than with conventional technology.

Photos: Gerhard Hagen

New Clinic Forchheim

2

3

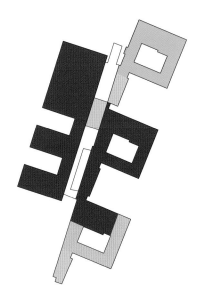

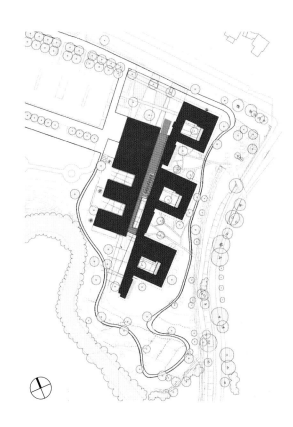

RRP Architekten + Ingenieure

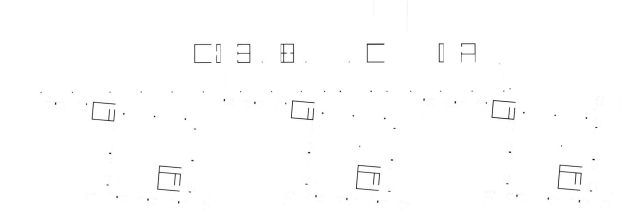

Second floor

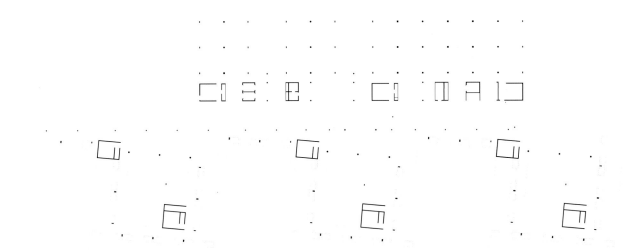

First floor

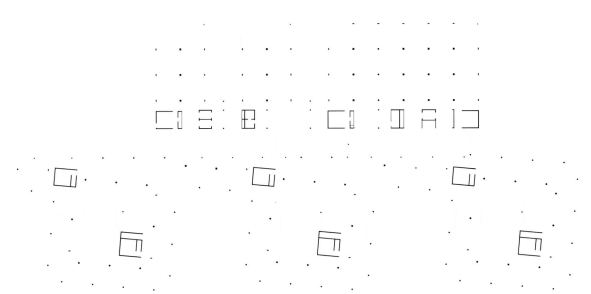

Ground floor

New Clinic Forchheim

4 Patient admission
5 Dining hall
6 Main entrance
7 Entrance hall

Plans without scale

Elevation east

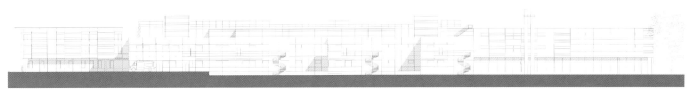

Elevation west

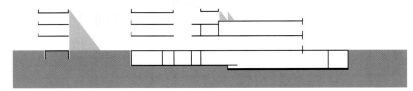

Section

Photos: Gerhard Hagen

New Clinic Forchheim

RRP Architekten + Ingenieure

Architect
Planungsring Dr. Pawlik + Co. Generalplanungsgesellschaft mbH

Operated by
Johanniter GmbH

Situated in
Johanniterstraße 1
14929 Treuenbrietzen

Planning | Construction
1994 | 1996–2005

Beds
300

Gross ground area
20,859 square metres

Gross volume
72,000 cubic metres

Usable area
8,900 square metres

Total cost
32m euros

1 Main entrance
2 South view
3 Café with outside area

✚ ● ○ Johanniter-Krankenhaus Fläming Treuenbrietzen

During the First World War, a complex was constructed consisting of four buildings which must be seen as the architectonic core of the present hospital. It was complemented by a tuberculosis centre in 1926–1927. This building is inspired by the then advanced idea of fresh-air therapy, providing not only spaciously separated loggia-buildings but also balconies in front of the patient rooms. The construction, a modern clinic for rheumatology, pneumonia and psychiatry, still profits today from these architectural merits; a nursing school is connected to it. The ensemble was renovated to reflect these new aims. It now houses, among other things, the lung clinic formerly housed in the Beelitzer Heilstätten. Moreover, a new ward block was constructed, along with intensive care and x-ray departments. The connection between the two main nursing buildings is a new, central entrance building for the entire hospital. The difference in height between the two clinics was visually overcome in this three-storey construction with lifts and bridges. The adjoining café with an exit to the garden is the social hub for patients and visitors. To complete the second building, a department for thorax surgery with a nursing area and operating department was constructed. The central sterilisation area is found under the operating rooms, with intensive care above it. The additions to the historical building was carried out in full respect of monumental-protective measures. The original colour concept from the 1920s was maintained, and the new parts of the building were erected in the style of the older buildings.

Photos: Archive Planungsring

Johanniter-Krankenhaus Fläming Treuenbrietzen

2

3

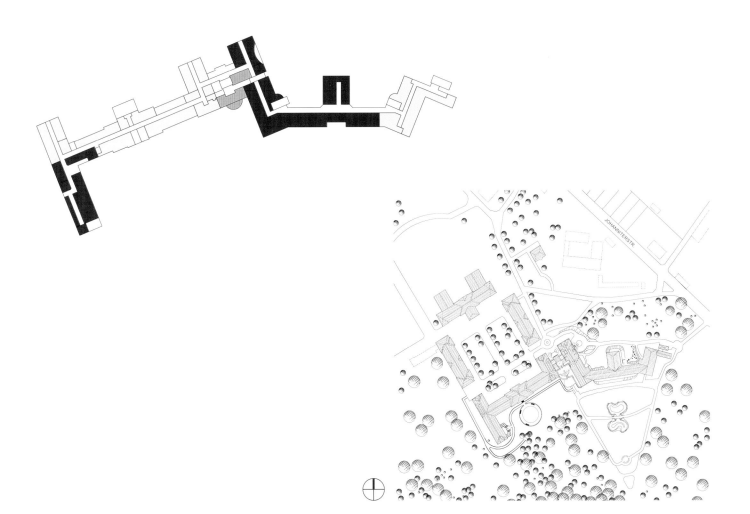

Planungsring Dr. Pawlik + Co. Generalplanungsgesellschaft mbH

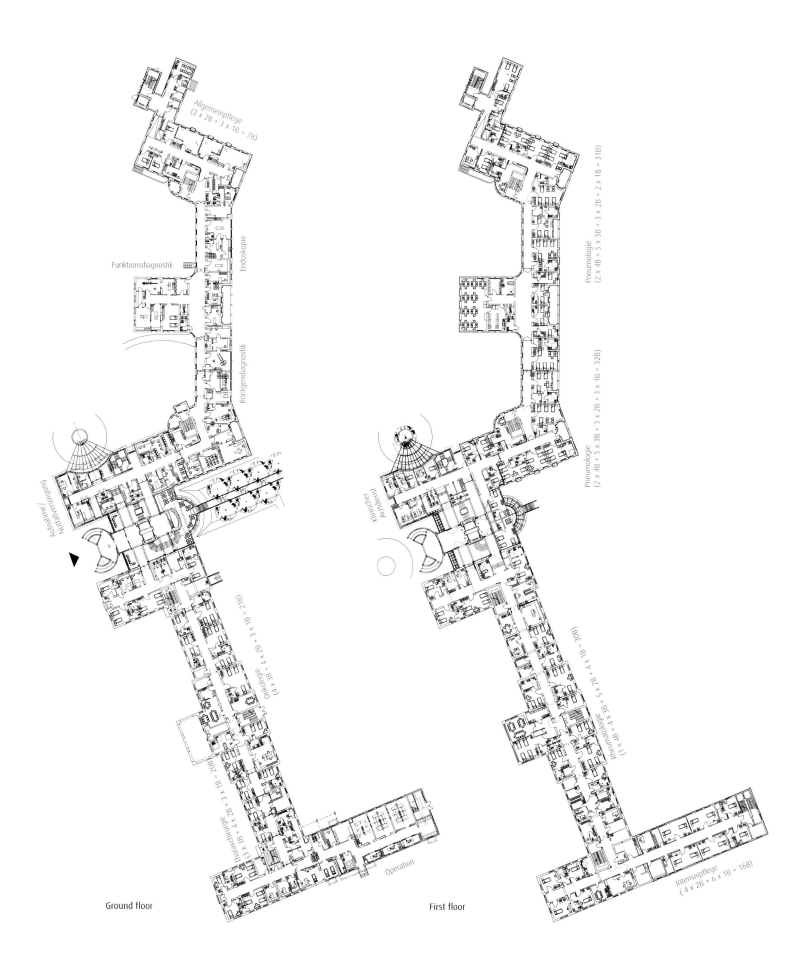

Ground floor — First floor

Johanniter-Krankenhaus Fläming Treuenbrietzen

4 Entrance hall
5 Cafeteria

Plans, scale 1:1,000

Elevation north-west

Section

Photos: Archive Planungsring

Planungsring Dr. Pawlik + Co. Generalplanungsgesellschaft mbH

General plan
Planungsring Dr. Pawlik + Co. Generalplanungsgesellschaft mbH

Architect
Prof. Gottfried Böhm/Friedrich Steigeweg, Rainer Goetsch
with Planungsring Dr. Pawlik + Partner, Dr. Peter R. Pawlik, Michael Mews

Operated by
Helios Klinikum Emil von Behring

Situated in
Gimpelsteig 9, 14165 Berlin

Planning | Construction
1989 | 1992–1995: first construction phase |
1995–1998: second construction phase

Beds
435

Gross ground area
28,326 square metres

Gross volume
134,838 cubic metres

Usable area | Access ways
15,600 square metres | 7,291 square metres

Total cost
62m euros

1 Covered inner courtyard with olive tress
2 Main entrance on Gimpelsteig
3 View from Gimpelsteig

✚ ● Emil-von-Behring-Clinic Berlin-Zehlendorf

This new construction of the Emil-von-Behrling Krankenhaus in south-west Berlin conveys the impression of a place of residence and services. The clear focus of attention is a graded cylindrical pilaster rising up out of steel and glass structure, which marks the entrance. On the side towards the covered courtyard, the cylinder reveals itself as the building's façade with bay-windows and balconies. Under the long, extended steel glass roof which unifies both patient wings like the nave of a cathedral, visitors stroll under olive trees, surrounded by façades as if in an Italian piazza. The patients, contributing to the vitality on the balconies and galleries, feel like residents of this little city. This conservative urbanity is reminiscent of motifs borrowed from shopping-malls. But what serves to stimulate, in the latter is dedicated to feeling at home here in a place where a stay is actually a state of emergency, something of which one senses very little. The covered courtyards of the two extended complexes are not only pleasant but also practical; they serve as clear orientation markers for the facilities inside the building and the bar that connects both wings with the main building in the middle. These new constructions bring together the nursing and treatment wards, which were formerly housed in different buildings, to form an economically feasible whole. The therapy zones are located on the ground floor of the ward block, with its bay-windows and rooms facing the sun. The ward block doctors' building and the main building, with its striking library tower, are the essential elements of this casualty and teaching hospital created to realise a small-town concept.

Photos: Archive Planungsring

Emil-von-Behring-Krankenhaus Berlin-Zehlendorf

2

3

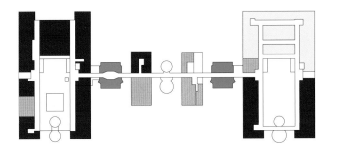

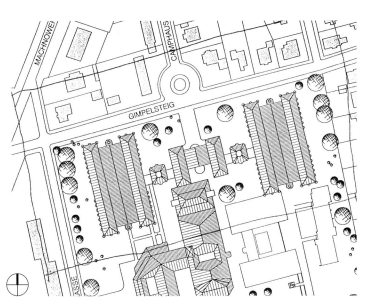

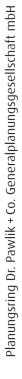
Planungsring Dr. Pawlik + Co. Generalplanungsgesellschaft mbH

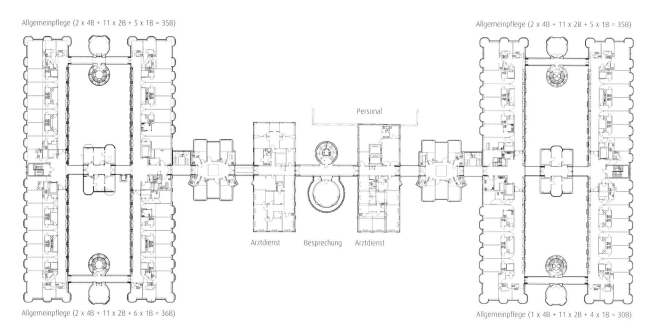

Second floor

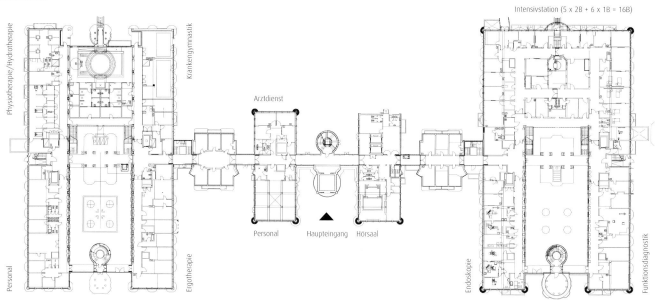

Ground floor

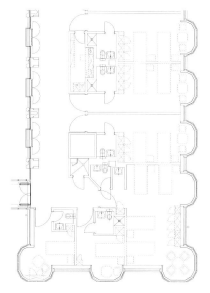

Nursing area

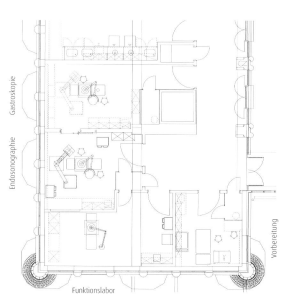

Endoscopies

Emil-von-Behring-Krankenhaus Berlin-Zehlendorf

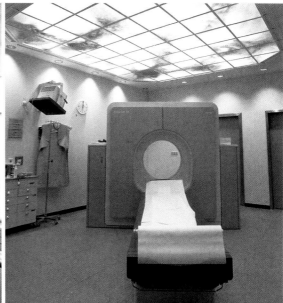

4
5

4 Sport ground
5 Computer tomography

Plans, scales 1:1,000 and 1:200

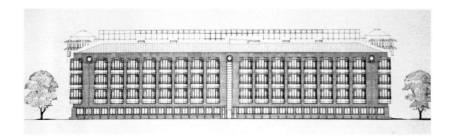

Elevation west

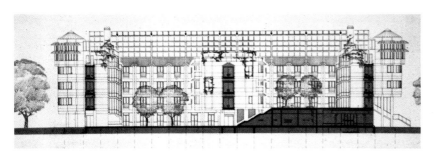

Section

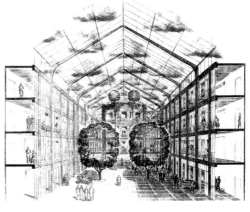

Isometry

Photos: Archive Planungsring

Planungsring Dr. Pawlik + Co. Generalplanungsgesellschaft mbH

Architect
Heinle, Wischer und Partner Freie Architekten

Operated by
Bundesministerium der Verteidigung

Situated in
Scharnhorststraße 13
10115 Berlin

Planning | Construction
1996 | 1997–1998

Beds
360

Gross ground area
2,800 square metres

Gross volume
15,000 cubic metres

Usable area | Access ways
1,636 square metres | 600 square metres

Total cost
9.1m euros

1 Main entrance
2 Night view, entrance hall and ward block

Bundeswehrkrankenhaus, Specialised Medical Examination Posts Berlin

The Bundeswehrkrankenhaus (Army Hospital) is officially the first new large scale military construction in Berlin since Reunification. The environment north of the new Berlin central rail station, characterised largely by industry and railways, will change a great deal the coming years. The new construction of the specialised medical examination posts for the Bundeswehrkrankenhaus, located on the site of a former military garrison hospital, is also coming to terms with this situation. The building rests upon a continuous foundation of stone. The formal execution is based upon a thoroughly clear module pattern and the façade is marked by horizontal window-bands. Unlike the former military hospital, the Bundeswehrkrankenhaus is also an urban address for medical care. The new construction of the specialised medical examination departments which resembles a polyclinic, is located directly on the street, so shielding the ward block from noise. Together with the first two storeys of this older building, which have been incorporated into the spatial concept, the new construction forms a structural unity, helping to separate the ambulatory from the stationary streams of patients. The new entrance hall is a striking architectonic connecting link between street and hospital, acting as a service centre for patients and staff. The building also has green inner courtyards directly accessible from the hall. From the street one especially notices the adjoining corridor with waiting zones; the functional rooms are situated towards the quiet inner courtyard benefiting from fresh air and plenty of daylight.

Photos: Rudi Meisel

Specialised Medical Examination Posts, Bundeswehrkrankenhaus Berlin

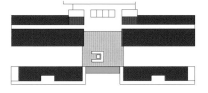

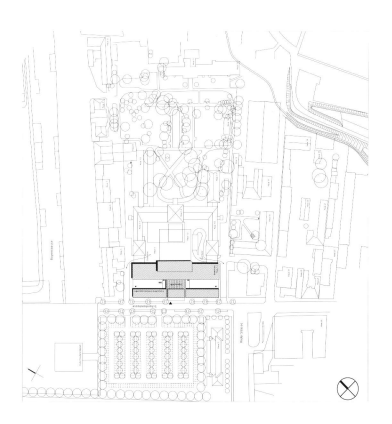

Heinle, Wischer und Partner Freie Architekten

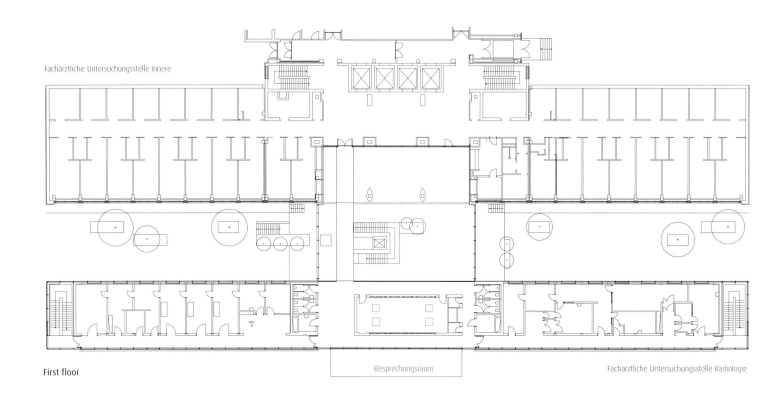

First floor

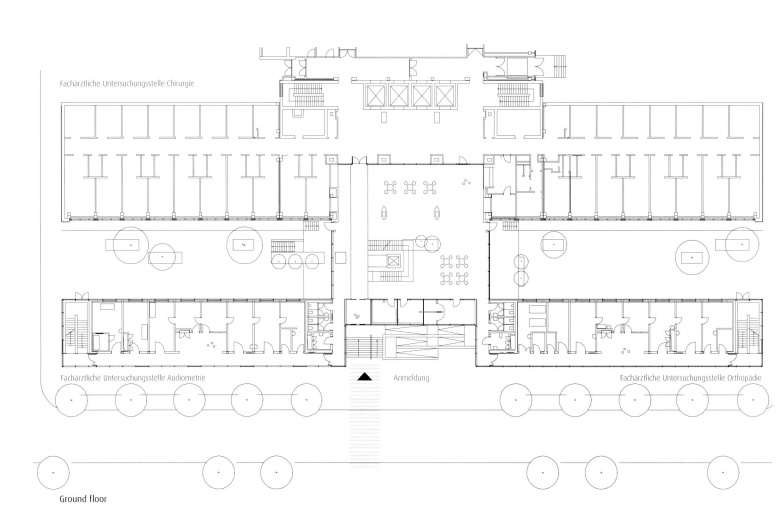

Ground floor

Specialised Medical Examination Posts, Bundeswehrkrankenhaus Berlin

3 Sitting room, entrance hall
4 Stairwell

Plans, scale 1:500

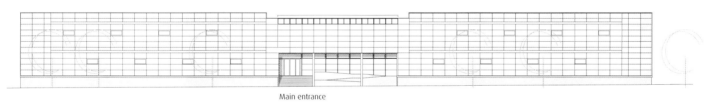

Elevation — Main entrance

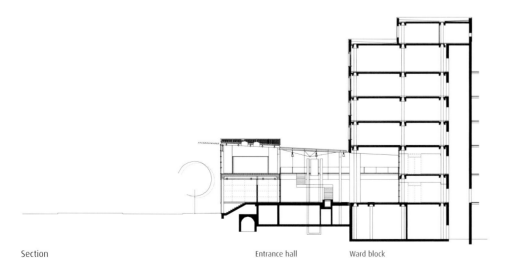

Section — Entrance hall — Ward block

Heinle, Wischer und Partner Freie Architekten

Photos: Rudi Meisel

Architect
Heinle, Wischer und Partner Freie Architekten

Operated by
Städtisches Klinikum Brandenburg GmbH

Situated in
Hochstraße 29
14770 Brandenburg/Havel

Planning | Construction
1998 | 2000–2003

Beds
128 (new construction west), 465 (total)

Gross ground area
11,177 square metres

Gross volume
57,229 cubic metres

Usable area | Access ways
7,169 square metres | 3,882 square metres

Total cost
69m euros

1 Main corridor and old building
2 New construction west, examination and treatment areas
3 Front driveway, accident and emergency

I ● Städtisches Klinikum Brandenburg/Havel

The Städtisches Krankenhaus in the inner city of Marienberg was founded in 1901. It grew from an institution of 150 beds into a district hospital with over 1,000 beds in the late 1970s. The main design concept behind the expansion of the Städtisches Krankenhaus is that of a cross overlapping the historical axis and the new glazed main corridor, connecting the old main building with the eastern and western new buildings on three levels. After the completion of the new west construction, which houses the examination, treatment, and general nursing wards, and is connected with the new helicopter landing area by a 50-metre-long bridge, another ward block will be built and the main building completely transformed in a second building phase. In the new surgical department, with ten operating rooms, the latest developments in anaesthesiology and hospital hygiene are put into practice.

Separate washrooms in front of each operating room are dispensed with in favour of a central working and service zone. The customary sterile corridor is also missing, thus making the ground plan more spacious and affording greater flexibility in operation management. The operating rooms border directly on the façade. The operating team can therefore work in plenty of daylight; one can see from the outside what is going on inside. In nursing, the concept of two wards with a maximum of 20 beds each is intended to encourage patients to take an active part in their recovery. Great value is placed upon direct and functional proximity between patient and staff. The room aesthetic also plays an important role as a psychological factor: the combination of warm colours, light and wood used here is intended to support and encourage the recovery process.

Photos: Bernadette Grimmenstein

Städtisches Klinikum Brandenburg/Havel

2

3

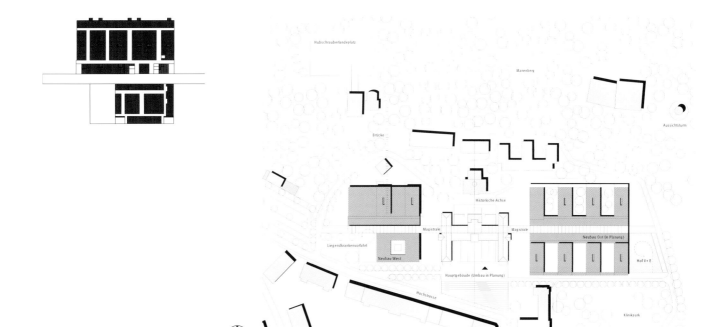

Heinle, Wischer und Partner Freie Architekten

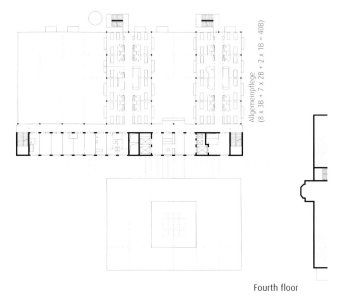

Fourth floor

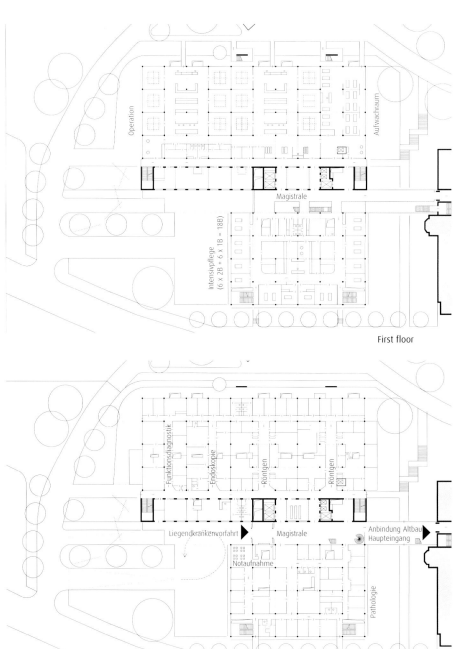

First floor

Ground floor

Städtisches Klinikum Brandenburg/Havel

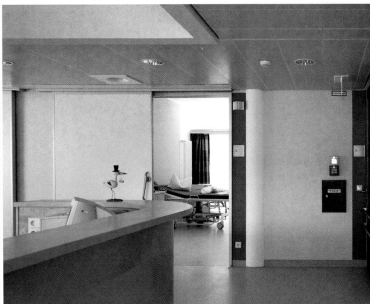

4
5

4 Patient lounge
5 Nursing station, maternity ward
6 Waiting area, accident and emergency
7 Operating theatre

Plans, scale 1:1,000

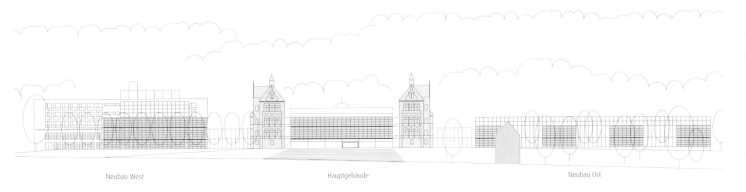

Neubau West Hauptgebäude Neubau Ost

Elevation, scale 1:1,250

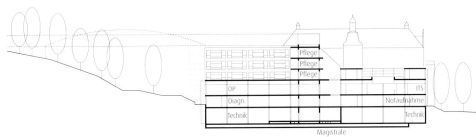

Section new construction west

Photos: Bernadette Grimmenstein

Heinle, Wischer und Partner Freie Architekten

Städtisches Klinikum Brandenburg/Havel

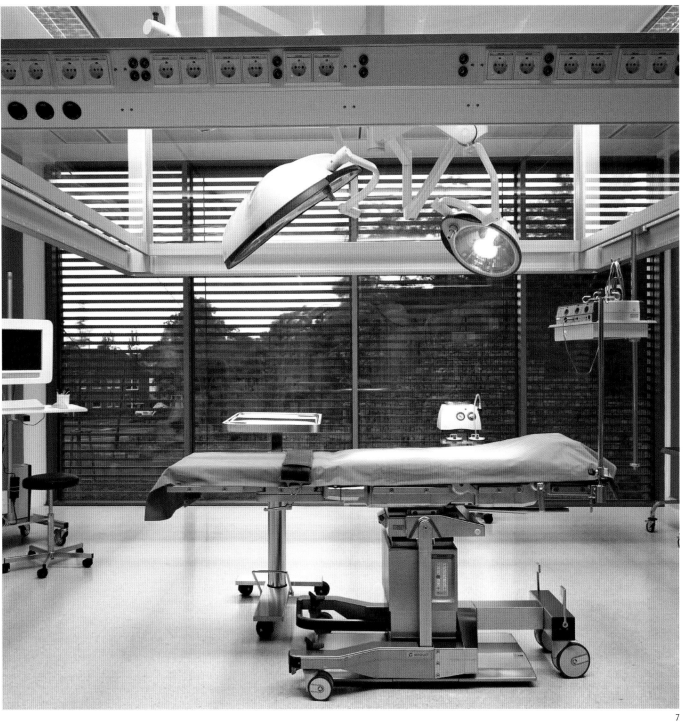

7

Heinle, Wischer und Partner Freie Architekten

Architect
Berg Planungsgesellschaft mbH & Co. KG

Operated by
Landkreis Bitterfeld

Situated in
Friedrich-Ludwig-Lahn-Straße 2
06749 Bitterfeld

Planning | Construction
1991 | 1994–1998: first construction phase |
2000–2004: second construction phase

Beds
525

Gross ground area
33,749 square metres

Gross volume
130,483 cubic metres

Usable area | Access ways
19,099 square metres | 8,296 square metres

Total cost
78.5m euros

1 Main hall
2 Main entrance
3 Entire complex

✚ ● ○ Kreiskrankenhaus Bitterfeld/Wolfen

The Kreiskrankenhaus was built in several phases at the Bitterfeld location whilst simultaneously using the Wolfen area. The new construction is surrounded by residential homes and undeveloped landscape, and is located on the grounds of the pavilion hospital originally built near the Bitterfeld city limits during the 1920s. Parts of the old building were incorporated into the new concept. The old east wing was connected to the new east wing by a glazed passageway and completed by two new portions. One reaches the beautifully landscaped inner courtyard through the old gate-house, which today houses the management. The glazed four-storey entrance hall can then be reached from there. This central hall strongly influences the appearance of the two-nave building, combining the functional areas of the east and west wings on all storeys. Level 0 is not only the central hub of the whole building, but also a place full of experiences for patients and visitors alike. The eye-catching glazed lift at the end of the hall serves to relieve the bed-lifts. From here the visitor reaches the lift and stairwell in the individual wards via the connecting passageway. The side rooms of the wards border on the atrium hall, by which they are lighted and ventilated. In the midpoint of the T-shaped wing we find the corresponding central service post, so that staff has to cover the shortest possible distance to reach their patients. The patient rooms are all orientated towards the outside, with views onto the city and its surroundings.

Photos: Christoph Petras, Berg Architekten

Kreiskrankenhaus Bitterfeld/Wolfen

2

3

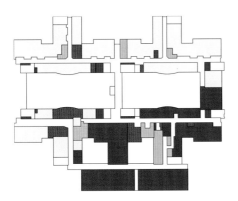

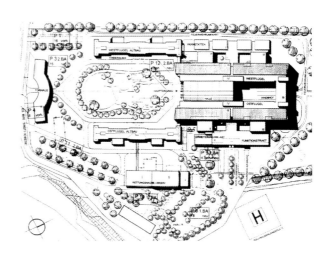

Berg Planungsgesellschaft mbH & Co. KG

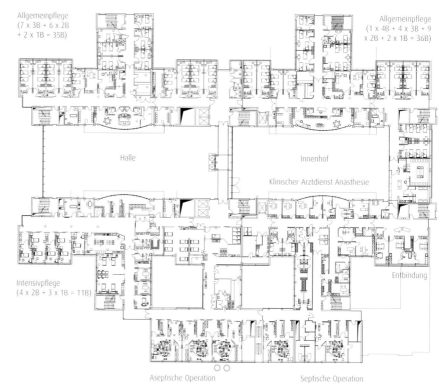

First floor

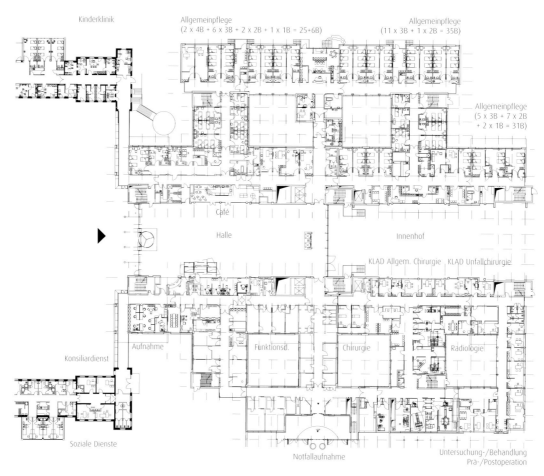

Ground floor

Kreiskrankenhaus Bitterfeld/Wolfen

4 Main hall, cafeteria
5 Courtyard

Plans, scale 1:1,000

Elevation

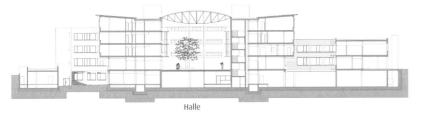

Section

Photos: Christoph Petras, Berg Architekten

Architect
Thomas Schindler Architekt BDA

Operated by
Focus Medical Klinikgesellschaft mbH, Berlin

Situated in
Klinik bei Waren/Müritz

Planning | Construction
1993 | 1995–1997

Beds
227 (nursing)

Gross ground area
20,500 square metres

Gross volume
72,500 cubic metres

Usable area | Access ways
12,768 square metres | 4,099 square metres

Total cost
29m euros

1 South façade
2 Aerial view
3 North façade

✚ ● Müritzklinik Waren/Müritz

The main concept of the Müritzklinik is that people should feel as well in this hospital as they would in a wellness-hotel. The colossus, placed like a sculpture in the woods, with 800 rooms over eight storeys, has a clear design, thanks to its façade covered with stripes alternating with windows. The gentle overlapping and ending of different building subdivisions and heights reflect the topography of the landscape, which is dominated by forests and lakes. Alongside glass, wood of different types is the predominant element, while the straight concrete parts lend stability to the complex. The smooth, clear transitions between facets of the design are continued in the interior. The centrepiece of the interior architectural arrangement is a bright, wide entrance hall furnished like a hotel lobby, along with a main staircase. A large glass opening in the roof allows light to stream down into the foyer. Wall niches and seating groups provide aesthetic touches in this orientation system, as do small foyers on the various storeys. The Müritzkinik is a building that eschews glaring colours and other eye-catching design devises. Exercise and the best nutrition guarantee a healthy lifestyle – something that the open layout of the kitchen and the gymnastics rooms demonstrably communicate to the patients in this clinic, which specialises in recovery from organ transplants, heart, rheumatic, and circulatory illnesses. The furniture, too, plays an important role in determining the high-quality – yet relaxed character – of this hospital building in the patient rooms as well, emphasising function, space, and architecture.

Müritzklinik Waren/Müritz

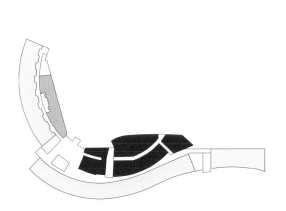

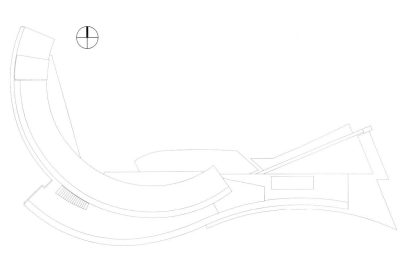

Thomas Schindler Architekt BDA

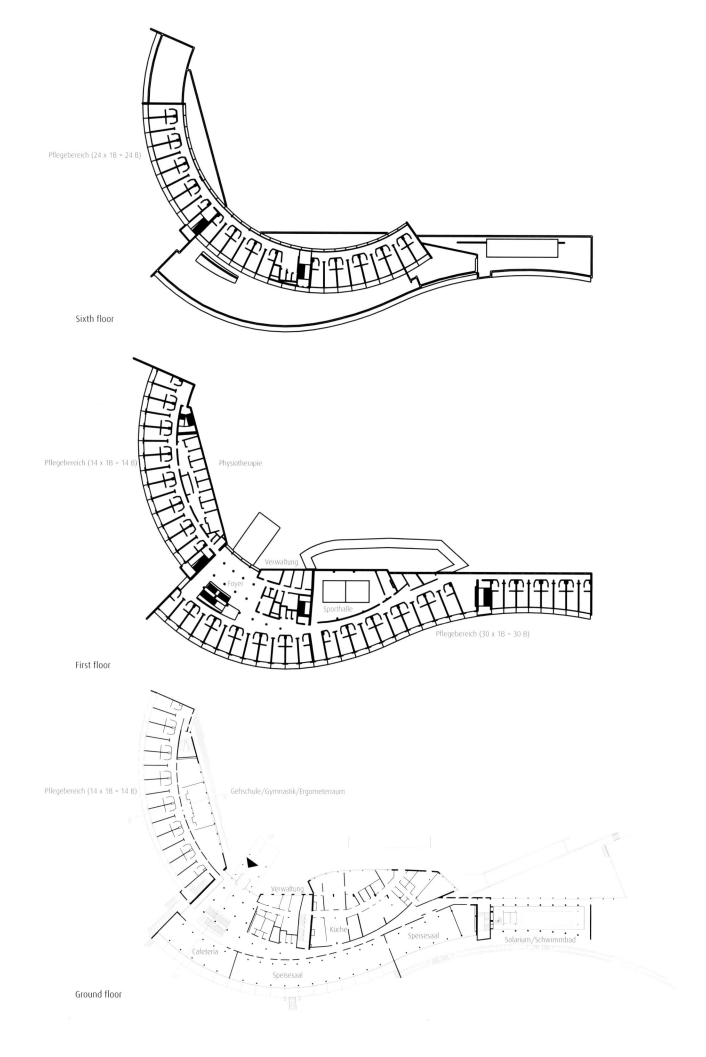

Sixth floor

First floor

Ground floor

Müritzklinik Waren/Müritz

4 Patient room
5 Entrance hall

Plans, scales 1:1,000 and 1:200

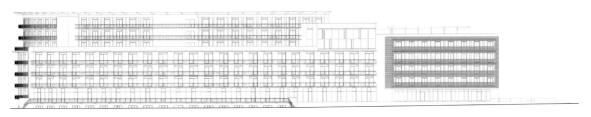

Elevation east

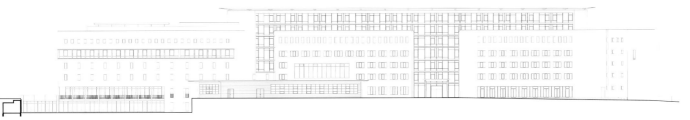

Elevation west

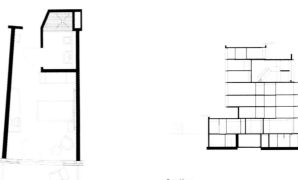

Floor plan, patient's room   Section

Thomas Schindler Architekt BDA

Architect
Steffen + Peter
Dr. Ribbert Saalmann

Operated by
Städtisches Klinikum Magdeburg

Situated in
Birkenallee 34, 39130 Magdeburg

Planning | Construction
1998 | 2000–2006

Beds
700

Gross ground area
56,861 square metres

Gross volume
193,369 cubic metres

Usable area | Access ways
28,694 square metres | 14,016 square metres

Total cost
117.2m euros

1 Ward block, façade detail
2 Ward block
3 Functional building

✢ ● ○ ∞ Städtisches Klinikum Magdeburg

The Städtisches Klinikum is located between the old village of Olvenstedt and the city of the same name, which was built over 30 years ago in the tradition of the Magdeburg garden settlements designed during the first third of the previous century. Somewhat old fashion in terms of organisation and practice, the former hospitals of Altstadt and Olvenstedt have been united in the new clinic. The result is a specialised hospital with fewer beds, in which operating costs have been reduced through the integration of essential equipment for diagnosis and surgery, as well as shorter distances; the nursing equipment is used much more efficiently as well. The good road and rail connections, plus the helicopter landing pad and the large available area, which offers space for further expansion, were good reasons for deciding to build on this site on the outskirts of Magdeburg. The hospital in Olvenstedt was renovated, expanded and newly structured in six building phases that lasted until late 2006. The functional building contains surgical areas with ten operating rooms, sterilisation facilities, and an intensive-care ward with 24 beds. The two newly built ward blocks incorporate surgical nursing wards, an emergency room, and psychiatry for children, adolescents, and adults. In the renovated third bed block concentrates the medical nursing ward, labour and delivery ward, pharmacy and supply storage. In the northern part the central laboratory, with the partially renovated x-ray department and pathology are found; the southern part houses urology, x-ray, physiotherapy, and the doctors' service.

Städtisches Klinikum Magdeburg

2

3

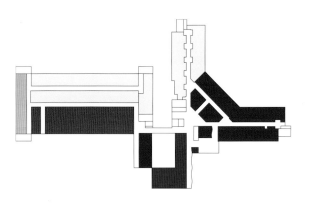

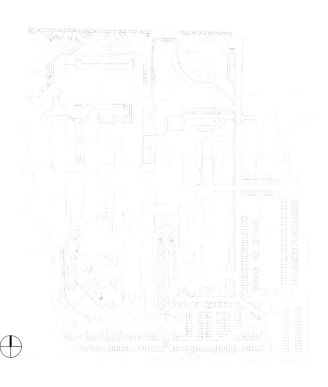

Steffen + Peter | Dr. Ribbert Saalmann

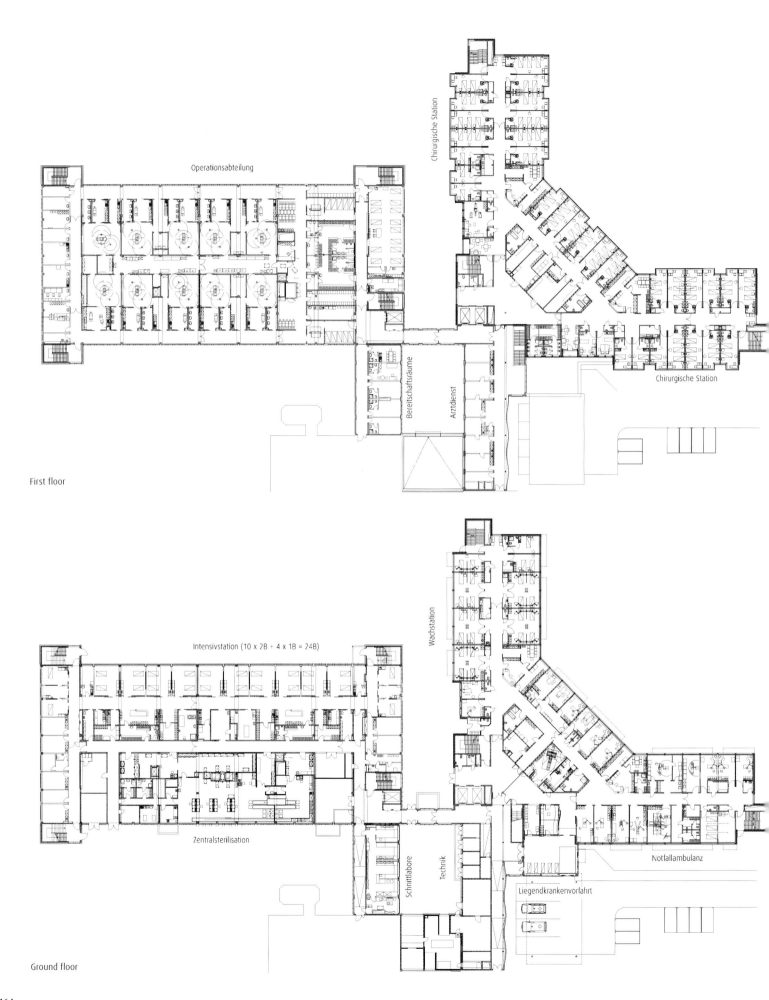

First floor

Ground floor

Städtisches Klinikum Magdeburg

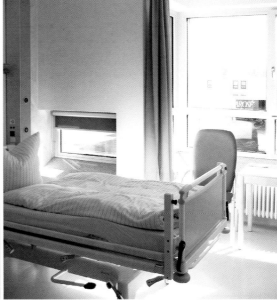

4 Entrance hall
5 Patient's room

Plans without scale

Main entrance

Elevation south

Architect
IAP Isin Architekten Generalplaner GmbH

Operated by
Ostalb-Klinikum Aalen

Situated in
Im Kälblesrain 1, 73430 Aalen

Planning | Construction
2001 | 2002–2004

Beds
426

Gross ground area
8,984 square metres

Gross volume
49,764 cubic metres

Usable area | Access ways
7,680 square metres | 2,310 square metres

Total cost
ca. 11m euros

1 Entrance driveway from multi-storey car park
2 Medical service centre
3 Entrance medical service centre

— ● ∞ Medical Service Centre, Ostalb-Klinikum Aalen

A person's mental state is a decisive factor in how he feels. Correspondingly, the Ostalb-Klinikum has changed its image from hospital to health centre. Visible signs of this are found in its architecture. The new entrance forum was designed so as to serve equally as reception, a lounge for conversation and a place for organised events all at the same time. Through a spacious design and the conscious use of colour and material, a space was created which is reminiscent of an urban square, thus conveying a comfortable atmosphere instead of the usual oppressive one. The entrance has the character of a reception area, which is achieved by the insertion of a long, open counter, resembling a hotel reception with its information and advice function. This bright, friendly atmosphere permeates the entire architecture and is emphasised by a practical aesthetic. In the spacious foyer, the cafeteria has been inserted as a clearly perceptible box. The patient transport from the helicopter landing area is in the basement, out of sight of visitors. The medical service centre, in the form of five bright, luminous cubes linked by inner courtyards, is connected to the entrance forum. It houses doctors' practices, speech therapy and ergotherapy, coiffeur and ambulance bay, counselling and office areas for out-patient care service, a psychosomatic day clinic, a nursing hotel of the German Red Cross, Health Ministry examination and treatment rooms as well as clinic administration offices. The four inner courtyards provide ample sunlight for all building units. Technical provision is accomplished through an installation core in each cube.

Photos: Ankenbrand/Photostudio Spectrum (2, 3)

Medical Service Centre, Ostalb-Klinikum Aalen

2

3

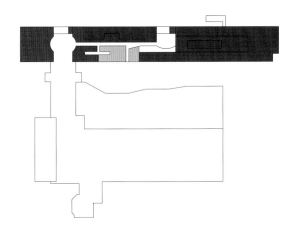

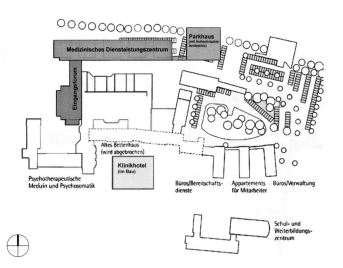

IAP Isin Architekten Generalplaner GmbH

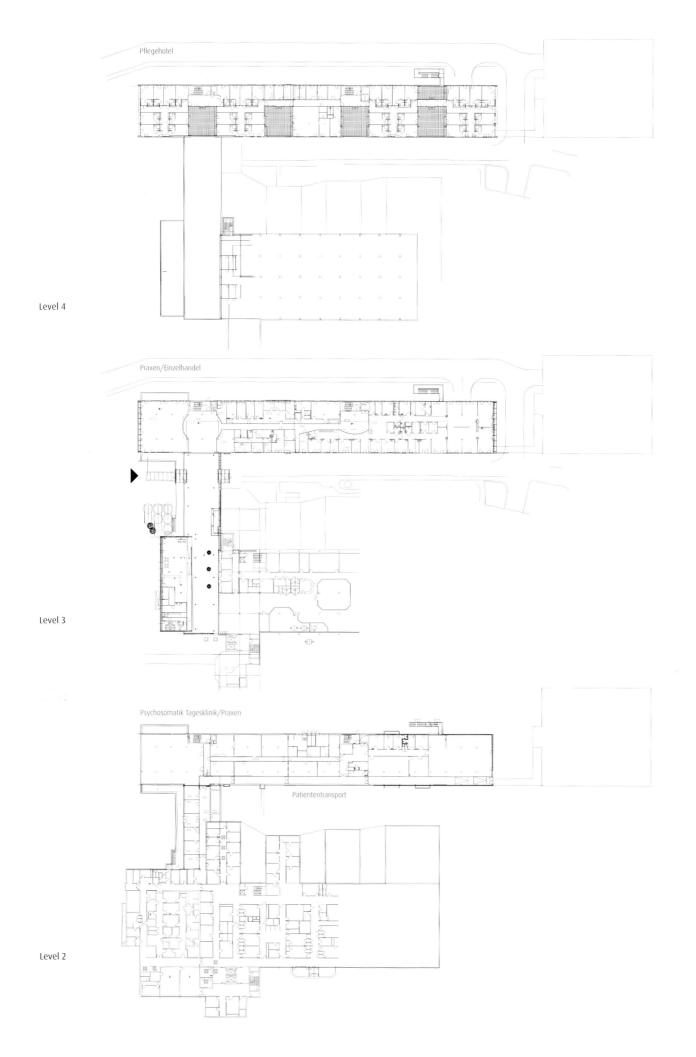

Medical Service Centre, Ostalb-Klinikum Aalen

4 Guding principle, German Red Cross
5 Patient lounge
6 Colour design, stairwell
7 Colour design, passageway

Plans, scale 1:1,250

Section

Elevation, south

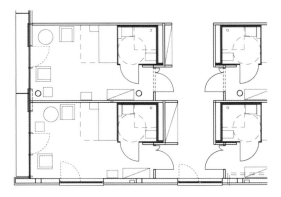

Extract room

Photos: Ankenbrand/Photostudio Spectrum

IAP Isin Architekten Generalplaner

Medical Service Centre, Ostalb-Klinikum Aalen

IAP Isin Architekten, Generalplaner

Architect
TMK Architekten Ingenieure
Thiede Messthaler Klösges

Operated by
Stiftung St. Johann Nepomuk Erfurt

Situated in
Haarbergstraße 72, 99097 Erfurt

Planning | Construction
1997 | 1999–2003

Beds
420

Gross ground area
43,000 square metres

Gross volume
172,000 cubic metres

Usable area | Access ways
18,500 square metres | 11,650 square metres

Total cost
92.5m euros

1 Main entrance
2 Ward block, park side
3 North side, main entrance

— ● Katholisches Krankenhaus St. Johann Nepomuk Erfurt

The hospital, located on the promenade of a lake which acts as a rain-water reservoir basin in this bishopric in south-eastern Thüringen, radiates a holiday atmosphere with its white architecture. With a hotel-like new building on the 10-hectare grounds, the building founded in 1735 with a nursing school to accommodate 80 people, together with staff apartments located above have entered upon a new future, in terms of both their organisation and their technical equipment. The new building, parallel to the street, is arranged in a compact, three-storey treatment wing which also houses the institution's IT services. The four buildings leading to a health park are nursing wings with single and double rooms. On the ground floor of the examination and treatment wing directed towards the street, we find patient admissions with emergency room, x-ray diagnostics, as well as radiology for bed-bound patients accessed via a separate entrance area. The link between these building parts and the ward block is a main corridor, glazed and located above a two-storey entrance hall and extending over four storeys, with open staircases and visitors' and patients' lifts situated towards the front. Along this main passageway, the cafeteria and the hospital chapel are integrated into the main entrance. The psychiatry department with out-patient department is on the ground floor. Amongst other things, the clinic doctors' service linked together with special out-patient departments are on the second storey of the entrance hall. The surgical department with intensive care, sterilisation and labour and delivery are found on the third storey.

Photos: Jochen Stüber

Katholisches Krankenhaus St. Johann Nepomuk Erfurt

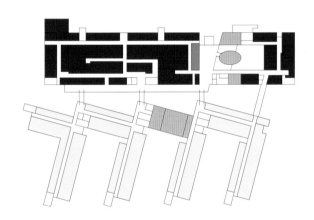

TMK Architekten Ingenieure

First floor

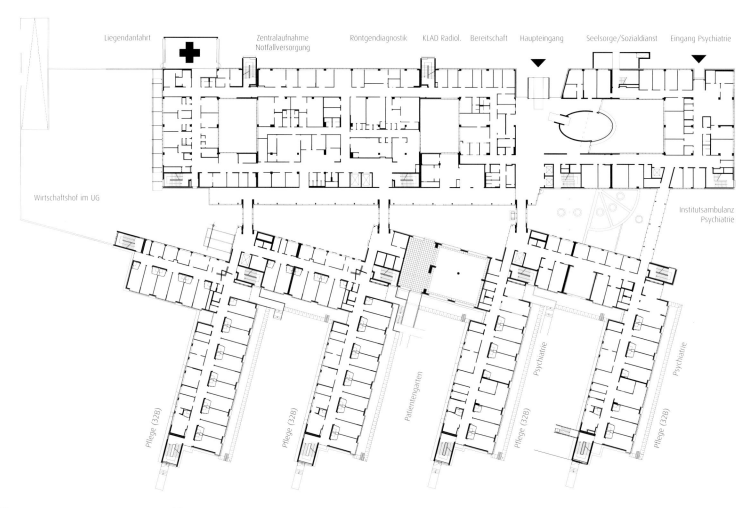

Ground floor

Katholisches Krankenhaus St. Johann Nepomuk Erfurt

4
5

4 Entrance hall with chapel
5 Main hall
6 Park side, ward block
7 Inner courtyard, ward block

Plans, scale 1:1,000

Main entrance Entrance hall Ward block

Section

Photos: Jochen Stüber

TMK Architekten Ingenieure

Katholisches Krankenhaus St. Johann Nepomuk Erfurt

TMK Architekten Ingenieure

Architect
Arcass Freie Architekten BDA

Operated by
Diakonissenanstalt Stuttgart
Stiftung Paulinenhilfe Stuttgart

Situated in
Rosenbergstraße 38, 70176 Stuttgart

Planning | Construction
1998 | 2001–2008

Beds
479

Gross ground area
46,000 square metres

Gross volume
177,000 cubic metres

Usable area
23,250 square metres

Total cost
132m euros

1 Green inner courtyard
2 Main entrance with old and new clinic buildings
3 Street-crossing Seidenstraße and Rosenbergstraße

✢ ● ○ ∞ Diakonie-Klinikum Stuttgart

The Evangelisches Diakonissenkrankenhaus and the Orthopädische Klinik Paulinenhilfe were formerly independent hospitals located in different areas in the western part of Stuttgart. They merged under one roof in the year 2000 on the site of the Diakonissenkrankenhaus to form a clinic capable of coping with the demands of the future. This new Diakonie-Klinikum forms a harmonious block-edge construction added to the part of town built during its formative period. A green, quiet inner courtyard is surrounded by buildings made of transparent constructional elements, creating brightly lit rooms and an atmosphere of comfort for the recovering patients. Most of the patient rooms are oriented towards the garden in the inner courtyard. An impressive view of Stuttgart can be had from the upper storeys of the service rooms directed towards the city and its public areas. The main entrance lies on the dividing line between old and new architecture, i.e. in the middle of the new, completed complex. A two-storey, angular entrance hall connects the old and new buildings, generating hotel-like atmosphere as do all the publicly accessible areas. Supply and waste disposal are carried out by means of an underground housekeeping department. The building ensemble, made of historic monument-protected building materials and sharply defined contemporary constructions makes an important contribution to this part of the city. The architecture combines tradition, medical progress and technological innovation in a clearly visible way.

Photos: Siegfried Gragnato

Diakonie-Klinikum Stuttgart

2

3

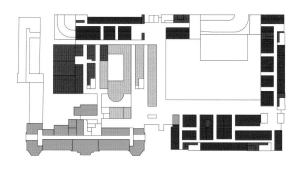

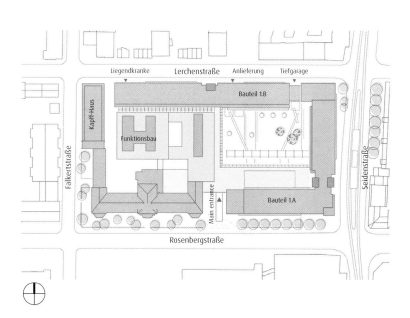

Arcass Freie Architekten BDA

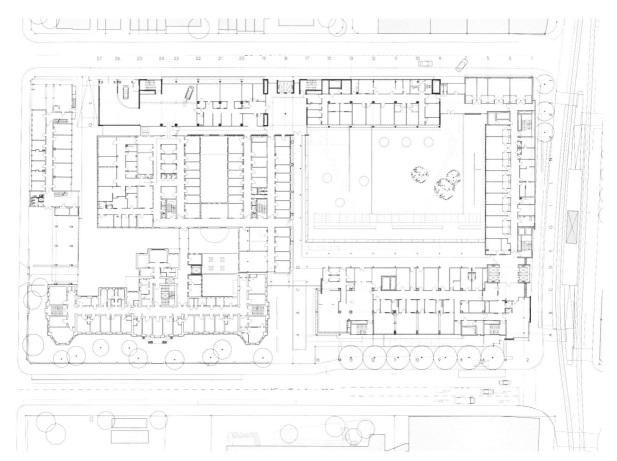
First floor

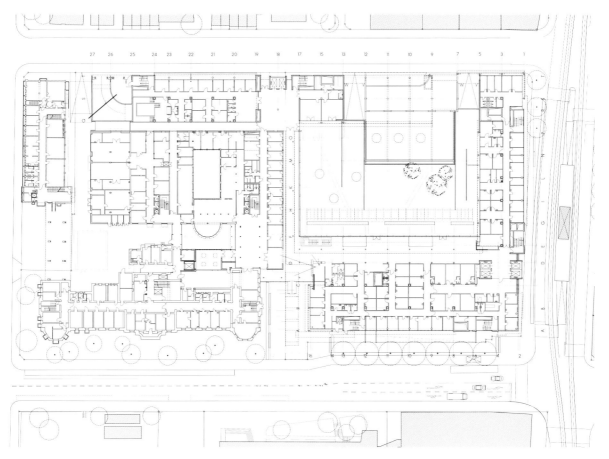
Ground floor

Diakonie-Klinikum Stuttgart

4 Double room
5 Access to patients' rooms
6 Entrance hall
7 Corridor operation area

Plans, scale 1:800

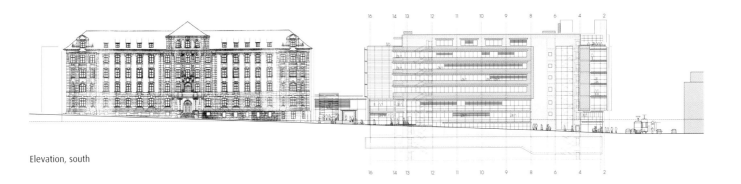

Elevation, south

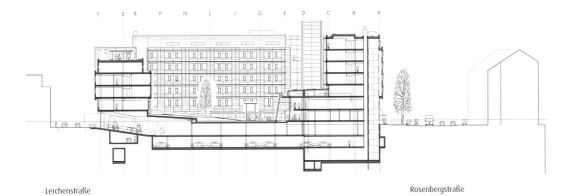

Lerchenstraße   Rosenbergstraße

Section

Photos: Siegfried Gragnato

Arcass Freie Architekten BDA

Diakonie-Klinikum Stuttgart

Arcass Freie Architekten BDA

Architect
Arcass Planungsgesellschaft mbH

Operated by
Universitätsklinikum Heidelberg

Situated in
Im Neuenheimer Feld 410, 69120 Heidelberg

Planning | Construction
1992 | 2001–2004

Beds
319

Gross ground area
53,200 square metres

Gross volume
304,000 cubic metres

Usable area | Access ways
25,000 square metres

Total cost
178m euros

1 Transitional hall to the ward block
2 U-shaped ward block
3 Aerial view

— ● University Medical Centre Heidelberg

The new construction for the Medizinische Klinik follows the idea of an integrated complete clinic. Situated on the university grounds in the Neuenheimer Feld near the Neckar river, it is planned to gradually link nearly all areas of specialisation into one architectural ensemble. This clinic is an important contribution to establishing the best conditions for research, education and patient care. The three-storey entrance hall is therefore located on the edge of the clinic, in order to serve as a connecting module for further extension. The overall compact new building is opened up by inner courtyards offering a wide variety of views towards the outside.

Clear lines and spacious forms in a simultaneously timeless, objective architectural language set the tone for the facilities, both inside and out. The character of the surrounding flora is important. A botanical garden with old trees lends the entrance area an unmistakable character. The new parking facility in the horseshoe-shaped inner courtyard of the hotel-like ward block guarantees a pleasant atmosphere with its visually integrated environment. All areas of patient care are clearly and recognisably connected by the patient passageway leading from a mall-like entrance hall. Despite the large dimensions, everything is clearly organised.

Photos: Dietmar Strauß

University Medical Centre Heidelberg

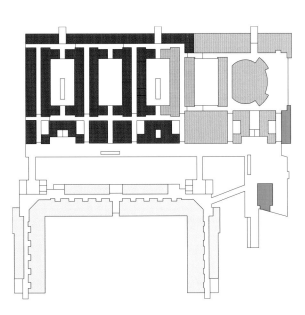

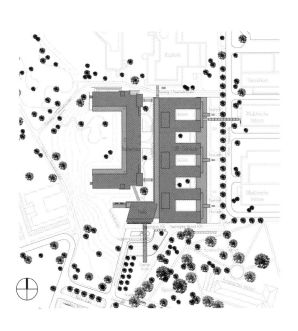

Arcass Planungsgesellschaft mbH

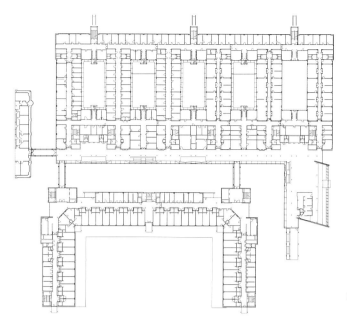

Examination and treatment area

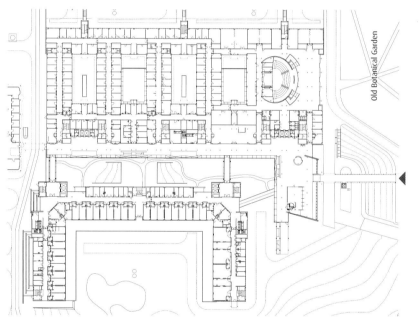

Main entrance level

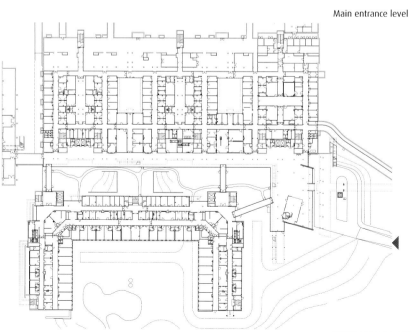

Stretcher access, entrance level

University Medical Centre Heidelberg

4
5

4 View in the patient passageway
5 Atmosphere, patient passageway
6 Corridor design, nursing area
7 Corridor design, examination area

Plans, scale 1:1,750

Elevation east

Elevation, east. Inner courtyard

Photos: Dietmar Strauß

Arcass Planungsgesellschaft mbH

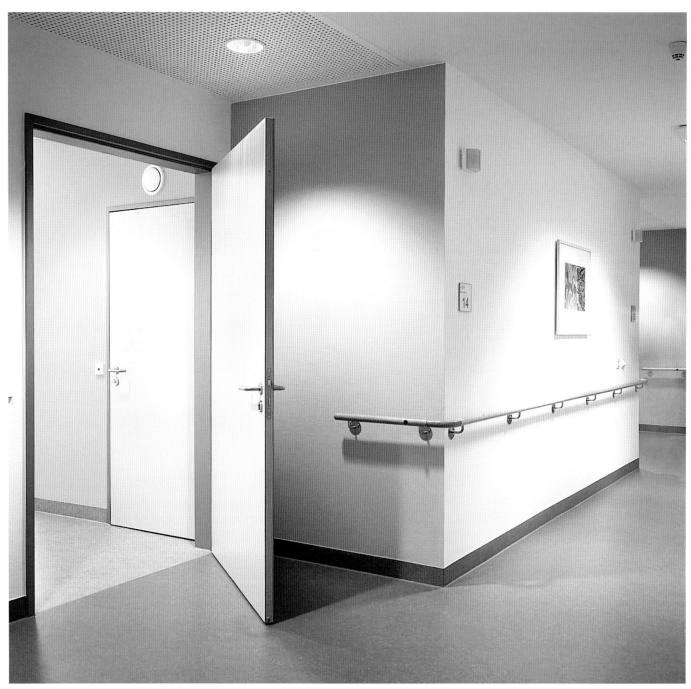

University Medical Centre Heidelberg

Architect
Arcass Freie Architekten BDA
Prof. Joachim Schürmann (entrance area)

Operated by
Robert-Bosch-Krankenhaus GmbH

Situated in
Auerbachstraße 110, 70376 Stuttgart

Planning | Construction
2000 | 2002–2009

Beds
521

Gross ground area
72,000 square metres

Gross volume
240,000 cubic metres

Usable area | Access ways
33,000 square metres | 16,700 square metres

Total cost
182m euros

1 Exit, garden side
2 Model of the complete facilities
3 Ward block, garden side

I ● O ∞ Robert-Bosch-Krankenhaus Stuttgart

The Robert-Bosch-Krankenhaus, which began operation in 1973 as a new construction, lies on the edge of the inner city on a rounded hilltop on the transition to still existing vineyards and green fields. The cruciform, five-storey building contains nursing areas on its upper storeys, complemented by in-patient treatment, surgical department, and labour and delivery ward. On the extended ground floor are located the general facilities as well as diagnostics, examination and treatment; supply and waste disposal are on the free-standing ground floor on the crest side. Corresponding to today's standards in function, operation and appearance, the overall renovation and further development of the building are being undertaken whilst the hospital is fully operational in three phases of 38 units altogether.

The new structure is the result of new construction, adding storeys and renovation, as well of prefabricated building modules for nursing and care with 120 beds. Alongside state of the art medical and research facilities, economic concerns are given a specially high priority. Moreover, the approach of the founder, Robert Bosch, required special attention to be paid to both the internal and external appearance of the hospital. The heavily exposed concrete façade, showing wear and tear typical of its ear, was replaced by a new, light, filigree façade in the area of nursing and care. The examination and treatment areas on the lower floors received band-shaped plaster façades, adjustable to fit different room sizes. The entrance facilities are located in glazed pavilions.

Photos: Dietmar Strauß

Robert-Bosch-Krankenhaus Stuttgart

2

3

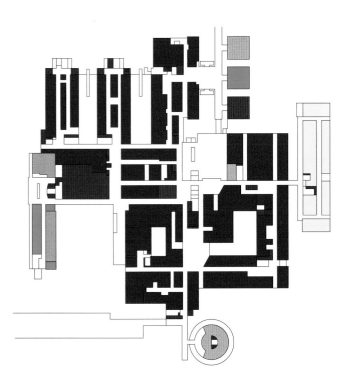

Arcass Freie Architekten BDA | Prof. Joachim Schürmann

Ground floor

Robert-Bosch-Krankenhaus Stuttgart

4
5

4 General care, corridor, first to fourth floor
5 General care, corridor, fifth floor
6 Delivery room
7 Patient room

Plans, scale 1:850

Elevation, north-west

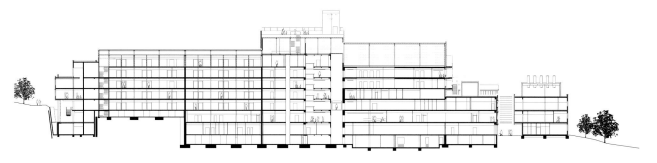

Section, west-east

Photos: Dietmar Strauß

Arcass Freie Architekten BDA | Prof. Joachim Schürmann

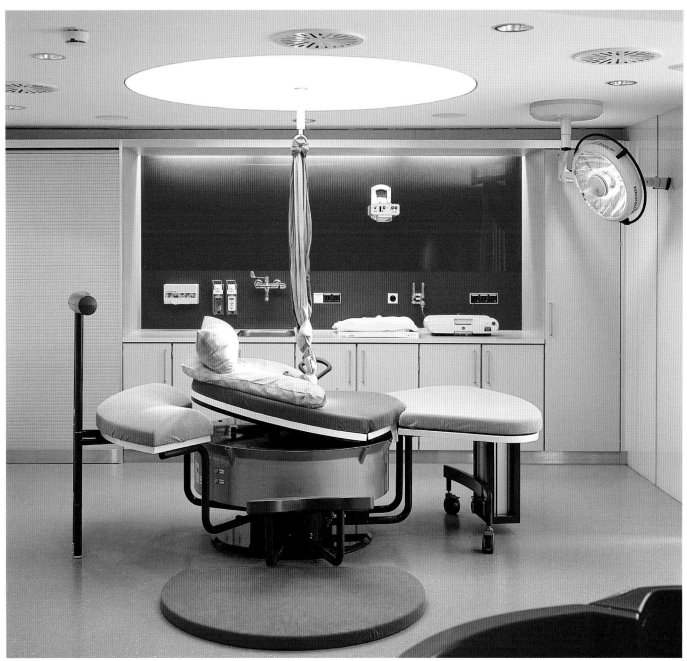

Robert-Bosch-Krankenhaus Stuttgart

Arcass Freie Architekten BDA | Prof. Joachim Schürmann

Medical Practices

Philipp Meuser
Architecture as a Factor of Quality

Franz Labryga
Planning Medical Practices: the German Principle

Medical Practices

Philipp Meuser

Architecture as a Factor of Quality

The German weekly magazine "Der Spiegel" dubbed him the "corpse manufacturer", producing human bodies as if on a serial assembly line and preparing them for the afterlife, by arranging them in ordinary, real-life poses as dancers, teachers, a rider astride the body of a horse prepared in the same way – just as if they had become fixed in the position of their final moments before death. In their second lives, these beings have taken on a different form and being turned inside out, displaying bones, tendons, muscles and internal organs.

The corpse cabinet of Dr. Gunther von Hagens, the well-known anatomist, is the result of a process he himself developed. Known as plastination, it was patented in 1977. It takes only a glance to understand that the exhibitions he has staged showcase human beings both as marvels and as models for creation. As medical man and director of the Institute of Plastination, Dr. von Hagens intends his highly provocative exhibitions to instruct the general public in the fascinating world of the "human body, its functions, diseases and changes" all within the context of health – a goal successfully achieved.

Since 2000, von Hagens has attracted some 30 million visitors to his exhibitions of post-mortal artefacts known around the globe as Body Worlds. In this, the media-savvy doctor paves the way. Even before healthcare reform was being publicly debated in Germany, he set health education at the heart of this political initiative, whose implementation has altered the healthcare system in Germany like no other measure since 1945. It became the cornerstone of his thinking.

Awareness of one's personal health, prevention of illness and the quest for knowledge of how the body works have long been preoccupations of advanced cultures. But even though this line of thought is not new, Dr. Gunther von Hagens' museum-type displays are definitely a novel form of education. Such is his skill that he not only popularises medicine – he also markets his own genius as an inventor. But anatomist and Renaissance man von Hagens, who, with his hat, resembles the artist and marketing supremo Joseph Beuys, is also his own publicist: "Wherever there is a microphone or a camera, the taxidermist is immediately in evidence," writes "Der Spiegel". His consummate marketing skills exploit the modern media – with an internet presence, targeted media work and educationally outstanding exhibitions. The recipient of international patents and awards combining the roles of entrepreneur and adviser in the field of healthcare in a variety of different ways. Born in 1945 in Poland a self-proclaimed internationalist and intellectual adventurer, Dr. von Hagens combines the very characteristics essential to best promote his medical expertise in the healthcare market of the future. In this respect, too, he is a role model for his colleagues.

The privatisation of health

In Germany the drive for privatisation has turned health into a market subject to competition. The mere fact of being a doctor is simply not enough and by no means a guarantee of a state protected status or of automatic commercial success. Accredited medical associations insure around 135,000 GPs, specialists and therapists, most of them independent practitioners [figure for 2009]. Then there are approximately 24,000 non-medical practitioners and holistic therapists and providers of purely non-medicinal treatments centred on "wellness", all serving a vast and paying clientele. In recent years, these professional groups have radically changed public perception of the body, sickness and health. Anyone running a practice these days must not only succeed as a businessman but must also face the challenge of a fundamentally different attitude to quality of life. As regards the competition, it helps to have a strong personal profile, easily visible to everyone – and as for the change in outlook, the approach required is an ability to rethink one's orientation and adjust accordingly. Both personal profile and public perception affect the identity of the enterprise and the health of the doctor's business. Health care reform their own contributions towards the cost of healthcare, has made people wary of consulting

The popularisation of medicine: Male body with torso cut twice vertically on show in the *Body Worlds* exhibition which has so far attracted over 37 million visitors around the world (source: Körperwelten)

their doctor. The state approach means that people must assume at least some responsibility for their own sickness prevention. In the wake of such change, statutory health sickness schemes have followed with their own powerful publicity campaign to create a new image for their business. They are now called health insurance schemes or life insurance companies like DAK. This system has long since transformed the visitor to the doctor's surgery from "patient" [from the Latin "patiens" = sufferer, or someone who endures] to "client" [from Latin "cliens" = person to whom protection is due], that is customer. While it is true that the doctor is still the healer that he always has been, where nothing else is possible nowadays he tends to be cast in a new role, that of advisor seeking to advise his customers on matters of prevention. People who go to the doctor do not necessarily suffer from some acute illness. Today, patients are also motivated by prophylactic concerns, are health and beauty conscious or are simply looking for advice on how to find relief from the stresses of modern living.

This role of the doctor as health adviser is the end result of the healthcare reform and has never so far been taken into account in monetary terms. But it can only be a matter of time before the standard wording on drugs packaging "For the risks and side effects, consult your doctor or pharmacist' features in some different format as a service on the bill for the health insurance provider. Doctors are therefore subject to competition not only other doctors, but also in their role as businessmen vying for customers. For a doctor to stand out from his colleagues he must at the same time make himself visible to the customers. The solution is advertising. Doctors in Germany were not allowed to advertise until 13 July 2005, when the German Federal Constitutional Court, ruled on a constitutional suit contesting the prohibition on advertising brought by the Association of Statutory Health Insurance Physicians. The ruling overturned the ban and allows doctors a form of advertising which enables them to emphasise the image of and trust in doctors that patients have acquired through the services and care received. By law, doctors are now also permitted to publicise these aspects of their work by advertising in newspapers, magazines, on the internet and television. Even the size of the nameplate outside the practice is no longer specified.

In Germany, for doctors, this switch from a self-imposed ban on advertising presents both an opportunity and a challenge. Since not all his colleagues have the media savvy of a Gunther von Hagens, they need expert advice on how to project themselves. Moreover, 80 per cent of outpatient medical facilities are single-doctor practices with the remainder being largely two-doctor practices. Most doctors have total control of their business, shouldering all the risks and responsibilities including organisation of the practice – staffing, technology, finance, purchasing, marketing – at the same time as further training. They have no time to think about the next advertisement; besides, it costs a vast amount to advertise in newspapers, magazines and as well on television.

However much the German associations of statutory physicians regulations are required to allow doctors to advertise, the remuneration system negotiated with the health insurance schemes conflicts sharply with all the rules of competition. This is because the system does not allow the "commercially particularly able, qualitatively good and popular doctors ... to grow too strongly". Marketing experts Felix Cornelius and Wolfram Otto write that in consequence of the physicians' guild's code, "popular doctors don't become less popular because of it. But they no longer have any incentive to expand their practices." As a consequence of this switch, customers, that is patients, registered with less popular practices and who would like to change to another doctor, must accept longer waiting times; in the worst case scenarios, they have no opportunity to change practice at all because the popular doctors are not taking on any new customers. Ultimately, there is no financial incentive to extend consulting hours and the range of services.

Dental practice "Kids Docs" in Berlin-Steglitz, architects: planbar 3 [formerly BHZ planning office], guidance system: 3 für Formgebung, branding: metome.design

Above: Reception area with a desk in the form of a plane cockpit
Left: Treatment room with the child-friendly name "Elf room"

In short, the association of statutory physicians, guild system prevents payment for services that would clearly disadvantage those doctors with fewer patient visits. So the system effectively prevents doctors from advertising their services commercially and from using this facility to make themselves stand out from their colleagues. This means that the guild's code penalises the successful medical service providers and actually deprives the customers, that is the patients, of their right to choose their doctor freely.

Quality through brand creation and identity
One solution is to make quality visible through brand creation and identity. Nothing makes a service provider's self-image more effective and more cost efficiently visible over the long term, than a "corporate design' which answers the questions "Who am I? What am I? What do I have to offer?" And such a design is on business cards, the internet page and even through the design of the practice premises.

For these are the questions that the customer expects answers to, answers that he will consciously analyse for their meaning and on which he makes the decision as to whether or not he will even enter the premises, enter once and never again, or do so repeatedly. Naturally what is covered by the company's image must match its core values, including the attitude and appearance of the staff. This is vital because doctors all too easily forget that each visit by a patient to a practice constitutes an intrusion into the patient's privacy. After all, consulting a doctor is an intimate affair. It is a very personal matter, whether it is to do with the body, the mind or both. Essentially, this "advance payment' of trust is made intuitively by people who nowadays are self-assured and increasingly sceptical about health-related matters. The decision whether – if at all – to go to which doctor is made far away from the practice's front door. The appearance of the doctor's business card or home page, which the patient will have consulted in advance, may well be decisive. Whatever feelings and expectations are aroused and whatever trust is engendered in the potential customer by these virtual visits, they need to be confirmed by the real environment of the practice: by the appearance, manner and conduct of the staff, even the architecture and design of the practice. It all comes down to the business's identity, which must chime with the character and service provision of the practice in terms of treatment, facilities and the very look of the building and with the customers themselves. The architecture and design serve to enhance the service provision, its quality and the practice's identity, which are experienced as a whole and viewed over the long term. According to tradition, the architect of a medical practice has an obligation to fulfil as the creator of an artificial world of health. Artist, architect and polymath Leon Battista Alberti [1404–1472] provided expert advice for a healthy life in his "Della famiglia" [1434–1441]. In his work "De re aedificatoria" [1452], he described the climate, air, the locality and even the direction of the wind as essential influences affecting human well-being. "The environment provided by inhabited houses must be at the right temperature – even animals should feel at ease," he says there. The humanist Alberti concludes his holistic vision of cause and effect in a healthy environment with an apposite picture in which he depicts the town as a large house and the house as a tiny town. He goes in great detail into the minutiae hygiene technology in the water supply system. Not the only one to address architecture in practical and theoretical terms, Alberti was, alongside Andrea Palladio, one of the major scholars of his time. Knowledge was gained by taking a panoramic view of things and was further enhanced by referring back to ancient sources. Health and architecture are intimately connected together in advanced cultures. The right ambient temperature and atmosphere are part of a business's self-representation and the development of its unique brand. As already noted, this ranges from the business card, the headed notepaper and the internet

Dental practice "Kids Docs" in Berlin-Steglitz,
architects: planbar 3 [formerly BHZ planning office], colour design,
guidance system: 3 für Formgebung, branding: metome.design
Above: "Boarding card" for young patients
Top right: "Kids Docs" business equipment
Right page: Postcards

presence to the design of the practice's premises. A doctor's professional expertise is not the sole deciding factor in his commercial success. Such expertise tends to be perceived subjectively anyway and well ahead of any appointment. The intangible influences include the architecture of a practice, that is a practical room layout which reflects the practice's internal organisation. This knowledge acquired by architects and schooled in the humanist tradition is essential because they are able to integrate all the constituent elements, including an assessment of the doctor's clientele. They will be able to use exactly this knowledge to develop a finite corporate design in combination with other media-related brand markers as a single-source solution.

Marketing and branding

Since in Germany a doctor is one in over 135,000, to stand out, it is very important that he make visible the company's unique identity. An easy matter? Hardly. Yet the answers to the questions on a unique company identity are primarily linked to personality. The very first step is for a doctor to view his own medical practice as a business. Next it helps to imagine this company as a human being with a body, mind and soul. This is because a company identity, that is the corporate identity, can be compared with the identity of a person.

In order to run the company [that is practice] in a market-orientated fashion, it is important for the doctor to visualise the customers he already has or would like to attract in the future. The answers to such questions will determine everything that needs to be done to define the practice as a brand. It is precisely the customers who determine the practice's particular approach and profile and so also its corporate identity. The architecture of the practice, "hardware" as it were, is an essential ingredient in enabling visualisation of the company identity. Taken together with the media used to present the practice, the architecture creates harmonious and memorable picture of the practice. In sum, all the elements used in self-presentation, including the architecture should be unified in character in terms of form and colour and combine to make the practice's identity visible – and so you have the corporate design. An implicit logo which readily identifies the practice for everyone is extremely helpful here. This is all the more vital when the medical practice is part of a joint practice or a medical centre. The media-friendly corporate design as part of the company identity serves to ensure that a positive image of the practice is generated in the subconscious of the potential client well before they have been to the practice. Corporate design is what triggers a sense of recognition in someone entering the practice for the first time. And quite apart from the visual impression, the conduct of the staff, how they communicate, their professional expertise and even the internal organisation of the practice all have a decisive role to play. Patients as customers have a finely tuned sense for whether there is a match between the style – which renders the design [on every level] visible as

"hardware" – and the staff, that is the "software" of the practice. To be sure there is no mismatch, it may well be helpful to have six to eight weeks' training in how to run the practice.

As well as the code of conduct, which makes palpable the special character of a medical practice, having a dress code for the staff makes it easier for the visitor to know who's who in the hierarchy. This is especially advisable in large or joint practices and in practices offering a wide range of services. Moreover, the actual location of the practice is important. This starts with the area surrounding the building and the location of the practice within the building. It is at this point that it is especially important to consult the architect as qualified adviser.

"Tell me where you live and I'll tell you who you are!" This aphorism applies to the location of a company as well as, in this case, a medical practice. The impression given by the exterior influences the well-being of the visitor. It begins with the ring tone of the telephone, includes the typeface of the headed notepaper and extends right up to the architecture and design. If the choice of colours for the logo reflects the colour philosophy behind the practice's design that is definitely an advantage. Because it all comes down to the recognition effect.

Space and room planning

Architecture for medical practices generally means designs for the interior of a building. It rarely happens that the opening of a new practice involves the construction of a new purpose-built building or extension. Medical practices are generally located in residential, office or former factory buildings. This means that the design must accommodate the existing ground plan whether in the rented floor, office or residential unit. And this is were any problems can start – which can rapidly turn into a disaster before the ink is dry on the lease. So it is astonishing that the most time-consuming activity is walking to and fro, that is the distances covered every day in the practice. In general, the female auxiliary staff walk almost ten kilometres on an average day. Where work processes are made unnecessarily complicated, the layout of the premises generates additional costs if nothing else. And such instances are not rare, they are true of most cases. This is because nine out of ten practices are planned without architects and operate as outlined above. In other words, the poor design of nine out of ten practices can significantly distort the owner's balance sheet. The remedy is a rational, architecturally designed layout achieved with a qualified planner – before the lease is signed. Indeed, at the heart of any well-organised practice is a space and room design that has been properly thought through – and this means for the practice as a whole and each room individually.

But this can't be done until both doctor and architect know what services the doctor actually provides, who the doctor's customers are, now and in the future. All this dictates the choice of location, the design of the building, the details of the furnishings and fittings plus

the specific needs in terms of floor space and the technical equipment to serve the rooms. During negotiations between landlord and tenant, the architect can be particularly helpful and valuable – for example with regard to the connection points needed for the service supply cables and lines, or issues and any consents needed from the authorities. A practice may be in a former apartment, but here too, the answers to the following questions are of crucial importance. Do the potential premises satisfy the building regulations? Can work place and hygiene regulations be complied with here? And, first and foremost, – is building consent necessary or will a change of use application have to be made?

The easiest way to settle all these points is to consult an experienced architect. He knows how to cast an expert's eye over the vision and expectations the doctor has of his practice, over the requirements and the site under consideration. As a qualified planner, taking into account the ground plan and standard of the potential premises, the technical requirements, the estimated building costs and the doctor's budget, the architect can judge whether the location being considered by the doctor is suitable or whether a conversion would be worthwhile.

The same applies to any structural work contemplated for an existing practice – in other words, would conversion be an option? If the reception desk is off to one side, if the distances are too great, if the staff keep getting in each other's way as they do their work and if that sort of fatal flaw is characteristic of the business, then a new practice design is needed. Even converting small medical practices into spacious, functioning healthcare businesses provides ample scope for the skill and expertise of an architect.

Space allocation

The structure of a medical practice can often say much more about a doctor's working methods than he would like or is even aware of. Clear workflows, rationally planned offices and the kind of benefits inherent in a peaceful atmosphere all depend largely on how the space is allocated. A well thought-out space allocation design will facilitate rational workflows, separate the work areas from the public areas and reflect the practice's workflows. This means everything, from reception to the treatment rooms, administration and even the waiting areas. The purpose of space allocation is clarity, in the form of a logical subdivision of available space for the different roles that make up the practice's workflow hierarchy. Allocating the practice's spaces in this way ensures that everything runs smoothly.

A well thought out-space design not only helps to cut down on walking distances and save time, and therefore reduce costs, but is also just as much an essential ingredient of the corporate identity as the choice of headed notepaper. The details of the practice's configurations express this identity non-verbally, that is visually. The space concept, taken as the template for the interior design, is also subject to the basic precept that the form must be fit for purpose. The reception area is the point from which the visitor should be able to grasp the layout of the practice. Reception should be sited at some central point. At the very least it should be clearly visible from the entrance and both welcome the visitor and be a credit to the practice – like a prestigious executive suite or a hotel lobby.

One should also bear in mind that the reception is more than just a place for visitors to report to. It is also the place for initial conversations, exchanging information, making arrangements and appointments for clients by staff. All important client details are forwarded from here straight to the next points in the system. If we compare the practice with human anatomy, then reception is to the practice as the heart is to the body. It is the point from which the whole of the practice's circulatory system is controlled and kept alive.

As regards the practice's other workflows, one could compare them with the reception desk in a hotel. It is as if the patient receives his "key" here, that is access to all the services available in the practice. Reception is the control point or heart of the practice. And just as

Medical Practices

the heart oversees the workings of the whole body, the practice's receptionist should be able to see all the doors so that she can keep the practice's workflows running smoothly. If the practice is logically structured, then the waiting rooms, treatment rooms and other rooms will all be grouped around the reception area. An alternative approach, however, is to have the rooms ranged along a corridor running from the reception lounge. In this case, the corridor also functions as guide to the practice's rooms. Space allocation permits of numerous variations. The point is to design workflows as effectively, and therefore as cost efficiently, as possible.

The watchword here is "rationalisation" which, according to Bruno Leo Friton, himself a doctor, applies to all measures taken together as a whole. According to Friton, such measures "are geared to designing a work process such that the greatest output is achieved with the least amount of effort. This is achieved by means of the configuration of a work place, observation of working patterns, fit-for-purpose preparation of a work process, simplification of the necessary work and elimination of superfluous work and by the design of special instruments and appliances." With reference to this precept, in 1961 Friton was already invoking the cultural philosopher Jacob Burckhardt and his optimistic description of a social authority: "I would wish that everything that can be mechanised would be mechanised as completely and as soon as possible so that the human spirit would have all the more power and leisure to do the rest."

With regard to healthcare reform, the Deutsche Ärzteblatt [the German Medical Association's journal] recently warned that the doctor with a stethoscope around his neck caring for his patients and wearing a white coat soiled with blood or other body fluids had been replaced by the doctor wearing a suit. Instead of a stethoscope around his neck, he'd be holding a time-and-motion chart in his hand and would most likely know the manual of process management by heart. Any marks on his clothes – if any – would be from the ink of his fountain pen or photocopier. Technology would be dictating to the human spirit and not the other way round. At least, everyone will have detected a pervasive hiatus , one which turns Friton's plausible statements on rationalisation into a template for Germany's medical practices.

The first thing a medical practice needs to install is an efficient intercom system. This type of technology does not need to be impersonal. The medium used to conduct a conversation does not define its character, rather, if anything does define it, it's the speakers themselves. The fact remains that a good intercom system saves two full working hours per day or at least two months' salary per year. It also makes no sense for the practice's staff to let patients into the waiting rooms themselves and then, having opened the door for them, to show them to the relevant treatment room. Neither auxiliary staff, nor doctors nor customers are small children. But it is supremely important and sensible to call the next patient from the reception area and tell him with a friendly smile where he is supposed to go. And that, of course, is only possible if the space allocation has been organised properly. This involves locating the waiting area as far forward as possible in the practice, in the area to which the public has direct access. In doing so the public area is clearly demarcated from the working area and workflows are not disrupted. Sensible space allocation in harmony with workflows also requires an interior design which functions as a nonverbal way finding system from floor to ceiling in terms of choice of materials, colour, and lighting.

All this constitutes a "top-level cultural support system by means of technology" [Friton] because its implementation enables time to be spent in concentrating on the patient on learning from the patient and increasing scientific knowledge. Indeed, the medical vocation is still a craft. This preparatory work at the interface between theory and practice also means that the doctor and architect are excellent professional partners with shared goals service to their fellow human beings.

How colourful a "non-colour" can be: White is the colour of ice and snow. It is a symbol of purity and clarity, serenity and innocence. It also works as a symbol of inapproachability, sensitivity, and cool reserve.

From left to right:
- Home in the Altai Mountains
- Surfboarder in Hawaii
- Crystals of ice on a flower bud
- Marble palaces in Turkmenistan
- Gravestones in Sarajevo
- Swans on the Caspian Sea
- Kaftan in Saudi Arabia

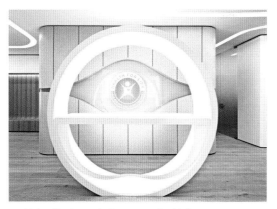

Waiting areas often serve the purpose of a business card for the company

Health for Life, Vienna, design: smartvollarchitekten

Interior design

Light, colour and the form of the materials for walls, floors, ceilings and furniture define the character of the medical practice. These elements should convey the company identity of a medical practice and at the same time harmonise with the professional bias of the doctor. It does not matter whether the practice is functional and cool with few nuances of colour, whether it is bright and transparent or has forceful design features. Nor does it matter whether it looks like an observatory or is defined by contrasts between old and new, or whether powerful organic forms and well accentuated photographic motifs or art predominate. The important thing is that the fittings can be seen to reflect a personality as well to convey more than that. It is just the same when you first meet someone. The first ten seconds after entering a practice are crucial for the patient's feelings of liking or disliking the experience – and hence crucial for the success of the doctor. The first impression counts. And the first impression will be of the area surrounding the practice, the building in which it is housed, the doctor's nameplate, the ring tone and finally the reception area, the atmosphere.

In other words, we are dealing here with a combination of many factors which together exert an intuitive and unconscious effect. It is the same with a person. We do not register a person selectively. Instead, we take the person as a constellation of many features – how he moves, the type and colour of the clothes he is wearing and the surroundings we encounter him in. The intuitive overall impression creates an image in the mind that then determines whether we like him or not. In this connection, again, it is impossible to overemphasise that its architecture and design are indisputably at the heart of the corporate design of a medical practice.

This merging of several factors in the total experience determines perception and whether or not a client feels at ease in an environment. Even so, it is also true that it is impossible to please everyone. What one person finds unpleasant excites and stimulates another. It is a natural law that antipathy and sympathy determine whether something is attractive, to be kept at a distance or embraced. But it is also beyond dispute that the changing role of the patient from someone who suffers and tolerates to someone in need of protection, has dawned on only a few doctors and been incorporated by them into the design of their premises. Only rarely in German treatment rooms and on the premises of therapists practising symptom-orientated academic medicine does one come across a practice with a contemporary, customer-centred design.

This is extremely unfortunate as the first impression also determines the tenor of the verbal messages. And in this way, the corporate design, the visual impression of the company's identity defined in human terms, continues to have its effect, as a form of recommendation, far beyond its own material limits. Recommendation by clients, however, is the most cost-efficient, effective and most enduring form of advertising for any businessman including doctors. Especially today. The design of the practice is therefore a factor in the doctor's self-presentation with a far-reaching effect as publicity.

For this commercial effect to occur, both form and function, as basic design factors, must be complemented by a relaxing and pleasant atmosphere for the client. Harmony is what drives everything. And it is significant that this harmony is generated by the combination of materials, form, colour and lighting.

If in doubt, remember that less is more when trying to get the message across to clients. This means that the choice of materials for the floor, wall and ceiling together with the lighting and any pictorial subject matter have a substantial effect on mood. Light not only represents vitality but also underpins the whole image and the perception of things in one location.

So, to create a successful overall impression, it is absolutely essential to create a colour scheme that is implemented faithfully throughout the premises, from the furniture

sporthopaedicum, Berlin, design: Meuser Architekten

Praxis im Olympiadorf, Munich, design: Harald Stricker
(photo: Quirin Leppert)

to the fittings. Here, both the practice's environment and the specialist bias of the doctor or the extremely unfortunate interests of a GP can influence design. Also, choices made for the design of the space, along with the surface areas, depend crucially on whether the practice is a single-doctor or joint practice or perhaps a medical centre accommodating doctors with different specialities practising independently of one another. It may well be the case that a design leitmotiv is adopted to govern the colour, lighting, form and material scheme for the internet presence, business cards and architectural design. Although the medical services are configured at a particular location, it is important to express the individual character of the practice with everything – business cards, internet presence and design of the practice. In joint practices, the skill lies in making each individual doctor stand out visually from his colleagues.

Irrespective of the size and specialist bias of the practice, access to the building should be easy to see and find and free of obstructions. Particularly, if the practice is sited in a residential or business block, it is important to have an obvious entrance area to distinguish the medical practice from its neighbours. A practice nameplate is not enough on its own. Instead, what is needed is a guidance and way finding system using visual prompts to guide the visitor from outside the practice into and throughout the practice. The basic elements of good architecture are colour, light and materials. In his standard work on the art of colour, published in 1961, the painter and art teacher Johannes Itten writes about the radiational power of colour. According to Itten, colours are "energies" which exert either a positive or negative influence on us, whether we are aware of it or not. We have the Catholic liberation theologian Dom Hélder Pessoa Câmara to thank for the following equally apposite insight about light: "Light transforms the things it falls on." For a professionally designed medical practice, this means giving due consideration to the effect of light entering from outside and the way colours and materials appear in artificial light – and making the appropriate choices. When using colours and materials, the way we combine them determines the quality of both the atmosphere in the rooms and the visitor guidance system. Not only floor and wall materials, but also colourful and eye-catching features and pictograms are simple resources to guide visitors as they move about the practice, or, in the case of a joint practice or medical centre, to draw attention at the same time to the variety of available services. The more sparingly the resources are used, the greater the effect. The interior design should serve the purpose of a well functioning, non-verbal guidance system, complemented for example by a digital signage and way finding system, all of which are flexible and extremely easy.

Whatever specialist bias the healthcare business has, at its heart is always the reception area, as already stated. This is the visitor's point of arrival. First impressions are made here, impressions that decide whether the client feels secure, accepted and welcomed as a guest in the practice. Perhaps the patient even senses his fears, tensions, uncertainty and unease ebbing away when a friendly smile from the person at the reception desk, itself the centre of a soothing atmosphere, conveys the first sensations of healing. As in a hotel, the reception desk is the first point of call for communication and at the same time acts as a business card for the enterprise – in this case a medical practice – and the control centre for its internal operations. The visitor registers its appearance and activities subliminally and his unconscious begins to draw unseen conclusions about the whole enterprise. As regards customer relations, reception is the nerve centre of a practice.

The expression "Report here" should never be used, not in print nor as a verbal message. This expression, which smacks of the barracks and still leads a shadowy life in some official premises, should be avoided here at all costs. A business such as a medical practice is no place to issue orders. This is a place where services are rendered. Here, the patient is king and not subject – and is always a guest. As terminology can change attitudes, it should be clear that the style and atmosphere of a practice's reception must match that of a five-star hotel in every respect. Bright, light, calm, relaxed and dependable – that is the impression to be created by the choice of materials and the lighting which should give the visitor an immediate grasp of where he is to go. As everyone knows, the first impression isn't selective but general and takes in the whole space at once. A prominent, unobscured position for the reception desk is ideal. If reception is set at a slightly higher level, like a cockpit, this will also create the visual impression that the friendly staff has healthcare totally under control from here and throughout the practice.

Reception should be sited with a clear view of the consulting room and give clients enough room to move and avoid any feeling of being hemmed in to counteract the nervousness and apprehension that visitors to the doctor undoubtedly experience. Like this, the customer is able to gain an impression of his environment. This has a relaxing effect, breeds confidence and in the best cases dissipates fears and fosters a feeling of ease. In small practices, the spatial impression can be extended by arranging the reception area such that it looks out over some charming countryside view or an interesting building. Otherwise, bright colours, a skilfully stage-managed lighting scheme and mirrors as well as bright materials – but only those that look light, all help to make the practice and reception look larger than they actually are. A wall covered with a photo reflecting the character of the practice or a mounted artwork can catch the eye and expand the dimensions of the practice – or at least divert attention and dispel nerves.

The combination of light and transparent materials and forms creates visions for the space. Round shapes make it easier, with little expense, to create a perception of depth, even in small practices, and so also reduce the sensation of being constricted. Round and undulating forms for the fittings and walls are relaxing and also generate a sense of well-being. Moreover, undulating wall surfaces can make it easier to find one's way around a medical practice. After hygiene, finding one's way around is the most important aspect of a practice. Both a rational room scheme and design features have a very important role to play. The interior design can be organic or associative. The shapes may refer to means of transportation such as a train, a ship, a car or a plane. The common feature of all these experience is a clear hierarchy and logical sequence of spaces. Creativity can be employed to objectify and reproduce this for the spatial design of a medical practice and gear it to the needs of the particular target group, making sure at the same time to find an original way to engender a positive atmosphere.

Nor should one overlook the role of flooring as a guidance system. As well as real wood flooring and industrial-grade parquet, there have long been other options such as rubberised flooring or floor coverings which combine acoustic and visual advantages in that they look like textiles and have the comfortable, warm and calm feel of wall-to-wall carpeting without its disadvantages.

This type of flooring can be cut to whatever size you like and is easy to fit so that even round, flowing transitional spaces, for example between the working and waiting areas, can be marked out to great effect.

No one with an appointment at a company goes straight in to see the boss. A previous appointment may overrun or something may have happened to upset the scheduling. That is the normal state of affairs in a medical practice. You simply have to wait. And so there is a direct link between reception and a waiting room or – speaking generally – the waiting area or zone. On average, the time spent waiting in medical practices in Germany is 45 minutes. This means that the waiting area is where visiting patients spend more time out of the total time on a doctor's premises than anywhere else.

Whether or not a patient can stand this wait time will depend on whether the atmosphere is inviting and relaxing. To an extent, there needs to be something to take

Medical Practices

the patient's mind off the waiting – something which will differ from client to client. A practice which is geared mainly to children will highlight different aspects from one serving mainly older people. And a practice with a clientele spanning many generations ought to have different waiting areas geared to the different ages and tailored to their individual needs. The idea is to avoid conflict. Because one thing is clear – no matter how attractive the atmosphere of a medical practice, no one will ever associate it with the pleasures of going to a bar or staying in a hotel. After all, consulting a doctor is hardly a leisure time pursuit, but rather an unpleasant and unusual necessity. Germans are afraid of going to their doctor – they feel it in their bones, to varying degrees. A current survey conducted by Forsa, the public opinion research institute, on behalf of the Techniker Krankenkasse, the technicians' health insurance scheme, found that one in five people is afraid of going to the dentist. This means that a place where a suspected or actual illness is diagnosed needs to be able to reduce fear and avoid tension. Waiting rooms are unsuitable for this because a waiting room gives the visitor the feeling of being shut out. Expressions such as waiting area or waiting zone convey a sense of openness. Interconnected rooms should also meet this need for openness. As soon as the patient enters the foyer of a practice, it should be made clear to him how free he is to make up his own mind about his visit. This greatly relaxes the patient's mood. Depending on the nature of the clientele, a combined bar and lounge feel can lighten the atmosphere for people as they wait. A play area for children designed so that a parent can either play with their children or have something to do while they wait would be a helpful addition. Apart from reading matter and something to drink, the choice of materials is important to create the right atmosphere.

Light is important in the waiting area. Subdued lighting creates a warm and comforting atmosphere. The other thing is colour. The positive energy emanating from both, achieved by the right choice of colours and the right technology, is put to the test in the waiting zone, as it stands in for the whole practice. The mood the patient feels here decides how things proceed from this point. Hygiene being the underlying theme of a practice's fixtures and fittings this implies the use of robust, elegant and natural materials. In this case, elegant also means resistant to rough handling. Nothing is worse than threadbare upholstery, scratched chairs or a worn floor. One recommendation would be to have comfortable armchairs suitable for reading. Where there is activity, robust wooden benches upholstered in some easy-to-care-for material would be more suitable.

Apart from the reception desk, the waiting area is the nerve centre of a doctor's practice. This is the place where a doctor's self-representation is expressed in the interior design and the decision is made whether the doctor gains the customer as a patient or loses him.

Further reading
Bergdolt, Klaus: Leib und Seele.
Eine Kulturgeschichte des gesunden Lebens.
Munich 1999.

Damaschke, Sabine | Scheffer, Bernadette | Schossig, Elmar: Arztpraxen. Planungsgrundlagen und Architekturbeispiele.
Leinfelden-Echterdingen ²2003.

Meuser, Philipp: Arztpraxen. Handbuch und Planungshilfe. Berlin ²2016.

Meuser, Philipp | Pogade, Daniela | Tobolla, Jennifer: Construction and Design Manual. Accessibility and Wayfinding. Berlin ³2018.

Thill, Klaus-Dieter: Marketing in der Arztpraxis. Cologne 2005.

Franz Labryga

Planning Medical Practices: the German Principle

Anyone interested in planning medical practices or similar medical facilities must take a very broad view in order to grasp the extremely complex parameters and to draw the right conclusions so as to arrive at the best of all possible results.

UNDERLYING HEALTH CARE POLICY CONDITIONS

Medical care mandate
The German medical profession has a statutory care mandate. To meet this responsibility, there were, on 31 December 2008, 319,697 male and female doctors. Of these, 181,400 were involved in inpatient work and with authorities or corporations, and 138,300 worked in outpatient care. This last figure breaks down as follows: 119,800 doctors affiliated to a statutory health insurance scheme [including doctors in partnerships], 12,600 doctors and practice assistants in employment and 5,900 doctors in private practice. The figure for affiliated doctors is made up of 58,500 practitioners and 61,300 specialist doctors.[1] In 2007, a total of 662,000 people were employed in German medical practices. Non-medical employees includes male and female medical assistants, nurses, medical technical staff, and employees in social and other professions; approximately 79 %[2] of the non-medical employees were women.

GENERAL PRACTITIONERS | Nowadays the first port of call for patients with medical problems is still the general practitioner. Between the GP and the patient there is normally a relationship of trust and consequently he often fulfils the role of a family doctor.
The general practitioner usually works on a self-employed basis [a registered doctor]; sometimes he is employed as a doctor. He is generally designated professionally as a doctor specialising in general medicine or also in internal medicine. In complex cases, he refers patients for diagnosis and treatment to an appropriately trained specialist physician.

General practitioners usually work in their own practice but if the patient is seriously ill, the doctor will visit him at home. A relatively recent innovation, there is also the concept of a "general practitioner in the hospital" who accompanies his patient, even in the clinic, as a sort of pilot. He will also provide care both before and after treatment, that is during the whole course of the illness.

SPECIALIST DOCTORS | If a doctor has decided to follow a particular branch of medicine, has completed several years of continuous training in his subject and has graduated with an appropriate specialist qualification, he may be known in Germany as a specialist.
Qualification recognition is the prerogative of the Medical Councils or is subject to the medical profession laws of the German Länder and the continuous training codes of the different regional Medical Councils.

USE OF THE DOCTOR'S SERVICES | In Germany, many people need to consult their doctor relatively often: approximately 18 times per year. Compared with this, Norwegians only go to their doctor three times a year.[3]

INTEGRATED CARE | It is useful, when planning facilities for outpatient care, to familiarise oneself with a new cross-sector care strategy, which has been developing for some time and which aims to foster improvements in the quality of patient care. It implies stronger links between general practitioners and specialists, hospitals and other institutions. A variety of healthcare reforms has laid the foundations for this evolution. It involves reaching joint-working agreements which are intended to facilitate cooperation between various institutions in a newly emerging care scenario and to avoid unnecessary costs.

RESOURCE SPENDING | In the year 2006, over 245 billion euros spent on healthcare in Germany[4]. Analysis of the various types of service shows that doctors accounted for the main expense. They examine and treat, they prescribe, they place orders and so essentially control the extent to which healthcare services are being used [Fig. 3]. Analysis of the healthcare expenditure of the different institutions is also instructive. Such analysis clearly shows that the expenditure of the outpatient facilities – at approximately 120 billion euros – is some 30 billion euros higher than that of the inpatient and partly inpatient facilities. A good indicator of the value of health is expenditure. In 2008, it amounted to 2,970 euros, that is 10.6 % of GDP [gross domestic product]. In Europe, only France [11.0 %] and Switzerland [11.3 %] spend more, though the highest expenditure is in the United States [15.3 %]. International comparative studies, which include additional efficiency indicators, show that Germany is roughly in the middle.

RESULTS OF SURVEYS | The increasing significance of medical care is reflected in the growing number of scientific studies. They often supply important indicators on the efficacy of healthcare policy measures, reveal important connections and produce relevant data for decision-making. Some examples [see below] should make this clear:
A study by the Institute for Quality and Efficiency in Health care shows that a doctor in Germany currently spends an average of 7.8 minutes per patient; the equivalent figures for the UK are 11.1 minutes and for the USA 19 minutes. It is quite clear that simple conversations between patient and doctor are inadequately remunerated. Financially speaking it is only worth treating a patient if equipment as costly as is technically possible is used for diagnostics and therapy.
According to the results of the Ernst & Young 2009 Health Barometer, 85 % of Germans give healthcare in their region a positive rating; 90 % of those questioned rated the medical care provided by general practitioners as "good or very good"; whereas specialist doctors scored 88 % and doctors in hospitals 84 %.
The results of the survey make it clear that patients like a "familiar face and individual care". In future, healthcare approval will tend to decline because of the increasing rise in costs. Patients do not want "high-tech medicine tailored to cost efficiency."[5]

Federal Medical Councils
At the head of the medical self-regulatory system in Germany stands the Federal Medical Council, which acts as an umbrella organisation for the 17 German Medical Councils. It takes an "active part in forming opinion on healthcare policy issues in society and develops perspectives for a healthcare and social policy that addresses the concerns of citizens and emphasises personal responsibility".[1] The continued training code has direct consequences for planning medical practices because it defines the designations of specialist doctors, main areas of interest [specialisations within a specialised field] and additional designations. The head office of the Federal Medical Council is in Germany's capital Berlin.

Association of Statutory Health Insurance Physicians in Germany
The Federal Association of Statutory Health Insurance Physicians is the umbrella organisation which covers all such existing Associations [German acronym – KV] in each of the German Länder. Under § 75 [1] of the Social Code V[6], it is the responsibility of this body to guarantee the medical care of insured persons by affiliated doctors.
Registered doctors and psychotherapists, employed doctors and authorised hospital doctors ensure the "right of patients to adequate, fit-for-purpose and economic care having due regard to the generally accepted state of medical knowledge."[7] The Associations of Statutory Physicians rule on licences to practise and needs planning. If needs are already covered, a licence will only be granted if an existing practice is taken over.

Health insurance schemes
In Germany the cost of medical care is borne largely by two types of insurance schemes, as set out below:

STATUTORY HEALTH INSURANCE [SHI – Gesetzliche Krankenversicherung – GKV] | This was launched by Bismarck. Today, approximately 87 % of all insured people belong to this scheme.[2] Social Code V limits the available scope of services to those that are "economic, adequate, necessary and fit-for-purpose". Following the solidarity principle, all those insured pay the same, irrespective of whether they are healthy or sick, young or old.

PRIVATE HEALTH INSURANCE [PHI – Private Krankenversicherung – PKV] | After the launch of SHI, private enterprises created PHI for those on higher incomes. A number of different companies provide these services. Approximately 13 % of insured persons belong to PHI schemes[2]. Each insured person is charged a separate amount under the cost reimbursement principle that reflects their age, sex, state of health and individual requirements.

The Institute for Quality and Efficiency in Health care
Like many of today's social developments, medical care also relies on the support and findings of scientific research. 2004 saw the foundation of the "technically independent, incorporated, scientific Institute for Quality and Efficiency in Health care" [Institut für Qualität und Wirtschaftlichkeit im Gesundheitswesen]. Its headquarters are in Cologne and it is enshrined in Social Code V. It operates under a mandate from the Federal Ministry of Health and the Joint Federal Committee.
Its main remit is to improve the quality of patient care.

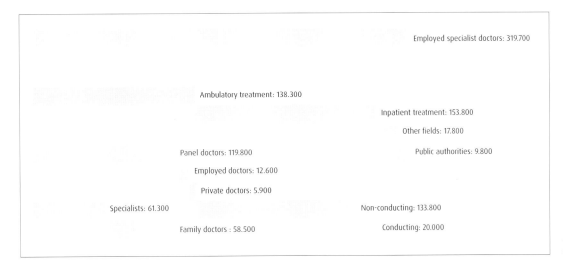

Medical profession in Germany, 2018
(www.bundesaerztekammer.de)

Other tasks are the evaluation of the cost-benefit ratio of drugs, assessment of guidelines for the treatment of important illnesses and provision of information on the efficiency and quality of healthcare services.

The institute sees itself as the advanced guard for evidence-based medicine [EBM] which aligns diagnosis and therapeutic measures with the current state of medical knowledge and where, in specific cases, the experience of the doctor is also taken into account[8].

The institute is fully funded by the public healthcare system. It is to be expected that the institute and its work make a considerable contribution to the most beneficial allocation of the rising costs of healthcare.

LAWS, REGULATIONS AND OTHER PROVISIONS

When planning medical practices it is useful to know the most important provisions governing this very complex area of work; such knowledge may actually lead to the development of facilities ideally suited to their functions.

Social Code

Social law in Germany is set out in the Social Code. So far the Code comprises twelve volumes the fifth of which deals with the organisation, mandatory insurance and services of the statutory health insurance schemes and their legal relations with doctors, dentists and pharmacists.[6] Outpatient care under the health insurance scheme is regulated in § 71 whereas § 77 lays down the main task of the Associations of Statutory Physicians as meeting the responsibilities assigned to them by Social Code volume V.

SHI [German – GKV] Modernisation Act

In 2003, reform of the German healthcare system was initiated by the Act on the Modernisation of Statutory Health Insurance[9]. Its main purpose is to reduce the contributions made to statutory health insurance . The following rules need mentioning:
– The provision of a general practitioner system as the first point of call for patients who are then referred to the relevant specialist doctors, should lead to better care.
– Subject to financial sanctions, a duty to undertake continued training will be incumbent on doctors.
– An independent institute will be created to evaluate the quality and economic efficiency of the treatment of certain illnesses.
– The ban by which pharmacists are prevented from owning more than one pharmacy will be lifted. In future, up to three branch pharmacies will be permitted in addition to the main pharmacy.

Affiliated Physicians Amendment Act

Since 2007 the effect of this Act has been to relax the conditions under which registered doctors practise. This Act allows a doctor to network with other medical practices, medical care centres and even hospitals so that patients with particular diseases can be cared for jointly.[10] The Act is intended to cut out bottlenecks in outpatient care, especially in the former East Germany.

Hospital Funding Reform Act

The financial situation of hospitals has seriously deteriorated in recent years. The same is also true of the clinical medical practices under consideration here. The reason is the inexorable rise of staffing and equipment costs.

This is further fuelled by the growing reluctance of the Länder to meet their statutory obligation, as enshrined in the Hospital Funding Act [German acronym – KHG] [11], to fund hospital investment. This has led to a situation in which the supporting authorities have had to fund part of the necessary investments from the resources for patient care, a purpose for which they are not intended.

The Hospital Funding Reform Act[12], passed by the Federal Parliament in the Spring of 2009, is supposed to improve the financial situation of hospitals. It seek to lay down a stable financial foundation giving hospitals a long-term financial perspective. The intention is to create a modern system of investment funding based on performance-related fixed sum investments for necessary building work. This will be in addition to investments for wage and salary increases.

Other provisions
Nowadays, standards are laid down for the delivery of quality in outpatient medicine. This means that medical practices are supervised and inspected like state authorities. The Association of Statutory Health Insurance Physicians has provided an overview of the statutory apparatus entitled "Supervision and Inspection of Medical Practices by Authorities"[13]. Other than that, the following acts, regulations and guidelines should be followed:

- Protection against Infection Act
  § 16 Action by responsibility authorities
  § 36 Maintenance of hygiene to prevent infection
- Medical Devices Act
  § 26 Carrying out supervision
- Medical Devices Operator Regulation
  § 6 Safety checks
- Medical Devices Safety Scheme Regulation
  § 3 Mandatory reporting
- Hazardous Substance Regulation
  §20 Official exceptions, orders and empowerments
- Biosubstance Regulation
  § 16 Notification of the authority
- Health and Safety at Work Act
  § 21 Responsible authorities; Cooperation with accident insurance funding bodies
- Youth Employment Protection Act
  § 51 Supervisory authority, inspection law and reporting obligations
- Maternity Protection Act
  § 20 Supervisory authority
- X-ray Regulation
  § 17 Quality assurance in different medical and dental facilities
- Health care Services Act
  § 11 Supervision of healthcare facilities

- Radiation Protection Regulation
- Provisions of the Occupational Health and Safety Association
- Recommendations of the Commission for Hospital Hygiene and Infection Prevention
- DIN provisions, for example DIN EN 554, DIN EN 285 and DIN EN 13060.

TYPES OF MEDICAL PRACTICES

Over the past ten years, developments in healthcare policy, in medicine and in the labour market have led to medical activities being carried out in a variety of forms. Doctors have adapted their practices to new demands by increasing the size of their usable floor space. Some doctors practising independently have decided to run their practices together with one or more other colleagues.

Financial imperatives and the standards set by the state have resulted in practices with quite different characters. The underlying conditions in which doctors work are subject to constant and increasing changes and this has resulted in numerous different types of specialist doctor practices and special forms.

Subdivision according to size
[taking single-doctor practices as an example]
In order to bring a little order into what may seem a rather confused picture and gain an initial overview, it would seem helpful to subdivide medical practices according to the size of their usable floor space. The subdivisions adopted below for small, medium and large single-doctor practices are taken from the properties showcased in this book.

Subdivision into three sizes gives us the opportunity to develop standard functional and spatial allocation plans suitable for the description and illustration of examples and allows comparisons within the different size groups.

SMALL SINGLE-DOCTOR PRACTICE | Small medical practices have a usable floor space of up to 125 square metres. Medical work should not be carried out in premises which are too small or have too few rooms. From 100 square metres upwards, they contain the rooms necessary for efficient operation; the surface area should be no smaller than this.

MEDIUM-SIZED SINGLE-DOCTOR PRACTICE | Medium-sized single-doctor practices always have a usable floor space of between 126 square metres and 174 square metres. They are furnished and fitted out and will have the necessary rooms at the required size.

Speciality	Number
General medicine	42.730
Internal medicine	41.722
Surgery	29.602
Anaesthesiology	18.327
Gynaecology and obstetrics	16.134
Paediatrics and youth medicine	11.973
Psychiatry and psychotherapy	7.856
Radiology	6.690
Ophthalmology	6.638
Otolaryngology	5.566
Skin and venereal diseases	5.180
Urology	5.040
Neurology	4.238
Psychosomatic medicine	3.890
Neuropsychology	3.746
Industrial medicine	2.728
Rehabilitation	1.684
Child and paediatric psychiatry	1.503
Neurosurgery	1.464
Oral and maxillofacial surgery	1.445
Pathology	1.405
Nuclear medicine	984
Laboratory medicine	977
Radiotherapy	933
Public health system	919
Microbiology, virology and epidemiology	666
Transfusion medicine	520
Pharmacology	452
Human genetics	251
Forensic medicine	210
Other speciality	631
Without speciality	93.553
Doctors in total	**319.697**

Employed specialist doctors in Germany, 2018 (www.bundesaerztekammer.de)

LARGE SINGLE-DOCTOR PRACTICE | Over 176 square metres, large single-doctor practices are regarded as being generously equipped practices with adequate usable floor spaces.

Subdivision into specialist areas
The original single-doctor practice was the province of the general practitioner who served and helped patients from his catchment area with all their health and hospital problems.

Subsequent developments in medicine have led to continued training for doctors who have specialised in certain medical areas.

GENERAL PRACTITIONER PRACTICE | The general practitioner fulfils an essential task, providing basic care for people especially in thinly populated and rural areas. He normally works in the fields of general or internal medicine and may also have a recognised specialist qualification.

SPECIALIST DOCTOR PRACTICE | The regional Medical Councils are responsible for the definition and remits of the specialist areas and their demarcation from each other. The register of the Federal Continuous Training Code of the Federal Medical Councils [as amended on 23 March 2008] provides the basis for the planning of specialist doctor practices[14]. It provides recommendations for the regional Medical Councils.

No distinction is drawn in this register between areas and specialist doctor competencies – both allow a specialist doctor practice to be set up the disciplines listed on the next page. In addition to the specialist medical areas, the Continuous Training Code currently includes a further 46 extra training options, among which are for example: acupuncture, allergology, diabetes, geriatrics, homoeopathy, intensive care, laboratory diagnostics, natural healing methods, emergency medicine, palliative care, physiotherapy and also balneology, sports medicine and tropical medicine.

Subdivision according to number of practices
As well as the option of enlarging a single-doctor practice by adding more usable floor space, a further option, and one taken with increasing frequency today, especially for financial reasons, is to expand by adding on further medical practices.

Specialist medical areas

- Anaesthesiology
- Anatomy
- Occupational medicine
- Ophthalmology
- Biochemistry
- General surgery
- Vascular surgery
- Cardiac surgery
- Paediatric surgery
- Orthopaedics and accident surgery
- Plastic and cosmetic surgery
- Thoracic surgery
- Internal surgery
- Gynaecology and obstetrics
- Ear, nose and throat medicine
- Speech and childhood hearing impairments
- Skin and sexual diseases
- Human genetics
- Hygiene and environmental medicine
- Internal and general medicine
- Internal medicine
- Internal medicine and angiology
- Internal medicine and endocrinology and diabetes
- Internal medicine and gastroenterology
- Internal medicine and haematology and oncology
- Internal medicine and cardiology
- Internal medicine and nephrology
- Internal medicine and pneumology
- Internal medicine and rheumatology
- Paediatric and adolescent medicine
- Paediatric and adolescent psychiatry and psychotherapy
- Laboratory medicine
- Microbiology, virology and infectious disease epidemiology
- Maxillofacial surgery
- Neurosurgery
- Neurology
- Neuropathology
- Nuclear medicine
- Public healthcare
- Pathology
- Pharmacology
- Pharmacology and toxicology
- Physical and rehabilitation medicine
- Physiology
- Psychiatry and psychotherapy
- Psychosomatic medicine and psychotherapy
- Radiology
- Forensic medicine
- Radiotherapy
- Transfusion medicine
- Urology

TWO-DOCTOR PRACTICE | The two doctors in a two-doctor practice have the choice of working as a professional team [also joint practice or partnership] in the same specialist field or, deliberately, in two different fields.

Working in the same specialist field has the particular advantage that each can stand in for the other when one is taking continuous training or is on holiday or sick and, of course, if one is pregnant or is on maternity/paternity leave. Working in the same specialist field allows the highest possible degree of synergy because the doctors will use the same staff, the same rooms and equipment. However, offering different specialist fields is of particular benefit to patients because this expands the range of available services and may allow for further specialist treatment in the same practice. Specialists in two different fields working together broaden the range of diagnostic and treatment options. The use of staff, the same rooms and appliances by two professionals make a two-doctor practice more cost-effective.

MULTI-DOCTOR PRACTICE | The advantages cited above of a two-doctor practice are increased if there are more than two specialist doctors. So far there is no discernible upper limit on the number of doctors that can work together. Combinations of up to approximately 40 medical practices working together deliver a perfectly efficient patient care service.

Subdivision into business models

Naturally, commercial aspects play a major role in the running of medical practices. This is especially true of patient accounting systems and the sharing out of the income from the facilities.

JOINT PRACTICE | In a business described as a joint practice with two or more specialist doctor practices, the principle is that they should be regarded as a single commercial entity. This cooperation model is a commercial and organisational merger to enable the joint practice of a profession. The Federal Medical Council's latest term for this is "professional community".

PRACTICE COMMUNITY | a practice community uses the same rooms, yet the practices remain independent of each other from a legal point of view – so there is no joint accounting system.

APPLIANCE SHARING SCHEME | In an appliance sharing scheme, a special way of sharing the use of expensive medical appliances; the actual use of the appliances is paid for under a jointly agreed accounting system.

MEDICAL CENTRE | One could take as the model here antique shops grouped together in the same street and offering customers as broad a range of items for sale as possible without any individual shop feeling threatened by competition from the others. A medical centre is similar, generally with doctors specialising in different fields, often situated in a multi-storey building. The doctors advertise their presence with an attractive sign that makes a deeper impression on patients than it would if they practised alone. Frequently, they are able to assist each other and there may be the joint use of ancillary facilities, for example a health-oriented café. A pharmacist and other healthcare facilities are glad to have premises there, too, either in the medical centre itself or nearby.

OUTPATIENTS CLINIC | The concept of cross-disciplinary practices in which to some extent the use of staff, rooms and medical equipment is shared, is not an invention of the former German Democratic Republic, but was rather the way in which most outpatient care was organised. Patients attending outpatient clinics can be referred to other practices in the same building and can continue to be treated there without having to travel long distances and with little loss of time. However, because of their size, outpatients clinics are only financially viable in central locations so that, in fact, many patients do have to travel very long distances. A further disadvantage occurs when employed doctors are not sufficiently involved in the clinic's turnover. After German reunification, doctors remained in the old buildings. Single-doctor practices emerged that regrouped under the roof of a medical centre.

Today, there are still outpatients clinics in most of the former socialist states, though there are some in Sweden, too. In South Africa, there are even mobile outpatients clinics.

MULTI-DOCTOR PRACTICE WITH CLINIC | As distinct from the multi-doctor practice already described, in which there is only outpatient care, in this sort of practice complemented by a clinic, patients can also be cared for after invasive inpatient treatment. This primarily applies to practices that perform operations.

An inpatient stay lasts in general from one to three days. The functional and spatial allocation plan usually required for inpatient care does not apply here. All that is needed are wards [with two beds or preferably one bed and a bathroom] and the side rooms for care-related purposes. The essential feature is the assured delivery of care, even at night. If a longer inpatient stay is needed, the patient ought to be moved to the inpatient department of a hospital.

The type of joint practice described here with its similarities to a clinic is often designated as a practice clinic. However, the term is sometimes used for practices that can only provide patients with a room and couch on which to lie down for a few hours post-operatively.

Medical Practices

HOSPITAL OUTPATIENTS DEPARTMENT | In years gone by, the Germany healthcare industry made a sharp distinction between outpatient and inpatient care. The care of outpatients lies in the hands of the registered doctors and inpatient care was the province of the hospital. Only hospitals with a teaching remit were allowed to provide care for outpatients as well because that was the only way of ensuring that medical students received a broad enough training. In these institutions the term outpatient clinic is still in use today.

As a consequence of the developments set out above, hospitals generally have a "medical service" which is in charge of examining and treating hospital patients.
The lifting of the distinction described above means that the "medical service" can also treat outpatients in hospitals. German hospital outpatient departments can now also be run in direct cooperation with registered doctors.

MEDICAL CARE CENTRE | On January 2004, the Statutory Health Insurance Modernisation Act came into force. This Act licensed medical care centres to provide care by affiliated doctors for patients insured under the statutory and private health insurance schemes. A medical care centre [MCC, German acronym – MVZ] is a cross-disciplinary facility run by a doctor and in which doctors work either as owners or employees. A MCC can be set up by pharmacists, preventive medicine and rehabilitation organisations and hospitals, not only by doctors and psychotherapists, provided they are involved in the medical care of insured patients by virtue of a licence, authority or contract. For patients, it is important to be aware that they are not entitled to personal treatment by one of the doctors in the MCC.

HEALTH CENTRE | The breakdown of the rigid distinction between outpatient and inpatient examinations and treatment has lent wings to the imagination of service providers and led to the creation of an institution called the health centre. This can deliver all the healthcare services for the population – almost without limitation.

In the health centre, a full, networked healthcare service is available for statutory health scheme and private patients. The integration in an electronic network of the medical staff can prevent long waiting times, save unnecessary duplicated examinations and enable doctors to discuss treatment methods and the prescription of drugs. There is practically no limit to the ways in which different healthcare institutions can cooperate and provide "integrated care". Many patients wish to have all health facilities grouped together in one place. As examples of how to satisfy this requirement, health information desks, sports medicine facilities, a pharmacy or a healthcare product sales outlet can be added as extra features.

SPECIAL FACILITIES FOR THE PROVISION OF MEDICAL CARE | Quite apart from the medical care facilities and institutions discussed above, there are numerous other special medical service provisions, for example:
- Obstetric clinic
- Rehabilitation centre
- Day clinic
- Drug outpatients department
- Geriatric centre
- Hospice

OPERATIONAL DATA

Before planning ideas for the rigid of outpatient care can be implemented, it is essential to discuss how the institution is to be run. This requires a vision or a corporate strategy – architects speak of business objective planning. This will form the basis for subsequent concept planning and conversion into the actual building. [See below for illustrations of important operational aspects relevant for the planning of medical practices.]

Business objectives
In their definitive book "Unternehmen Arztpraxis"[15], Schurr, Kunhardt and Dumont present a detailed analysis of many of the business aspects of a medical practice concept that need to be addressed:

Management
- Ability of the practice owner as business manager
- Accuracy of the accounting system
- Quality control system
- Planning and forecasting ability
- Staff management and social competence
- Information technology and policy
- Communication procedure with the bank

Human resources
- Doctor's qualifications
- Integration of key personnel
- Knowledge retention
- Working atmosphere

Patient demand | Specialist area
- Patient potential, volume and growth
- Development of the specialist field
- Dependencies
- Competition
- Health care services | key areas
- Patient benefit
- Service standard
- Marketing strategy
- Patient relations
- Transparency
- Communication of information

Commercial conditions
- Finance management
- Assessment of the annual accounts
- Total asset situation
- Liquidity control
- Funding policy
- Risk management

Practice development
- Developments since the last annual accounts
- Corporate planning
- Earnings planning and ability to service credit
- Special company risks
- Analysis of strengths and weaknesses

Once the individual objectives have been worked out, a business process management scheme needs to be developed, which will enable the individual objectives to be measured. This makes it possible to evaluate and compare alternatives, the result of which will be informed decisions.

Catchment area
The population profile of the catchment area and the type of patients have an importance for the commercial development of the practice which cannot be overstated. Successful medical practices take account of such data and use them to tailor their publicity strategy.

The size of the catchment area supplying the majority of the patients is a good indicator of the prominence and importance of a medical practice. Quite a number of medical practices have an extensive catchment area because their location, for example within a rural area, is relatively sparsely populated.

If patients have to travel long distances to the medical practice, either on foot, by public transport or by car, the doctors need to think about what welcoming facilities they can offer them. Things to consider are places to rest, especially for old and handicapped patients, and refreshments, for instance a mineral water dispenser.

Location
The criteria that apply to the choice of a good home also apply to medical practices. The first three main criteria are location, location, and location. For this reason, it is of paramount importance to make a thorough analysis of the location before drawing up customised plans in order to discover the most appropriate place for the future medical practice. The following criteria can be helpful:
- places with a high number of passers-by, perhaps in main thoroughfares, pedestrian zones or in buildings where the passers-by are constantly changing, such as department stores or shopping centres, railways stations and airports;
- places in or near other healthcare facilities. The best places are hospitals, health businesses or centres, gym and physiotherapy practices, care support points and podiatry practices, cosmetic studios and wellness centres. The following are not helpful:
- locations too close to existing medical practices, unless patient numbers make this competition acceptable and unless this competitive situation turns out to be advantageous;
- locations not easily accessible by the local transport. In general, locations further than 200 metres from a public transport stop are not acceptable.

General and specialist doctor' services
Volume V of the Social Code contains a breakdown of medical services. The main groups are the following:
- Services to prevent illness and exacerbation of existing illness
- Services to aid the early detection of illness
- Services for the treatment of illness
- Services for medical rehabilitation

The services are delivered by general practitioners and specialist doctors. The above breakdown does not take into account the location in which services are delivered. Following the lifting of the former strict demarcations between outpatient and inpatient care, all medical services can in principle be delivered anywhere. This applies, for example, even to operations and the technically expensive services of nuclear medicine. Increasingly, there are doctors who provide their services both in medical practices and in facilities intended mainly for inpatient care.

SPECIAL DOCTORS' SERVICES | In Germany, the financial situation of registered doctors has become significantly harsher and this has forced them to keep their range of service under constant review and to expand it. Imagination and creativity have led to the creation of numerous new services to help enhance the appeal of the practice concerned. There are many fields in which medical practices have become decidedly proactive:

EXPANSION OF THE RANGE OF SERVICES |
- Following continuous training opportunities in fields which have a strong appeal for patients, for example acupuncture, natural healing methods, allergology, diabetology, geriatrics, sleep medicine and sports medicine.
- Provision of individual healthcare services. Consideration needs to be given to the sorts of patients willing, in the future, to pay for such services out of their own pockets.
- Stronger networking with company doctors.

EXERCISE OF HEALTH CARE TASKS |
– Information and education campaigns
– Prophylactic measures
– Early detection of illness.

EXPANSION OF SERVICES |
– Information by email
– Supply of medication to patients' homes
– Collection of cost estimates from hospitals
– Acceptance of orders by telephone, fax and email
– Organisation of self-help groups
– Setting up health-oriented cafés

RUNNING COURSES AND SEMINARS |
– Courses on healthy eating
– Lectures on particular disease profiles, for example diabetes
– Readings on subjects around health promotion
– Seminars on dietetics and healthy eating
– Provision of informative brochures.

The planner ought to be aware of the type and scope of any special services since these activities have consequences for the type and number of rooms and their use.

Operational organisation

TREATMENT AREAS | The treatment areas laid down in the medical practice's objectives need to be kept under constant review because they can vary with the changing healthcare policy standards. Schurr, Kunhardt and Dumont have collected the factors which determine the treatment areas[15]:
– Training conditions
– Previous and planned medical concept
– Patient demand
– Competing services in the region
– Fixtures and fittings in the practice
– Legal situation
– Financial situation of the medical practice and practice owner
– Cooperation opportunities or cooperation necessities
– Medical innovations

TREATMENT PATHWAYS | The subject of the development and application of treatment pathways, which are also termed guidelines or standards, is increasingly the object of intense debate at specialist medical congresses. The call for evidence-based medicine [EBM], which is to be as internationally valid as possible, is intensifying, even in relation to quality assurance for medical services, nowadays universally regarded as essential. Treatment pathways have the following important characteristics:
– Description of the objective of a treatment
– References to other treatment pathways/ EBM sources
– Aids for decision-making
– Instructions for doctors' conduct
– Support of the process-oriented programme
– Overview of the required personnel and physical resources
– Information source for all patients.

After application of the treatment pathway, maintaining a detailed record should help keep the individual elements of the pathway developing.

QUALITY ASSURANCE | Improvement and assurance of the quality of the medical activities is one of the key roles of patient-driven, need-driven and commercial care. Transparency about the quality of treatment results and a good degree of quality in patient care are the main goals of quality assurance.

The Social Code [§ 135a and § 137][6] is the legal basis for quality assurance. Numerous institutions work in the field of quality assurance, for example the Medical Councils, the Associations of Statutory Physicians, the Joint Federal Committee, the Institute for Quality and Efficiency in Health care, the hospital companies and the medical service providers of the health insurance schemes.

QUALITY MANAGEMENT | The standard of quality necessary in medical care institutions is achieved through a range of measures. These include optimum work process design, improvement of communication structures, increased patient and staff satisfaction, measures to raise patient safety, development of joint care formats and quality indicators, and the setting up of quality control systems as well as the fitting out and design of work rooms so as to be fit for purpose.

All activities, which are grouped together under the general heading of "Quality Management" [QM], nowadays form a part of business management. Continuous medical training is one of the core tasks. For example, Medical Council seminars promulgate the principles of QM. For outpatient care, the Medical Centre for Quality in Medicine published the QMA Compendium. Its third edition appeared in 2008.[16]

Among the features it contains are quality management benchmarks for practice, quality criteria and quality indicators, a perspective for the future of quality management in outpatient medical care and, finally, a patient questionnaire [equally useful for architects] for assessment of outpatient care.

USE OF STAFF | The success of a practice depends entirely on the professional expertise of the staff. This is why special attention should be given to staff selection. Staff management is no minor matter. Here, the practice's management has to decide between authoritarian, laissez-faire and democratic management styles.¹⁵ Finally, there needs to be a good working atmosphere in the practice that gives every member of staff the feeling of being safe, accepted and having room to develop. The type and especially the number of staff members are determining factors which relate to the size of the work space, staff rest room and changing room areas.

USE OF EDP | It is no longer possible to run a medical practice efficiently without using any computers. In any case, before suitable hardware and software are selected, professional advice should be taken if at all possible. Once the choices are made, they will have a major influence on the quality of all work activities, that means for example, appointment management, accounting, documentation, quality assurance, communication and market research.¹⁵

Problems arise when different software systems need to work together and when data are exchanged between different integrated care locations. According to estimates, 70 % of all transactions are still in paper form.¹⁵ The paperless medical practice simplifies data processing, but special care must be paid to data protection and data security. A 30-year retention period is recommended for treatment data, so thought must be given to finding suitable storage locations.

The Statutory Health Insurance Modernisation Act has created a new requirement in connection with the launch of the electronic health record card; it is intended to contribute transparency and quality in treatment and to improve efficiency. Here, the patient can decide which of his data is to be stored.

WASTE DISPOSAL | The Federal Ministry for the Environment has issued strict rules for disposal. Waste generated in medical practices is generally collected at regular intervals by waste disposal companies. Special attention must be paid to hazardous substances, including drugs and organic and inorganic chemicals and anaesthetics that can no longer be used. Disposal of these is governed by the Narcotics Act. A special regulation governs out-of-date drugs.

Practice logo
A carefully selected logo accurately reflecting the spirit and philosophy of the medical practice can make a definite contribution to the success of the business. A good logo must meet five requirements; it must be:¹⁵
– Easily recognisable and understood
– Unmistakable
– Memorable
– Reproducible.
The logo should not be limited to the symbol itself – in other words, it is helpful to repeat the colour and forms in the design of printed matter, whether it is notepaper, receipts, brochures or notices. The logo is rendered particularly memorable if its colours and forms are used in the design of the practice's rooms.

Operating costs
Finally, all the thinking about operational matters that has gone into the organisation of a patient-friendly and dynamic medical practice ultimately has the aim of keeping operating costs within economically acceptable limits.

STAFFING COSTS | These can account for up to two thirds of the overheads. This is why careful thought should be devoted to using staff efficiently. Efficient staffing ratios are dependent on having streamlined the medical practice's essential work processes as rationally as possible.

MATERIALS COSTS | These are composed mainly of funding costs, depreciation, publicity costs, vehicle and travel costs, office and administration costs, insurance, rents and utilities costs. A good operating result is dependent on cutting out any wastage in these areas.

Supervision of medical practices by the authorities
A variety of laws and regulations lays down measures to ensure that quality standards for medical care are met. Examples of these are the Protection against Infection Act, the recommendations of the Robert Koch Institute Commission for Hospital Hygiene and the Prevention of Infection, the Medical Devices Act, the Health and Safety at Work Act and the provisions of the Occupational Health and Safety Association. Different institutions are tasked with supervisory roles, especially the healthcare offices as bodies within the public healthcare service. They conduct inspections with or without prior notice. If deficiencies are detected, fines are normally imposed immediately, though in particularly serious instances, restrictions may be imposed on the practice's activities or the practice may even be closed. In 2005, the Association of Statutory Health Insurance Physicians published a special paper entitled "Supervision and Inspection of Medical Practices by Authorities"¹³. Among the reports included are accounts of the numerous statutory bases and methods for inspecting medical practices subject to the procedures of Joint Self-Government. The publication's appendices contain helpful examples of checklists for medical practice inspections.

CONSTRUCTION DATA | Once the details of the practice's proposed operations have been worked out, construction data can be calculated and collated. Both data packages form the basis for further action and decisions on the design and form of a medical practice.

Medical Practices

Town-planning conditions
SITE | If a new building is planned, points to consider when choosing the site include the size and any existing conditions impinging on development potential as well as cost per square metres. A far-sighted property developer will seek to acquire a site large enough to allow extensions later so that the practice can grow if necessary.

TRANSPORTATION LINKS | Good transportation links are very important for a medical practice if it is to thrive. This is particularly true of customers who come on foot and also for those using public transport who need the nearest stop to be close.

MONUMENT PROTECTION | a medical practice subject to a protection order because it is an old building of outstanding architectural merit is something to be treasured by both the owner and his clients, which is why people like going to such buildings. The authorities generally impose strict conditions on such buildings making it very difficult to make alterations with the result that the practice frequently has to accept a degree of functional inconvenience.
Measures intended to maintain the substance of the building can also be very expensive. Sometimes, consolation comes in the form of financial support from public funds and occasionally even from sponsors keen to maintain the historical heritage.

Usable floor space
Apart from the area accessible to vehicles, the floor area and technical function area, the usable floor space is the most important type of area defined in DIN 277[17] for the subdivision of the total floor area of a building. Unlike the other three types of areas, the usable floor space provides information on the particular remit of the project. This is the most important parameter of a ground plan for medical practices.
Since size is one of the key distinguishing features of medical practices, the usable floor spaces are used to differentiate between them. Evaluation of the usable floor spaces of the examples cited in this book and analysis of the existing developmental trends lead to the following main classifications for medical practices:

Single-doctor practices
– Small single-doctor practices:
   usable floor space of up to 125 sqm
– Medium-sized single-doctor practices: usable floor space of between 126 sqm and 175 sqm
– Large single-doctor practices:
   usable floor space over 175 sqm

Multi-doctor practices
– Small multi-doctor practices:
   usable floor space of up to 200 sqm
– Medium-sized multi-doctor practices: usable floor space of between 201 sqm and 400 sqm
– Large multi-doctor practices:
   usable floor space over 400 sqm

Functional and spatial allocation plans
The functional and spatial allocation plans define the usable floor space. Other areas only become apparent at the design phase. Details of rooms are given in terms of number, type and size in square metres. The light exposure required by the rooms should be stated, using the following symbols:

○ Natural daylight necessary
◐ Natural daylight desirable
● Natural daylight unnecessary

DIN 13080 "Division of Hospitals into Functional Areas and Functional Sections"[18], a standard which has been applied successfully for over 30 years in hospital planning both in Germany and beyond its borders, contains a recommendation for the subdivisions of functional and spatial allocation plans. According to the above standard, the rooms should be subdivided into four groups:

– Main rooms
– Ancillary rooms
– Communication rooms
– Staff rooms.

It is left to the discretion of the function and space planner whether to subdivide the spaces according to his functional and spatial allocation plan or to choose the mode of subdivision in terms of function groups described in the following section; the latter makes it possible to have major differences between the rooms. For this reason, this book will focus on function groups.

Subdivision into function groups
A variety of different types of rooms can be incorporated into the plans for a medical practice. To reduce the large number of different rooms, some with similar purposes have been grouped under a single heading.
It has proved useful for planning purposes to assign the approximately 80 different types of rooms left over after nomenclature standardisation to particular room units.
Because, unlike in hospitals, the room units in medical practices are relatively small, the classificatory term "function group" is used here. As in DIN 13080[18], colours are used additionally. These make it considerably easier both to grasp the organisation of the ground plan and to

analyse and compare plans. Not every medical practice will use the eight function groups illustrated briefly below. However, they are representative, depending on the type and size of the practice and its service profile.

PATIENT ROOMS | Every medical practice works primarily for the benefit of its patients. Accordingly, all the rooms and equipment should be planned with this in mind. Bearing in mind the need to have rooms with different functions, rooms can be reserved primarily for the use of patients – for example, a reception room, waiting room with a children's play corner, consultation room, changing cubicles, patient transfer hatchway, recovery room and sickroom. Special attention should be given to these patient rooms at design stage. This is the point where sensitive patients pick up on how they are valued by the practice. In the ground plans the patient room function group is marked in yellow.

EXAMINATION AND TREATMENT ROOMS | The rooms in which the basic medical services are carried out constitute the "production area" of the business. These are the rooms used for doctor-patient consultations, for taking patient histories, initial examinations and blood samples: the general examination and treatment rooms, even those that double as consultation rooms, individual examination rooms, individual treatment rooms and the necessary laboratory rooms with different specialised uses, for example, as a urine lab or a lab analysing blood samples. This function group is marked below in red.

SPECIALIST MEDICAL ROOMS | The "production areas" of a medical practice also consist of the rooms destined for specialist examinations and treatments. They comprise their own function group, which is the one with the most variety because the rooms are of such different types. They are not so different in size – there is a much greater difference in the way they are equipped. There are approximately 70 specialist fields catered for in the Continuous Training Codes across the Länder and this is reflected in the number and type of specialist examination and treatment rooms. Also, most specialist fields require extra rooms specific to their field. This function group is coloured pink.

ADMINISTRATION ROOMS | This relatively small function group covers the rooms required for the administration of the medical practice. Key features are an office or even several offices, designed to fulfil different administrative roles, an administration department or, in large medical practices, several such offices for the different specialist fields. Administration rooms are shown in the ground plans in green.

SERVICE AND STAFF ROOMS | Service, recovery, changing and sanitary areas intended for the staff constitute their own function group. It is useful to locate the rooms together in a quiet section of the practice. In a medical practice that attaches importance to social compact issues, these rooms will be given the necessary special attention. Provision of recovery and rest rooms and table tennis rooms will definitely remain the exception. Some larger practices will have service rooms for particular groups of staff. These rooms are marked in orange.

SUPPLY AND DISPOSAL ROOMS | The rooms required to meet the supply and disposal needs of medical practices include primarily the sterile and unsterile work rooms, stores and equipment rooms, a cleaner's room and a disposal room. The supply and waste disposal rooms are coloured brown.

TRAINING AND TEACHING ROOMS | Rarely, rooms are provided for various staff and patient training options. These will be lecture, teaching and seminar rooms, which may also be multipurpose rooms which can be reserved for future developments. These rooms are marked in violet.

TECHNICAL SERVICE ROOMS | The blue marking in the ground plans identifies the rooms for technical equipment, such as the control rooms for X-ray, CT and MRI scanning equipment, EDP rooms, development rooms, rooms to house heating and air-conditioning plant and mechanical service rooms.

Important individual rooms
Of the approximately 80 standard rooms in a medical practice, some are largely destined to everyday purposes. These are the rooms in the administration, service and social room function groups, some of the supply and disposal rooms and the training and teaching rooms. Some indications of special types will suffice. Some of the individual rooms typical of medical practices need to be looked at in closer detail here. The order in which the rooms are presented follows the sequence of function groups described in the previous section. In general it is true that these rooms do not necessarily have to be usable floor spaces completely enclosed by walls; they are often open or half open areas to create an overall impression of the greatest transparency.

RECEPTION ROOM | Just as a well designed business card gives an indication of the nature of the owner, the reception area also conveys an initial impression of the whole medical practice. Natural daylight immediately by the entrance is highly desirable. Here there should at least be some preferably daylight exposure to create a soothing ambiance, a generous subdivision of the space, suitable, well designed furniture and, most important, an understanding, friendly person who listens to requests patiently, gives accurate information and can accompany the patient into the waiting room.

WAITING ROOM | It would be best for patients if the practice could manage without a waiting room. Unfortunately, however, even the best organisations have been unable to arrange for patients' appointments to be kept so promptly that waiting times and therefore a waiting room prove unnecessary. Crowded waiting rooms are not generally a sign of a well run appointment system. The following factors should be taken into account as regards the size of the waiting room: the number of doctors, the practice's hours of business, the average duration of treatments, the type of appointment system and average number of patients without appointments and emergency cases. The last group can ruin any scheduling system, no matter how precise. In any case, care should be taken that the number of waiting patients per doctor never exceeds five, otherwise, as patients arrive, they get the feeling of being part of a mass production system. This is why large practices have discovered the benefit of having a second waiting room which, ideally, should be in the immediate vicinity of the reception desk. This can also be used for patients at risk of infection. Mineral water should be available for patients in the waiting room. Having a range of current magazines makes the waiting time seem shorter and these should reflect the interests of patients. A clock and leafy plants also add to the feeling of wellbeing.

CHILDREN'S PLAY AREA | Since parents increasingly have to bring their small children with them to the medical practice, suitable toys should be available in the waiting room. It is helpful to set up the playing area in a corner because this keeps the children in one part of the waiting room where it is easier to keep an eye on them.

CONSULTATION ROOM | For a first consultation with the doctor or for longer, detailed discussions [perhaps with several relatives], large medical practices have found it beneficial to a special meeting room, which should be exposed to natural light if at all possible. This option frees up the doctor's examination and treatment room [consultation room]. It should be furnished with a group of four to six chairs arranged together. There should be a supply of fresh water.

CHANGING CUBICLE | Provision for changing cubicles needs to be arranged in consultation with the doctor in charge. They can be in the form of a curtained off space or a separate cubicle and should be located outside the examination and treatment rooms. These cubicles should have enough space to enable handicapped people to change, if necessary with a helper. Cubicles satisfy the wish for privacy.

PATIENT TRANSFER HATCHWAY | Some treatments require special protection against contamination. For these, the construction will need to make provision for a patient transfer passage. This applies mainly to invasive procedures carried out in an operating theatre. Patients enter the passage either on foot or lying down and are then transferred to an operating trolley or a mobile operating table. The route goes into the operating theatre via a holding area and then, after the operation, back into the patient transfer area. Now that outpatient operations are permitted in properly equipped specialist medical practices, there are more and more practices carrying out such work. The rule enshrined in the Robert Koch Institute guideline is crucial here, that is that outpatient operations must meet the same hygiene standards as for inpatient operations.[19]

RECOVERY ROOM | Peace and quiet may be necessary and beneficial for patients immediately after exhausting examinations or treatments. A simple room with a couch and somewhere to sit for an accompanying person is enough to meet this need. However, if a recovery room also has a soothing picture, some leafy plants and fresh water, then it becomes something more – an oasis of peace.

SINGLE-BED ROOM | Specialist practices with a clinic section have rooms, in addition to those necessary for outpatient care, that are available for a patient requiring an inpatient stay. The single-bed room is a place for a postoperative patient who cannot leave the practice on the same day or has to stay for observation, care and nighttime treatment. Essentially, the sickbays of a specialist medical practice do not differ from those of a hospital because the patients have the same needs. The only difference is that the length of stay in a specialist medical practice is considerably shorter than in a hospital.
A single-bed room should contain a bed [adjustable if at all possible], a bedside table with a drawer, a wardrobe, a table with two seats [one of which should be particularly comfortable], a television, telephone and an internet connection point for a computer. The sickbay should have an en suite bathroom complete with a wash basin, shower and lavatory. Relatively few practices go to the expense of having the larger areas needed for handicapped patients and the more costly equipment in the bathroom.

TWO-BED ROOM | Some patients prefer a two-bed room because they appreciate having someone with them for company, someone who can summon immediate help in an emergency. Otherwise, the requirements for two-bed rooms are the same as for a single-bed room. Having the same space around them and the same furniture, they present economic advantages. From the point of view of the staff, they are easier to take care of. This however should not be taken as reason for dispensing with the single-bed room, which in many cases is urgently needed.

EXAMINATION AND TREATMENT ROOM | The core of a medical practice is the general examination and treatment room. This is where the main activities of the practice take place – diagnosis and treatment involving the main "players", the patient and the doctor.
The examination and treatment room generally doubles as the doctor's working and consulting room where patient and doctor meet for initial consultations, the taking of patient histories and for conducting basic examinations. These lead to the specialist medical rooms where the specialist diagnostic and treatment procedures are carried out. The furniture of the usual examination and treatment room will consist of the doctor's desk and an examination couch. In addition, there will be a variety of cupboards and shelves for instruments and books and some appliances, for example, scales to measure weight and equipment to record physical dimensions. Depending on the doctor's personal inclinations, the room should radiate warmth, which may be objective, sober or homely. The doctor should pay special attention to this at design stage.

CONSULTATION ROOM | Some larger medical practices have a separate consultation room in addition to the examination and treatment room. It is an advantage if diagnostic and therapeutic procedures – which of course are commonly associated with small – can be conducted away from the doctor-patient consultation. As the space has no medical or technical equipment, the consultation room can take on a homely, almost private character. Sound insulation materials need to be considered when fitting out the room to ensure that the confidential nature of the consultation is not breached. The room should have adequate space to move around, comfortable seating, carefully chosen lighting and so on.

LABORATORY | Diagnostic examinations of the blood and other body fluids involve the taking of samples. This is generally carried out in a screened off part of the laboratory or in a small adjacent room with a hatch. Depending on the available appliances, the analyses are carried out in the same laboratory or in a central laboratory.

Numerous regulations govern the construction and fitting out of laboratories. The priority regulations to adhere to will be the Construction Code of the particular federal Land, the fire protection regulations and DIN 12924, section 4^{20}. These regulations govern, for example, the minimum size [12 square metres], require that walls, ceilings and floors are constructed of fire-resistant materials, specify a second escape route, an exhaust vent with a suction unit for work with flammable liquids and at least one fire extinguisher. Other requirements include good lighting, light that displays true colours, both natural and mechanical ventilation and surfaces that are easy to clean.

SPECIALIST EXAMINATION AND TREATMENT ROOM | Most medical practices are specialist practices with a specific medical remit; they have the space and appliances needed for that specialist field. Depending on the specialist field, there will be separate rooms for examination and for treatment and also rooms in which both are possible. The following are some examples of specialist examination fields: allergies, audiometry, blood pressure, computer tomography, echocardiography, electrocardiography, endoscopy, gynaecology, hearing tests, pulmonary function, magnetic resonance imaging [MRI], mammography, psychiatry, X-ray diagnostics, sonography and ultrasound. Examples of specialist treatments are: acupuncture, surgery, dermatology, dental hygiene, ergotherapy, infusions, inhalations, gymnastics, maxillary orthopaedics, cosmetics, laser treatment, massage, oral hygiene, nuclear medicine, prophylaxis, shock therapy, first aid and dentistry. Each of the necessary rooms will have its own ideal design in terms of size and equipment. For this reason, it is not possible to give a detailed description of the examination and treatment rooms in this account of the general fundamentals of planning. The specialist architect or specialist planner will need to work out the appropriate solution with the specialist doctor. Since flexibility of use is desirable, the rooms in the planning examples contained in this text are assumed to have a standard usable floor space of 14 square metres. Where the specialist examination and treatment room is also used as the doctor's consultation room, the usable floor space increases to 22 square metres.

PREPARATION ROOM | Before major interventions, operations or some complex diagnostic procedures, patient preparation time is needed. In larger medical practices a preparation room should be available for this. The siting of the room is determined by its main use. Its dimensions should be the same as those of the specialist examination and treatment rooms. In emergencies, this room can also be used for other purposes.

INTERVENTION ROOM | Most medical practices, generally those constituted as multi-doctor practices with a clinic but also health centres and hospitals contain their own intervention area in which patients with life-threatening symptoms are examined and treated. The room should be in the immediate vicinity of the entrance and it should have sufficient space to move about in and be well lit and ventilated. When operations are performed in the practice, the intervention room can also function as a preparation room or a holding area. The route from here to the operating unit should be as short and straight as possible.

OPERATING THEATRE | The fact that more and more outpatient operations are being performed in medical practices does not mean that such practices require the same spatial configurations as are usual in hospitals. Nevertheless, certain basic criteria must be met to ensure that medical interventions are performed under the proper conditions and especially that adherence to hygiene standards is guaranteed. The operating theatre must be located in a quiet area, screened off from the rest of the practice. The patient reaches this area via a patient transfer passage or, if applicable, via a holding area. Post-operatively, the patient is generally taken while still recumbent to a post-anaesthesia care unit, adjacent if at all possible, and he will remain here until he has fully regained consciousness. After that and depending on the severity of the intervention, he will stay in a single-bed or two-bed room for one or two days on average. The operating theatre should measure at least 36 square metres and, where feasible, be lit by natural daylight. Whether the room is ventilated – or better, air-conditioned – depends on the type of operation and also on the financial means of the operator. That applies also to the installation of a laminar flow unit.

POST-ANAESTHESIA CARE UNIT | After operations in which the patient has been anaesthetised, he should spend approximately two to five hours during the recovery period in a post-anaesthesia care unit. Here he will be subject to close observation and medical care. The post-anaesthesia care unit should have windows. As they come to, patients generally find it a wonderful experience to see natural daylight or even the sun. The number of beds in each case [at least two] depends on the number, type and duration of the operations. The post-anaesthesia care unit should be suitably fitted out with the necessary media and monitoring units.

ADMINISTRATION DEPARTMENT | To cope with the growing volume of secretarial duties, it may be useful, in large multi-doctor practices and hospitals, to set up one or several administration departments. The administration department supports the manager or managers of the unit and handles tasks which have to do with external relations, ordering, accounting, bookkeeping, prescription processing and correspondence. Obviously, the number of staff working in the administration department depends on the volume of work.

OFFICE | As an alternative or as an addition to the administration department, a separate office – or in larger medical practices two such offices – is a practical means of ensuring that the mounting load [despite EDP] of administrative work is carried out undisturbed. This keeps files from heaping up in the rooms intended for medical duties. The office should be in the vicinity of the reception area, be lit by natural daylight and appropriately furnished. A medical practice generates files that must be stored. Sufficient space for such storage [allowing for the years to come] needs to be included in the plans unless a separate archiving room is available. In large medical practices, separate offices can be included for essential administrative duties, for example for accounting and the correspondence service, which will also be responsible for writing medical reports.

ARCHIVE | An archive is useful in large medical practices to accommodate the swelling flood of files. This room can also house a photocopier, and any ladders, chairs and spare equipment that may be necessary.

STUDY ROOM | In major practices, but especially in hospital outpatient departments, study rooms can aid the performance of the organisation because this is the place for undisturbed, concentrated and creative work. Such rooms need not occupy a great deal of space. The important thing is that they are individually well designed. Study rooms are for medical directors, senior physicians, nurses and midwives.

STAFF CHANGING ROOM | According to the Work Place Regulation[21], each member of staff is entitled to and must have an individual cloakroom facility and a locker in which to keep valuables. However, small single-sex changing cubicles are preferable to meet the personal needs and hygiene standards of employees. A shower with direct access to the changing area is desirable. After all, a medical establishment has a particular duty to respect the hygiene and wellbeing of its staff.

STAFF REST ROOM | § 29 of the Work Place Regulation[21] lays down that employers with a staff of over ten must provide an easily accessible room for break periods. In a medical practice, this should be the rule even if there are fewer than ten employees because eating at the work place is inconsistent with maintaining hygiene standards. A fitted pantry will do in most cases. In larger medical practices this can become a separate room. A view out on to a green area would serve the needs of staff for relaxation particularly well.

STAFF LAVATORY | Under the Work Place Regulation, work forces of more than five are entitled to separate single-sex lavatory facilities. These must be for the exclusive use of the staff.

STERILISATION ROOM | There are various ways of organising the supply of sterile material for a medical practice. In any case, a storage facility for sterile material is essential. This must be available for disposable materials and goods sterilised outside the practice and, where applicable, for goods sterilised in the practice itself. If sterile goods are supplied from an external source, only storage space is essential, but additional work surfaces are useful for handling materials. However, if the materials are sterilised in the practice, whatever the quantity, a sterilisation room with the required appliances is essential. For hygiene reasons, unclean work before sterilisation in a sterilisation room must be carried out in a sterilisation room – unclean, whereas work after sterilisation should be carried out in a sterilisation room – clean.

WORK ROOM – CLEAN | In almost every medical establishment, lots of tasks must be carried out in a clean environment. In general, the staff use the examination and treatment room for this. However, the situation cannot be regarded as ideal with regard to hygiene. Better would be a separate room which is reserved solely for clean work. The room can be kept small. It requires a work surface with a water tap and shelves.

WORK ROOM – UNCLEAN | By way of contrast, the above remarks on the work room – clean are valid for this work room. On closer inspection, there is a stronger justification for setting up this room.

STORE | In Germany, it has long been the practice of suppliers of medical and office goods to deliver immediately on receipt of an order, ideally on the same day, so that goods no longer need to be kept in storage for long periods. However, there is still a need for medical practices to keep a small stock of materials on the premises. A properly constructed and managed store only needs a small area. Its size depends on that of the practice itself and its purchasing policy.

APPLIANCE ROOM | In general, medical and technical appliances are kept in the examination and in the treatment rooms when in use. An appliance room can be useful in case a newly acquired piece of equipment should fail or to store equipment used only on an occasional basis.

CLEANER'S ROOM | Many medical practices have not provided space for a cleaner's room. As a result, a cleaning trolley stands in a corner of a corridor, a vacuum cleaner is stowed in a cupboard not intended for this and cleaning materials are kept in various places. This situation is not acceptable if an efficient cleaning service is to operate. In the eyes of patients, this is a clear sign that the ground plan has been inadequately designed.

DISPOSAL ROOM | In future, it may well be that the waste of an affluent society won't be so highly valued – even so, the sorting of medical practice waste into different categories is a major priority for reasons of hygiene. A small room specifically set aside to accommodate different types of waste makes it unlikely that a medical practice's waste will pose any risks.

SEMINAR ROOM | Despite there being now almost far too much in the way of media contact, the direct exchange of information between doctor and patient will not lose its importance. In a large medical practice, a seminar room with information materials and a technical medical library provides good opportunities for face to face discussions, lectures with slide shows and any type of further training, also for ancillary staff. One should not forget that a practice enhances its image by having such a facility.

MULTIPURPOSE ROOM | Taking the longer view, it makes good sense to have a spare area with no specific intended purpose. Medical developments happen so quickly – anything can suddenly create the need for space. New medical standards can emerge overnight and require a rethinking of a previous strategy – with space implications. Such new developments cannot be accommodated completely but a room of this type can help get round many a bottleneck.

TECHNICAL ROOM | Often no attention is paid to the need for a utility room, the room in which all the utility supply services enter the premises. This feature may be omitted at design stage for lack of space and because there is inadequate coordination of the various technical work elements. However such a room makes sense for servicing work.

Structural characteristics

A range of quantity data is needed to make an assessment of a design in addition to any statement on quality-related characteristics.

GROSS FLOOR AREA | According to DIN 277[17], the GFA in square metres is the sum of the floor areas of all the ground plans of the building project. By relation to the usable floor spaces, which have been discussed above, this value gives an indication of the compactness of the building solution.

GROSS CUBIC VOLUME | Also according to DIN 277[17], the GCV in square metres is the volume of the body of the building enclosed by all the external boundary lines.

CONSTRUCTION COSTS | DIN 276[22] provides the framework for calculating the construction costs of building engineering[22]. The cost groups in question are 300 Structure-Construction designs and 400 Structure-Technical plant. Because it is desirable to be able to make comparisons the costs should always include VAT. The construction costs per cubic metre of gross cubic volume are often quoted as a value for purposes of comparison.

TOTAL COSTS | DIN 276[22] also applies here, specifically cost groups 200 to 700. These are total costs including VAT. The cost of the site itself is not included as it can vary enormously because it is dependent on local circumstances. The total costs are also known as investment costs. Together with the operating costs they are the most important criteria for decision-making on new buildings, extensions or conversions for a medical practice. The characteristic value of the total costs per square metre of usable floor space is a particularly potent figure.

Parties involved in planning
In view of the many aspects to be considered with regard to building work for the healthcare business, it is useful to form a planning support committee. This committee advises on all matters relating to planning and construction and attempts to find efficient solutions that all can agree on if possible. The main players in this group are the client, as an investor with no background in the subject, and | or the user[s] and manager[s] of the medical practice. The practice planners take on the coordinating and controlling responsibilities. In general, these are independent architects or interior designers, increasingly also representatives of construction companies specialising in building medical practices. If the client does not introduce a specialist customer, than the doctor managing or designated to manage the business must be a member of the planning team, backed up by a member of the medical staff. It is important that all questions on operational procedures, on the functional and spatial allocation plan, and on the execution planning are discussed in detail with these specialist individuals closely involved in the day to day running of the practice. The effort made here is worthwhile if it helps prevent constructional and organisational errors. Because the main business of the medical practice is the delivery of services to patients, it can only be beneficial to have their approval of the planning decisions. Therefore it is a good idea to find at least one interested patient to follow the planning procedure through with criticism and input. The results of a questionnaire conducted among the patients can be included here, either instead of or in addition to those of a single interested patient. It may be useful to invite representatives of the supervisory authorities to attend meetings of the planning committee.

ELEMENTS OF THE DESIGN

Conventional architecture firms are not normally commissioned to design medical practices. Experienced architects and facility planners have often specialised for years in design commissions of this sort. Thus, often the client will have to deal with architects for whom such a commission is their first and constitutes a challenge. Especially for these architects, the following suggestions should give ideas for the design. They are also suitable for potential clients and operators since they give an overview of the planning process.

Functions
Observation of the daily routine in a medical practice often makes clear that there are various inadequacies and opportunities for improvement. For this reason, it is worthwhile making a thorough analysis of the functional procedures and the relations between functions that this reveals before the planned construction work to ensure that operations are as smooth and effective as possible. The following illustrations will consider some aspects of design planning.

FUNCTIONAL RELATIONS BETWEEN THE FUNCTION GROUPS
Among the first things to consider at design stage is how to clarify the relations between the function groups. The matrix of functional relations will bring these out sharply. Three function groups lie at the heart of a medical practice: patient rooms, examination and treatment rooms and specialist medical rooms. The other functions are grouped around this core, depending on the size and service structure of the medical practices. Note should be taken of the strength of the relations between functions.

FUNCTIONALLY POSITIVE ARRANGEMENT OF THE FUNCTION GROUPS IN TWO-STOREY OR THREE-STOREY MEDICAL PRACTICES | If there are several floors, movements between the different floors will necessarily create extra work. Here, it makes a difference if the distances covered are via an easily accessible lift or have to be walked up and down narrow angled staircases or even spiral staircases.

The picture of the scheme shows an appropriate layout for the function groups for small, medium-sized and large practices spread over two or three floors. Always immediately adjacent are patient rooms, examination and treatment rooms and specialist medical rooms. If the rooms in the basement have sufficient natural daylight, the service and staff rooms can be located here. It is more convenient to have the supply and disposal rooms close to the examination and treatment rooms and specialist medical rooms. Alternatively, the upper floor can be used for the function group of the service and staff rooms.

INDIVIDUAL FUNCTIONS | Essentially, the rooms of a particular function group should be located as close together as possible. After all, this was the rationale behind setting up the function groups in this way. There are standards to be met by a sequence of operational functions which require close relations between function groups, for example between patient rooms and specialist medical rooms. There are some relations within function groups that should be adhered to – for example between reception | waiting room and operating theatre and post-anaesthesia care unit.

FORMING ZONES IN THE GROUND PLANS | The form of the ground plan of a medical practice depends largely on how the available floor space is divided up in what is generally an existing building. If there is any choice in the way the floor space is allocated or if a detached medical practice building is being planned, it is appropriate to think about zoning the ground plan. The type of zoning determines the type of construction the medical practice will have. The concern here is the arrangement of the rooms as a patient entering the premises will see them. A distinction is made between one-, two- and three-zone premises. The individual room zones can be directly adjacent to each other, but they can also be separated by corridors. In a one-zone ground plan, all the rooms are grouped together. In a two-zone plan the rooms are arranged one behind the other. The rooms in a three-zone ground plan are ranged even further back. In principle, these planning variants apply for small, medium-sized and large medical practices in one-, two- and three-storey buildings. Three storeys are not an option for small practices.

The zoning style and the number of storeys are determining factors for the efficiency of a ground plan of a medical practice. In general, it is true that single-storey, two and three-zone premises provide the best conditions to patients. Patients in such premises are not faced with having to move from storey to storey, an important consideration, and such practices are very compact, irrespective of their size. Also, the typology of the building and energy consumption profile are further advantages.

A measure of the compactness is the extent of vertical external surfaces [U]. This feature has the smallest values by far in the two recommended premise types.

Environmental conditions

With regard to the increased incidence of allergies and sensitivities among patients and staff, it is important, even in medical practices, to think seriously about measures for a healthier environment. This involves thinking critically about choice of site, decisions on materials for construction and equipment and about environmentally friendly measures for running the medical practice.

SITE SELECTION | It is best to avoid, if at all possible, being in the immediate vicinity of a petrol station, a dry cleaner's or industrial premises that release chemicals into the air, and having car parks close to the entrance.

CHOICE OF MATERIALS FOR CONSTRUCTION AND EQUIPMENT | Essentially, materials with a low gas emission potential should be selected. Hard surfaces are preferable because they present a less likelihood of deposits of harmful substances forming on them. Preferred media are tiles, stone, terrazzo, hard wood and linoleum as floor coverings, rough-textured wallpaper and curtains.

ENVIRONMENTALLY FRIENDLY OPERATIONAL MEASURES[23] | The following are some of numerous actions that can achieve environmental improvement in medical practices:
- Selection of suitable disinfectants [for example, without phenol]
- Strict avoidance of cleaning materials containing chemicals, of sprays, perfumes, ether oils, cigarette smoke and pesticides
- As far as possible, minimisation of electric smog [for patients who react to electric and electromagnetic fields]
- Air quality improvement [especially for hypersensitive patients]
- No daily newspapers in waiting rooms because of their high solvent content
- For sensitive patients, avoidance of cut flower because of pesticide contamination and leafy plants because of potential mould in the growing medium. A desirable homely effect can also be created with art works and objects made of safe materials.

Hygiene standards

In facilities for outpatient examination and treatment, care must be taken to ensure that patients and staff are protected against infectious pathogens. The guidelines for hospital hygiene and infection prevention of the Robert Koch Institute in Berlin set out the standard requirements for the prevention and containment of

Medical Practices

Matrix of the functional relations between the function groups

	Patient rooms	Examination and treatment rooms	Specialist rooms	Administrative rooms	Offices and staff rooms	Supply and waste disposal rooms	Training rooms	Plant rooms
Patient rooms		1	1	2	3	4	2	4
Examination and treatment rooms	1		1	3	2	2	3	3
Specialist rooms	1	1		3	2	2	3	2
Administrative rooms	2	3	3		3	3	2	4
Offices and staff rooms	3	2	2	3		3	3	3
Supply and waste disposal rooms	4	2	2	3	3		4	3
Training rooms	2	3	3	2	3	4		4
Plant rooms	4	3	2	4	3	3	4	

1 – Various functional relations

2 – Certain functional relations

3 – a few functional relations

4 – No functional relations

of infections. Particularly relevant is the appendix entitled "Hygiene Requirements for the Functional and Constructional Design of Hospital Facilities for the Care of Outpatients".[24] Even the rules for hospital outpatient departments, suitably adapted for medical practices and other outpatient care facilities, are definitely relevant, for example:
- If major diagnostic and therapeutic facilities are available, waiting room separation is preferable.
- Several patients should not be examined and treated at the same time in the same room.
- a separate waiting room should be provided for patients suspected of having an infection.

The guideline's requirements may well seem extravagant, but considering the growing worldwide hazard presented by new viral infections, they ought to take on increasing significance.

Natural light

In recent decades, the desire for natural light in rooms has grown significantly. Essentially, rooms in which people work for lengthy periods and rooms in which staff spend time must be lit by natural daylight. The German Länder have different regulations specifying the operative times. Solutions in which as far as possible all work and rest rooms have some natural daylight are regarded as forward-looking. With regard to the suggestions in the functional and spatial allocation plans about natural daylight, architects are enjoined to find ways, even with technical aids, that allow people working in medical practices to natural light as possible.

Wayfinding

The increasing overload of stimuli that people are exposed to is certainly responsible for the fact that some people, particularly senior citizens, have problems finding their way around buildings. Even the entrance often presents a psychological barrier. The pathway through a medical practice can seem like the entrance to a maze. Routes indicated by lighting and coloured markings on ceilings and walls, and the choice of materials and flooring can be of assistance here. But locating the built-in and moveable furnishings in clear and obvious positions helps as well.

Technical equipment

The sources of standards to be met by technical equipment are quoted in the chapter entitled "Laws, Regulations and other Provisions". For reasons of space, illustrations of technical equipment can only be referred to briefly here:

STATICS | If renovation work is to be undertaken, the planner must work with the existing structural characteristics for reasons of economy. For new buildings, the planner should design for statics that create the largest possible bearing distances. They create spacious areas not divided up by intrusive walls or supports and enable design freedom and flexible changes.

VENTILATION | The medical practice should appeal to the senses in many ways, with fragrant aromas for example, but care should be taken to ensure these are not

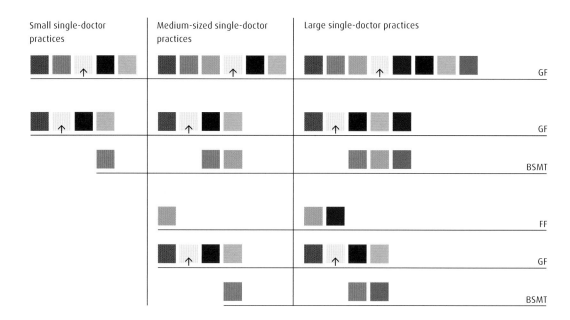

Arrangement of the function groups in one-, two- and three-storey buildings

overpowering because people's sensitivity to smell differs markedly. This is why it is necessary to install a ventilation unit that is effective and easy to control, one which creates a neutral atmosphere comparable with clean outside air. The time may come when further research is going to produce fragrances which can be used in ventilation units to generate a health-enhancing effect.

HEATING, COOLING, AIR HUMIDITY AND FILTERING | The Work Place Regulation[21] stipulates that work rooms should have a minimum temperature of 21 °C and a maximum temperature of 25 °C. This creates a pleasant ambient temperature for patients as well. For comfort, an agreeable degree of coolness and a pleasant level of air humidity are needed on hot days because dry air has a negative effect on the airways. Finally, filtration of incoming air is desirable for people suffering from allergies. Conventional heating systems, cooling appliances, air humidifiers and filters are available as solutions for these needs.

AIR CONDITIONING | The requirements cited in the foregoing section are all met by a modern air conditioning system. The decision to install such a system in new buildings must be made when planning starts; installing such facilities later is more costly. A good air conditioning system has the advantage of delivering an ambient atmosphere that can be adjusted at any time without draughts and disruptive noise. There will be no need for space for radiators. It is important to check in good time if the investment and operating costs can be kept within acceptable limits.

ACOUSTICS | Noise abatement is one of today's paramount concerns in the field of environmental protection. The Construction Codes of the Länder stipulate insulation to prevent noise carrying in buildings. DIN 4109[25] contains requirements to be met by rooms and components and the Work Place Regulation[21] governs noise levels in work and staff rooms. The requirements can be met by separating construction components, choosing the right floorings that reduce structure-borne and airborne noise, by installing ceilings with acoustic properties and noise-absorbing wall cladding and through the design of facilities and equipment and the use of noise-insulating doors. All this is highly desirable, especially with regard to the need to maintain patient privacy.

LIGHTING | Planning lighting equipment is a difficult task for planners because the field is so complicated. This should be undertaken by a qualified specialist because it is so important for the overall appearance of the medical practice. The chapter "Elements of the Design" deals with the requirements for exterior and interior lighting.

PROTECTION AGAINST BURGLARY | There is a heightened risk of burglary in medical practices, not only because of the material value of medical and office equipment but because of the lure presented by drugs on the premises. This risk cannot be eliminated completely but the use of technology can reduce it considerably. Roller shutters and grilles are the principal means used. Also used are high-resistance safety glass and motion detectors that trigger lighting and acoustic warnings. Video monitoring systems have proved effective

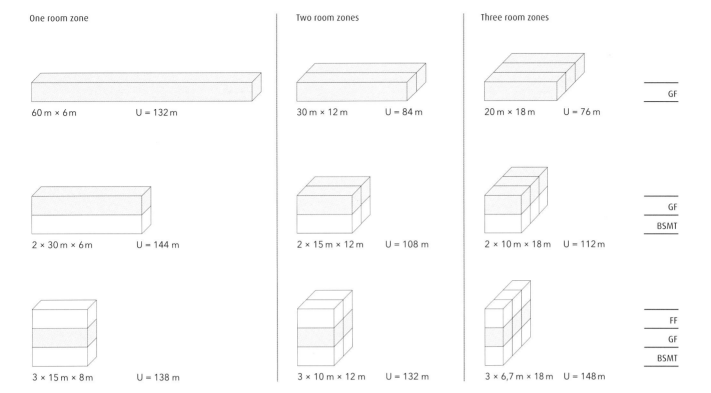

Schemes for one-, two- and three-storey medical practices with one to three room zones. The verticals are exaggerated in height for reasons of clarity. The converted room is the same size in all schemes. U means extent of the external areas in metres.

because they can help identify culprits. A radio link to nearby police stations can lead to early arrests. These also serve as strong deterrents.

FIRE PROTECTION | The Construction Codes of the Länder, the Work Place Regulation[21], DIN 4102[26], the Occupational Health and Safety Office and the Occupational Health and Safety Associations all attach great importance to preventing fires in medical practices, particularly in laboratories. There are requirements for the fire resistance classification of load-bearing and non load-bearing parts of buildings and the fire resistant properties of building materials. The length and number of escape routes are also laid down.

Special features of renovations

The fact that facilities and the substance of the building have become worn is the reasons for renovation work. Another reason is generally the need to make better use of the areas and spaces and to improve operational procedures. Recently, minor or major work can result from changes to technical equipment, for example the installation of a new lighting system, new heating, ventilation, cooling and filtration plant or even whole air conditioning systems. Whereas the erection of a new building is only subject to the need to finish by a fixed date, renovations, including extensions and conversions, are associated with significantly greater difficulties. Temporary closure of the medical practice certainly guarantees the quickest progress for building work. Generally, however, this route is not taken for commercial reasons so that the building work must proceed while the business is running. Dustproof partitioning will be necessary to protect patients and staff alike, but also mainly for hygiene reasons. It should at the same time help reduce building noise. Normally, it is also necessary to move some facilities or to clear some function groups. In such cases, careful advance consideration should be given to ways in which the necessary operations of the practice can continue during the individual phases of building work. Renovations entail inconvenience to everyone affected by them. Hence, it is advisable to apologise to the people concerned in good time.

STANDARD FUNCTIONAL AND SPATIAL ALLOCATION PLANS, FUNCTION SCHEMES AND STANDARD GROUND PLANS FOR SINGLE-DOCTOR PRACTICES

The functional and spatial allocation plans, function schemes and ground plans described below for small, medium-sized and large single-doctor and multi-doctor practices are intended to clarify, by means of a standard example, the basics of operation and construction. In this case, standard means a sample or model of minimum expense allowing for a functional and efficient solution.

The subdivision explained in the section entitled "Subdivision into Function Groups" applies to functional and spatial allocation plans. These plans also contain a suggestion for light exposure [see section entitled "Functional and Spatial Allocation Plans"]. Distinctions are made between individual rooms [functional elements] and rooms that need to be immediately adjacent [function units] which, taken together, always form function groups. The plan code prefixes [PCP – German acronym PKZ]

PCP	FUNCTION GROUP / FUNCTION UNIT / FUNCTIONAL ELEMENT	LIGHT EXPOSURE	USABLE FLOOR IN SQM
1	**Patient rooms**		**30**
1.1	Reception room	○	10
1.2	Waiting room with cloakroom	○	18
1.3	Patient lavatory [L + G]		
1.3.1	Anteroom	●	1
1.3.2	Lavatory	●	1
2	**Examination and treatment rooms**		**34**
2.1	Examination and wtreatment room [Consultation room]	○	22
2.2	Laboratory	○	12
3	**Specialist medical rooms**		**8**
3.1	Examination room – Electrocardiography	●	8
5	**Offices and staff rooms**		**16**
5.1	Staff rest room with pantry	○	8
5.2	Staff changing facilities with shower	◐	6
5.3	Staff lavatory [L + G]		
5.3.1	Anteroom	●	1
5.3.2	Lavatory	●	1
6	**Supply and waste disposal rooms**		**12**
6.1	Work room – Unclean	○	4
6.2	Stores	●	4
6.3	Cleaner's room	●	4
	Usable floor of a small single-doctor practice		**100**

Example of a functional and spatial allocation plan for a small single-doctor practice

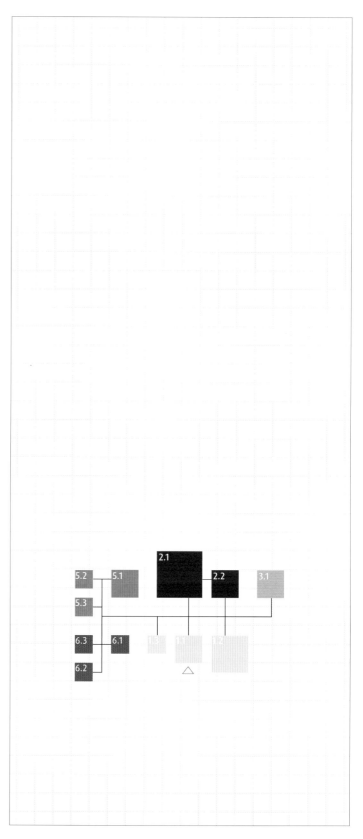

Function scheme of a small single-doctor practice

identify the function groups and then in succession the function units and functional elements. The function schemes show all the rooms and the functional relations. The different sized squares chosen to symbolise the rooms give an indication of the size of the usable floor spaces.

Conversion of the function scheme into a ground plan presupposes a ground plan format into which the individual rooms of the functional and spatial allocation plan must be fitted. This task can be compared to a jigsaw puzzle because the individual components need to be cut to size taking due account of the available area and of the functional and [potential] structural engineering demands. So it is understandable that with respect to a given functional and spatial allocation plan, minor "off-cuts" occur or occasionally a small area remains which can then be used as a box room or a technical room.

The standard ground plans are not intended to serve as templates for designs. They represent, as an example in the deliberately selected one-storey format, only a visualisation of the functional and spatial allocation plans. [The one-storey format has been chosen because it cuts out stairs and lifts and is therefore compact.] The standard ground plans are therefore evidence of feasibility and a guide in the initial planning considerations. Designs remain the province of the architects, interior designers and other medical practice planners, who work together creatively with the doctors and their colleagues to produce individual solutions.

Standards for a small single-doctor practice

In Germany, the number of small medical practices [up to 125 square metres usable floor space] out of the total number of such practices is showing definite signs of shrinking. This is with regard to the further development of outpatient care and also to the growing medical options and the raised expectations of patients. Taking commercial aspects into account, there is an accelerating trend towards larger medical practices, generally in the form of two-doctor or multi-doctor practices, because they benefit from the available synergies, giving a greater chance of survival. However, small single-doctor practices, especially those in rural and thinly populated areas, retain their raison d'être.

FUNCTIONAL AND SPATIAL ALLOCATION PLANS AND FUNCTION SCHEMES | The functional and spatial allocation plan presented as an example, at 100 square metres, lies in the lower range of benchmarks cited in the chapter entitled "Functional and Spatial Allocation Plans" where these benchmarks for small single-doctor practices range upwards to 125 square metres. In cases where the areas quoted here are not available some rooms must be reduced in size or abandoned. It may also be possible to combine some functions, for example the staff rest room and changing cubicle or work room – unclean area and cleaner's room, in order to save a few square metres of usable floor space. However, in this, the functional and spatial allocation plan approaches a lower limit,

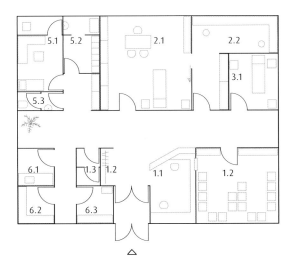

Standard ground plan of a small single-doctor practice, scale 1:200

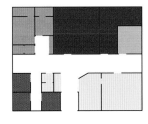

Colour scheme for the ground plan of a small single-doctor practice, scale 1:400

PCP	FUNCTION GROUP / FUNCTION UNIT / FUNCTIONAL ELEMENT	LIGHT EXPOSURE	USABLE FLOOR IN SQM
1	**Patient rooms**		**40**
1.1	Reception room	○	14
1.2	Waiting room with cloakroom	○	22
1.3	Patient lavatory [L]		
1.3.1	Anteroom	●	1
1.3.2	Lavatory	●	1
1.4	Patient lavatory [G]		
1.4.1	Anteroom	●	1
1.4.2	Lavatory	●	1
2	**Examination and treatment rooms**		**42**
2.1	Examination and treatment room [Consultation room]	○	22
2.2	Examination room and changing facilities	○	8
2.3	Laboratory	○	12
3	**Specialist medical rooms**		**8**
3.1	Examination room – Ultrasound	●	8
4	**Administration rooms**		**10**
4.1	Office	○	10
5	**Offices and staff rooms**		**32**
5.1	Staff rest room with pantry	○	12
5.2	Staff changing facilities – L		
5.2.1	Changing facilities	◐	8
5.2.2	Shower	●	2
5.3	Staff changing facilities – G		
5.3.1	Changing facilities	◐	4
5.3.2	Shower	●	2
5.4	Staff lavatory [L]		
5.4.1	Anteroom	●	1
5.4.2	Lavatory	●	1
5.5	Staff lavatory [G]		
5.5.1	Anteroom	●	1
5.5.2	Lavatory	●	1
6	**Supply and waste disposal rooms**		**18**
6.1	Work room – Unclean	○	6
6.2	Stores	●	8
6.3	Cleaner's room	●	4
Usable floor of a medium-sized single-doctor practice			**150**

Example of a functional and spatial allocation plan for a medium single-doctor practice

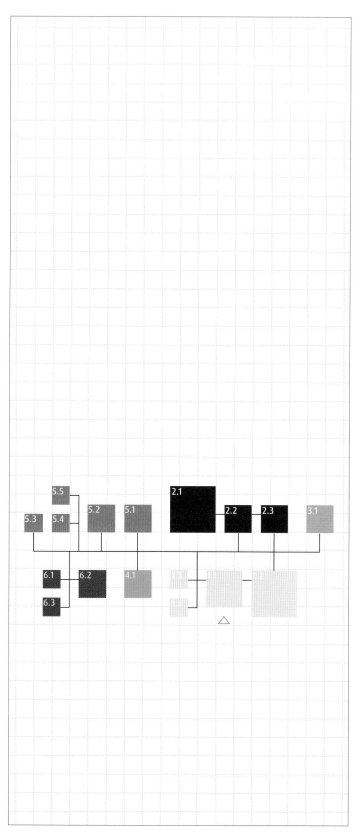

Function scheme of a medium single-doctor practice

below which it is no longer certain that a practice can be run on commercially sound lines. The function scheme emphasises the desirable direct relationship between the reception area and the waiting area, the adjacent site of the examination and treatment rooms and the specialist medical rooms and the combination of the service and staff rooms.

GROUND PLAN | The choice of a rectangle with two room zones for the schematic ground plan of the medical practice results from the view that the premises should feature a large measure of compactness and efficiency. The rooms are ranged along two sides of a straight central corridor so that they all receive natural daylight. The function groups of the patient rooms and the supply and disposal rooms are on the side of the entrance. On the opposite side are the examination and treatment rooms, the specialist medical rooms and the service and staff rooms. In a good position almost in the middle of the corridor there is a wind trap through which patients pass into the reception area, the adjacent waiting area and, immediately opposite, the examination and treatment room. An advantage of this ground plan solution is that the rooms are easy to find and reach.

Standards for a medium-sized single-doctor practice
The design principles which apply for small single-doctor practices also apply for the medium-sized version, with some modifications.

FUNCTIONAL AND SPATIAL ALLOCATION PLAN AND FUNCTION SCHEME | In addition to the larger size and different room types, the functional and spatial allocation plan for the small single-doctor practice has here been extended with an additional small examination room and an office. With 150 square metres of usable floor space, it occupies the middle zone of the benchmarks [125 square metres to 175 square metres] and has the size of medical practices frequently encountered today. Compared with the small single-doctor practice, all the function groups in the function scheme have been extended and furthermore the function group "Administration Rooms" is represented by one room located in the vicinity of the reception areas.

GROUND PLAN | The only change in the room plan, compared with a single-doctor practice, is the addition of an office opposite the reception area and the enlargement of the room group "Examination Rooms".

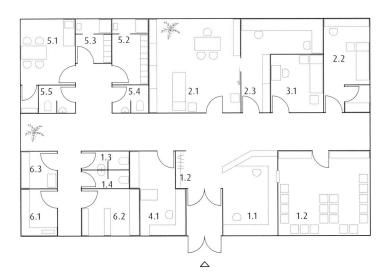

Standard ground plan of a medium single-doctor practice, scale 1:200

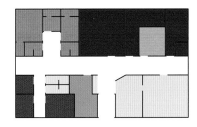

Colour scheme for the ground plan of a medium single-doctor practice, scale 1:400

PCP	FUNCTION GROUP / FUNCTION UNIT / FUNCTIONAL ELEMENT	LIGHT EXPOSURE	USABLE FLOOR IN SQM
1	**Patient rooms**		**50**
1.1	Reception room	○	16
1.2	Waiting room with cloakroom	○	26
1.3	Play room	○	4
1.4	Patient lavatory [L]		
1.4.1	Anteroom	●	1
1.4.2	Lavatory	●	1
1.5	Patient lavatory [G]		
1.5.1	Anteroom	●	1
1.5.2	Lavatory	●	1
2	**Examination and treatment rooms**		**50**
2.1	Examination and treatment room [Consultation room]	○	22
2.2	Examination room	○	8
2.3	Examination room	○	8
2.4	Laboratory	○	12
3	**Specialist medical rooms**		**8**
3.1	Examination room – Electrocardiography and Ultrasound	●	8
4	**Administration rooms**		**12**
4.1	Office	○	12
5	**Offices and staff rooms**		**40**
5.1	Staff rest room with pantry	○	16
5.2	Staff changing facilities – L		
5.2.1	Changing facilities	◉	10
5.2.2	Shower	●	2
5.3	Staff changing facilities – G		
5.3.1	Changing facilities	◉	6
5.3.2	Shower	●	2
5.4	Staff lavatory [L]		
5.4.1	Anteroom	●	1
5.4.2	Lavatory	●	1
5.5	Staff lavatory [G]		
5.5.1	Anteroom	●	1
5.5.2	Lavatory	●	1
6	**Supply and waste disposal rooms**		**20**
6.1	Work room – Unclean	○	6
6.2	Stores	●	10
6.3	Cleaner's room	●	4
7	**Training rooms**		**18**
7.1	Multipurpose room [Extension]	○	18
8	**Plant rooms**		**2**
8.1	Technical room	●	2

Usable floor a large single-doctor practice
200

Example of a functional and spatial allocation plan for a large single-doctor practice

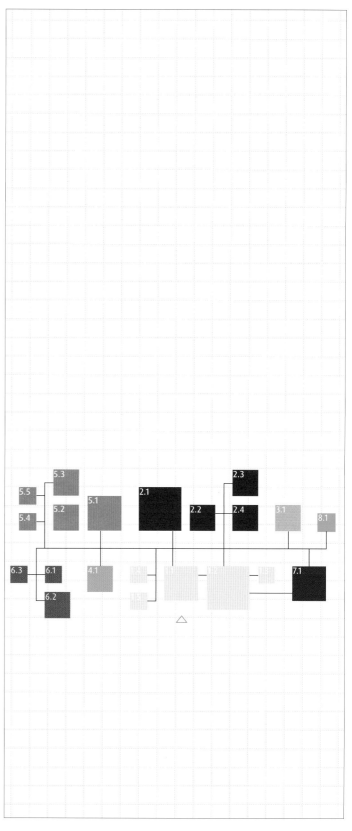

Function scheme of a large single-doctor practice

Medical Practices

Standards for a large single-doctor practice
The previously selected layout can be retained for large single-doctor practices.

FUNCTIONAL AND SPATIAL ALLOCATION PLAN AND FUNCTION SCHEME | Compared with the foregoing detailed description of single-doctor practices, some rooms have been enlarged further. Additional features are an examination and treatment room, a children's play corner and a room from the function group "Training and Teaching Rooms", which can be used as a multipurpose or spare room.

The size [200 square metres usable floor space] is somewhat above the benchmark for large single-doctor practices, which start at 175 square metres. There are numerous single-doctor practices in this size range. Their usable floor space rises to about 250 square metres.

The function group "Patient Rooms" has received an additional room with the function of children's play corner directly adjacent to the waiting area. Also, the function group "Examination and Treatment Rooms" has been enlarged by an additional examination room. A multipurpose room can be added to it, connected by a sliding door if possible. This functional and spatial allocation plan also has a technical room added to it.

GROUND PLAN | The central corridor solution with two room zones gives a large single-doctor practice sufficient development potential, even if more rooms have been added. Whatever the case, small dead end corridors must be added. It is essential to be aware at planning stage that the inward facing rooms that this will create will have to do without natural daylight.

Standard ground plan of a large single-doctor practice, scale 1:200

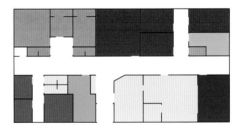

Colour scheme for the ground plan of a large single-doctor practice, scale 1:400

PCP	FUNCTION GROUP / FUNCTION UNIT / FUNCTIONAL ELEMENT	LIGHT EXPOSURE	USABLE FLOOR IN SQM
1	**Patient rooms**		**46**
1.1	Reception room	○	16
1.2	Waiting room with cloakroom	○	26
1.3	Patient lavatory [L]		
1.3.1	Anteroom	●	1
1.3.2	Lavatory	●	1
1.4	Patient lavatory [G]		
1.4.1	Anteroom	●	1
1.4.2	Lavatory	●	1
2	**Examination and treatment rooms**		**12**
2.1	Laboratory	○	12
3	**Specialist medical rooms**		**74**
3.1	Examination and treatment room [Consultation room] – Mouth, jaw and facial surgery 1	○	22
3.2	Examination and treatment room [Consultation room] – Mouth, jaw and facial surgery 2	○	22
3.3	Examination and treatment room [Consultation room] – Mouth, jaw and facial surgery 3	○	22
3.4	Examination room – X-ray	●	8
4	**Administration rooms**		**12**
4.1	Office	○	12
5	**Offices and staff rooms**		**40**
5.1	Staff rest room with pantry	○	16
5.2	Staff changing facilities – L		
5.2.1	Changing facilities	◐	10
5.2.2	Shower	●	2
5.3	Staff changing facilities – G		
5.3.1	Changing facilities	◐	6
5.3.2	Shower	●	2
5.4	Staff lavatory [L]		
5.4.1	Anteroom	●	1
5.4.2	Lavatory	●	1
5.5	Staff lavatory [G]		
5.5.1	Anteroom	●	1
5.5.2	Lavatory	●	1
6	**Supply and waste disposal rooms**		**16**
6.1	Work room – Unclean	○	6
6.2	Stores	●	6
6.3	Cleaner's room	●	4
	Usable floor of a small multi-doctor practice		**200**

Example of a functional and spatial allocation plan for a small multi-doctor practice

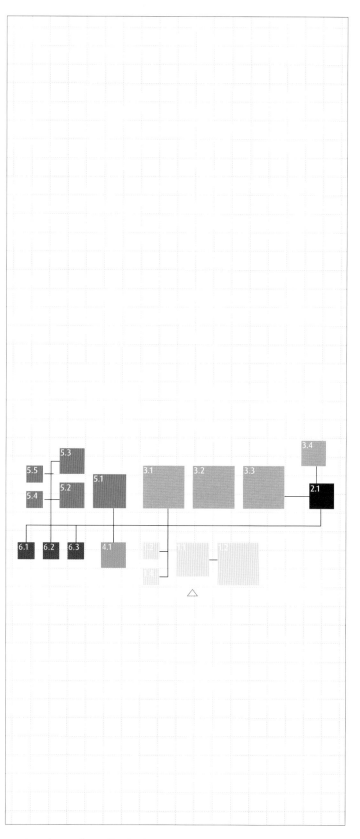

Function scheme of a small multi-doctor practice

Medical Practices

STANDARD FUNCTIONAL AND SPATIAL ALLOCATION PLANS, FUNCTION SCHEMES AND STANDARD GROUND PLANS FOR MULTI-DOCTOR PRACTICES

Standards for a small multi-doctor practice

As there is an increasing trend towards two-doctor and multi-doctor practices providing outpatient care, this type of practice will be described in greater detail with the aid of a few examples. The small multi-doctor practice is the most economic user of space.

FUNCTIONAL AND SPATIAL ALLOCATION PLAN AND FUNCTION SCHEME | The smallest workable size for a multi-doctor practice is around 150 square metres of usable floor space. For the standard example, 200 square metres has been chosen. This means that small multi-doctor practices have roughly the same space needs as large single-doctor practices.

The essential difference from single-doctor practices is in the form adopted here of a joint practice, that is in the interactions between at least three specialist fields whose managers each have their own examination and treatment rooms, using them also as consultation rooms; all the other rooms of the practice are being run jointly. As regards the basic structure, the main change is to the function group "Examination and Treatment Rooms". The specialist medical rooms replace several additional individual rooms. The examination and treatment rooms are not only used as consultation rooms, but also take on a specialist character, that is they contain the medical and technical facilities necessary to provide specialist medical services. The room allocation deducible from the function scheme corresponds largely to that of a large single-doctor practice. Only the general examination and treatment rooms are replaced by specialist medical rooms. In the small multi-doctor practice there is no space for a children's play corner, a multipurpose room and technical room.

GROUND PLAN | The layout of a small multi-doctor practice is similar to that of the large single-doctor practice. On the side opposite the entrance area, the available space allows the creation of a laboratory and X-ray room. These rooms are transferred to the entrance side and are accessible via the dead end corridor. All the rooms of the efficient rectangular ground plan receive natural daylight.

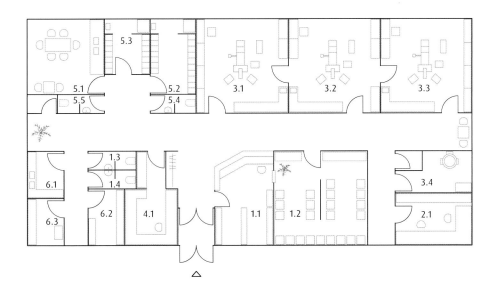

Standard ground plan of a small multi-doctor practice, scale 1:200

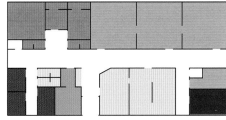

Colour scheme for the ground plan of a small multi-doctor practice, scale 1:400

PCP	FUNCTION GROUP / FUNCTION UNIT / FUNCTIONAL ELEMENT	LIGHT EXPOSURE	USABLE FLOOR IN SQM
1	**Patient rooms**		**54**
1.1	Reception room	○	20
1.2	Waiting room with cloakroom	○	26
1.3	Play room	○	4
1.4	Patient lavatory [L]	●	2
1.5	Patient lavatory [G]	●	2
2	**Examination and treatment rooms**		**12**
2.1	Laboratory	○	12
3	**Specialist medical rooms**		**118**
3.1	Examination and treatment room [Consultation room] – Gynaecology and obstetrics	○	22
3.2	Examination and treatment room [Consultation room] – Urology	○	22
3.3	Examination and treatment room [Consultation room] – Ophthalmology	○	22
3.4	Examination and treatment room [Consultation room] – Otolaryngology	○	22
3.5	Examination and treatment room [Consultation room] – Skin and venereal diseases	○	22
3.6	Examination room – X-ray	●	8
4	**Administration rooms**		**14**
4.1	Office	○	14
5	**Offices and staff rooms**		**42**
5.1	Staff rest room with pantry	○	18
5.2	Staff changing facilities – L		
5.2.1	Changing facilities	◐	10
5.2.2	Shower	●	2
5.3	Staff changing facilities – G		
5.3.1	Changing facilities	◐	6
5.3.2	Shower	●	2
5.4	Staff lavatory [L]	●	2
5.5	Staff lavatory [G]	●	2
6	**Supply and waste disposal rooms**		**38**
6.1	Sterilisation		
6.1.1	Sterilisation room	○	8
6.1.2	Stores – Sterile goods	●	6
6.2	Work room – Clean	○	6
6.3	Work room – Unclean	○	6
6.4	Stores	●	8
6.5	Cleaner's room	●	4
7	**Training rooms**		**18**
7.1	Multipurpose room [Extension]	○	18
8	**Plant rooms**		**4**
8.1	Terminal compartment	●	4
	Usable floor of a medium-sized multi-doctor practice		**300**

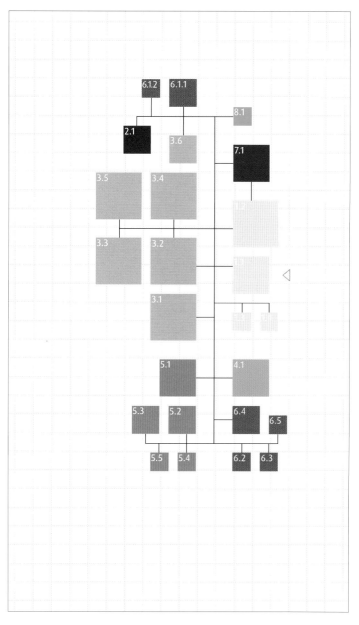

Function scheme of a medium-sized multi-doctor practice

Example of a functional and spatial allocation plan for a medium-sized multi-doctor practice

Standards for a medium-sized multi-doctor practice

Medium-sized multi-doctor practices are quite common. Obviously, their size is suitable for the duties they have to perform and synergy effects work particularly well.

FUNCTIONAL AND SPATIAL ALLOCATION PLAN AND FUNCTION SCHEME | The benchmarks for the usable floor spaces amount to between 250 and 400 square metres. The example chosen here is that of a medium-sized multi-doctor practice with 300 square metres of usable floor space which has rooms for five identical or different specialist fields. It should be noted that with different specialist fields, depending on the field, there will be an additional specialist examination and treatment room of one type or another. With this in mind, the 300 square metres of usable floor space must be regarded as the lower limit for a medium-sized multi-doctor practice with five specialist fields. The function scheme for the medium-sized multi-doctor practice includes additional rooms for the function groups Specialist Medical Rooms and Supply and Disposal Rooms. In addition there will also be a multipurpose room and a technical room.

GROUND PLAN | Retention of the previous ground plan structure shows that the limits of this form have been reached. A central corridor should not extend any further. A longer corridor is unacceptable, not only because of the long distances but also for design reasons. The alternative ground plan developed specifically for the medium-sized multi-doctor practice contains the same rooms, but needs 10 per cent more floor area after deducting the inner courtyard. The reason for the introduction of the inner courtyard in this schematic ground plan is the need to light all the practice rooms with natural daylight. Even so, the inner courtyard increases the overall quality of the design. Unlike the initial scheme, in which natural daylight penetrates the building both via the ends of corridors designed to allow the passage of light and via the central reception and waiting area, the alternative scheme seems to be flooded with light via the ends of four corridors designed to allow the passage of light, the similarly light-filled reception and waiting area and the inner courtyard.

A comparison of the distances to be walked shows, for example, that the distances between the furthest apart specialist medical rooms, measured from the middle of the door walls, are more or less the same [16.0 metres and 16.5 metres]. The central location of the staff rooms means that these distances are acceptable. Even if some distances should be a little longer, they would not pose a problem, as they would be balanced by the enhanced experience of the alternative.

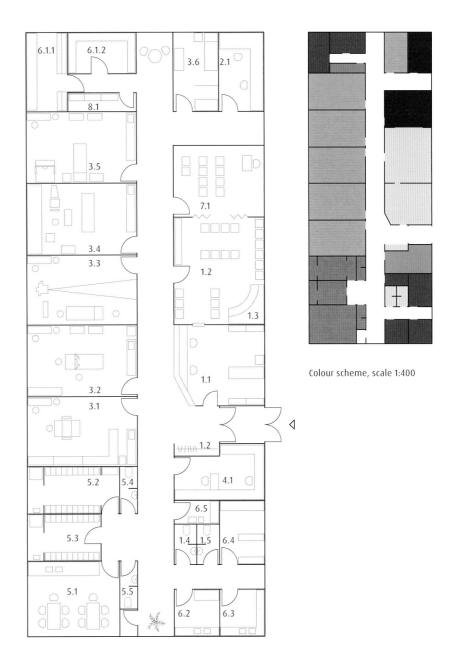

Colour scheme, scale 1:400

Standard ground plan of a medium-sized multi-doctor practice, scale 1:200

PCP	FUNCTION GROUP / FUNCTION UNIT / FUNCTIONAL ELEMENT	LIGHT EXPOSURE	USABLE FLOOR IN SQM
1	**Patient rooms**		**102**
1.1	Reception room	○	20
1.2	Waiting rooms		
1.2.1	Waiting room 1	○	28
1.2.2	Waiting room 2	○	14
1.2.3	Play room	○	6
1.2.4	Cloakroom	●	4
1.3	Meeting room	○	10
1.4	Hygiene room	◐	6
1.5	Patient lavatory [L]		
1.5.1	Anteroom	●	2
1.5.2	Lavatory	●	2
1.6	Patient lavatory [G]		
1.6.1	Anteroom	●	2
1.6.2	Lavatory	●	2
1.7	Lavatory for handicapped persons	●	6
2	**Examination and treatment rooms**		**14**
2.1	Laboratory	○	14
3	**Specialist medical rooms**		**194**
3.1	Examination and treatment room [Consultation room] – Mouth, jaw and facial surgery 1	○	22
3.2	Examination and treatment room [Consultation room] – Mouth, jaw and facial surgery 2	○	22
3.3	Examination and treatment room [Consultation room] – Neurology	○	22
3.4	Examination and treatment room [Consultation room] – Ear, nose and throat medicine	○	22
3.5	Examination and treatment room [Consultation room] – Urology	○	22
3.6	Examination and treatment room [Consultation room] – Gynaecology and obstetrics	○	22
3.7	Examination and treatment room [Consultation room] – Paediatrics and youth medicine	○	22
3.8	Examination room – X-ray	●	12
3.9	Examination room – Electrocardiography	○	14
3.10	Examination room – Ultrasound	○	14

PCP	FUNCTION GROUP / FUNCTION UNIT / FUNCTIONAL ELEMENT	LIGHT EXPOSURE	USABLE FLOOR IN SQM
4	**Administration rooms**		**32**
4.1	Office		
4.1.1	Office – Prescription processing	○	16
4.1.2	Office – Bookkeeping	○	16
5	**Offices and staff rooms**		**64**
5.1.1	Staff rest room	○	26
5.1.2	Pantry	○	8
5.2	Staff changing facilities – L		
5.2.1	Changing facilities	●	12
5.2.2	Shower	●	2
5.3	Staff changing facilities – G		
5.3.1	Changing facilities	●	6
5.3.2	Shower	●	2
5.4	Staff lavatory [L]		
5.4.1	Anteroom	●	2
5.4.2	Lavatory	●	2
5.5	Staff lavatory [G]		
5.5.1	Anteroom	●	2
5.5.2	Lavatory	●	2
6	**Supply and waste disposal rooms**		**54**
6.1	Sterilisation		
6.1.1	Sterilisation room	○	14
6.1.2	Stores – Sterile goods	◐	8
6.2	Work room – Clean	○	6
6.3	Work room – Unclean	○	6
6.4	Stores	●	16
6.5	Cleaner's room	●	4
7	**Training rooms**		**32**
7.1	Multipurpose room [Extension]	○	32
8	**Plant rooms**		**6**
8.1	Technical room	●	6
	Usable floor of a large multi-doctor practice		**500**

Example of a functional and spatial allocation plan for a large multi-doctor practice

Standards for a large multi-doctor practice

Recent developments in healthcare policy tend to concentrate outpatient care increasingly in large high-performance and economically efficient centres. The large multi-doctor practice is one of the first facilities of this type. Additionally, in Germany, hospital outpatient departments, medical centres, diagnostic centres, medical care centres and health centres have also sprung up. Essentially, these are also large multi-doctor practices but follow a different care approach.

FUNCTIONAL AND SPATIAL ALLOCATION PLAN AND FUNCTION SCHEME | The group of Multi-Doctor Practices begins at over 400 square metres of usable floor space. There is practically no upper space limit because these projects also include large medical buildings which can house up to 40 practices and more. Assuming an average minimum size of 60 square metres for a specialist medical practice, then the medical centre referred to above would have at least 2,400 square metres of usable floor space. The example chosen here has seven specialist medical practices with a usable floor space of 500 square metres. In comparison with the medium-sized multi-doctor practice, four specialist examination and treatment rooms have been added. The available space for the remaining rooms has been expanded and given different uses. With regard to the function groups Patients, Specialist Medical Rooms, Administration Rooms, and Service and Staff Rooms and compared with the previous model, the function scheme for large multi-doctor practices shows that some individual rooms have been enlarged and given different uses, whereas the remaining function groups have merely been given more usable floor space. The linear arrangement of the related function groups displays a high degree of clarity and order. As for the supply and disposal rooms, the sterilisation rooms are, for functional reasons, situated in the vicinity of the specialist medical rooms.

GROUND PLAN | The structure of the ground plan is the same as that of the medium-sized multi-doctor practice variant. Because of the enlargement and different uses of some rooms, the office and the multipurpose room must be accommodated by the inner courtyard. The two connections for the examination and treatment rooms have been lengthened by the addition of some specialist medical rooms. The ground plan incorporates the area of an almost equilateral rectangle, which gives the building geometry an ideal shape. Once again, however, this shows that the limits for this form have been reached. The second ground plan level should be considered for even bigger premises. The administration, service and staff rooms should be shifted first of all to this level. All of the rooms of the 500 square metres multi-doctor practice receive natural daylight with the exception of some ancillary rooms.

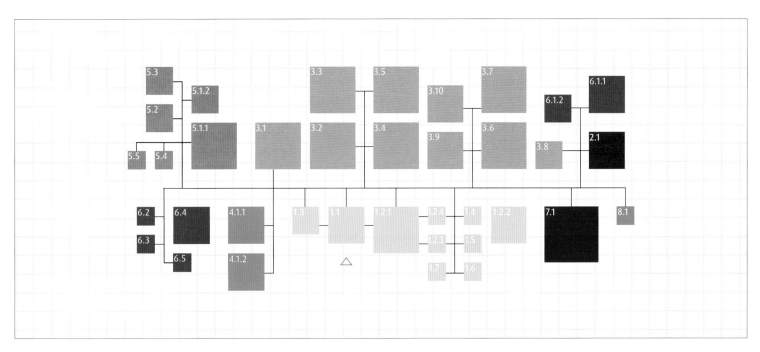

Function scheme of a large multi-doctor practice

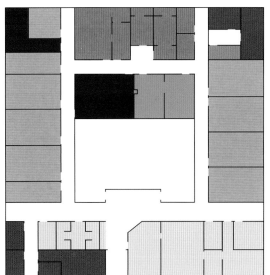

Colour scheme for the ground plan of a large multi-doctor practice, scale 1:400

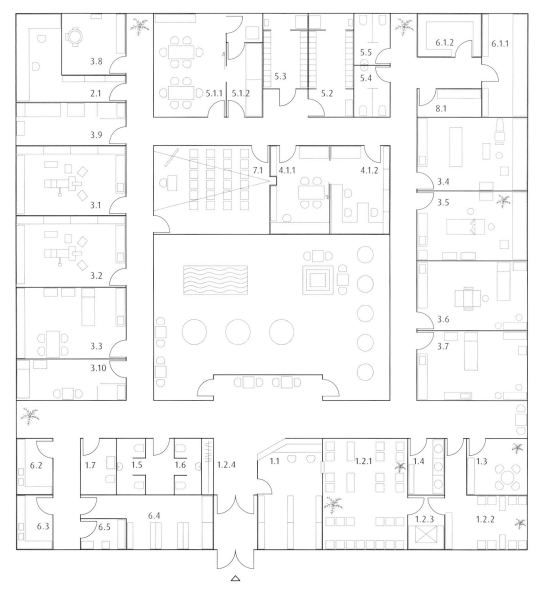

Standard ground plan of a large multi-doctor practice, scale 1:200

NOTES ON THE FUNCTIONAL AND SPATIAL ALLOCATION PLANS FOR SELECTED FACILITIES

Specialist medical practices
The highly diffuse nature of specialist medical fields and the effect this has on the design and layout of medical practices demands wide-ranging specialist knowledge. However, the architect can only acquire this know-how if he works closely with the doctors involved. The present author had the opportunity of working with doctors in this way, resulting in a planning manual for hospitals and care homes.[27] The listed suggestions include some that are applicable to hospital outpatient departments. Suitably adapted, they apply also to specialist medical practices.

Some years ago, Damaschke, Scheffer and Schossig published a book which examined in detail the different functional and spatial allocation plans of numerous medical practices. It also presented examples and illustrated functional, spatial and design aspects.[28] Now considered as a classic text, Teut and Nedeljkov's book "The Group Practice" [1973] also takes in experience gained in the German Democratic Republic. It provides information on technical planning principles, minimum space needs and functional inter-relations.[29]

OPHTHALMOLOGICAL SPECIALIST MEDICAL PRACTICE | Among the range of services provided here, the key areas are electrophysiological diagnostics, visual field diagnostics, laser therapy, eye training and ultrasound diagnostics. If operating facilities are available, resections and treatments for such disorders as cataracts, and drooping eyelids can be carried out.
Depending on the range of services provided, the main rooms are: examination room, eye training room, visual field checking room, laser coagulation room, sonography room, a room for fluorescence angiography and a room for electroretinography [dark room].

SURGICAL SPECIALIST MEDICAL PRACTICE | The following are some of the key areas with regard to examination and treatment: general surgical diagnostics and therapy, obesity treatment and follow-up care, hand examination and treatment, emergency examination and treatment, rectal examinations and treatments, soft-laser treatment, tumour follow up care, thyroid examination and treatment [nuclear medicine diagnostics and therapy related disciplines]. A separate room for rectal examinations should be provided in addition to general examination and treatment rooms. It is useful to site this room near to diagnostic facilities, for example radiology and endoscopy units and an operating unit.

ORTHOPAEDIC SPECIALIST MEDICAL PRACTICE | Listed here are some of the services delivered: examination and treatment of ailments of the feet, hips, knee, shoulders and spinal column, orthopaedic diseases of newborn babies and the delivery of oncological orthopaedics. The main rooms needed are examination and treatment rooms and rooms for ultrasound, plaster of Paris casts and infusions, and a recovery room is also needed as an extra. The waiting room must have comfortable armchairs with arm rests and armchairs with raised seats [for patients with sitting supports]. It is helpful if the medical practice is sited near to X-ray diagnostics and physical diagnostics facilities.

UROLOGICAL SPECIALIST MEDICAL PRACTICE | Among the range of services provided here are diagnostics and therapy for general urological ailments, fertility problems and endocrinology of the aging male, incontinence, also neurological diagnostics, stoma aftercare, pre-operative diagnostics and post-operative care for those suffering from tumours. Among the main rooms needed are examination and treatment rooms for cytoscopic examinations, a room for urodynamics and, if applicable, a lithotripsy room.

ORAL AND MAXILLOFACIAL SURGERY SPECIALIST MEDICAL PRACTICE | The following are some of the key areas with regard to examination and treatment: acute and pain therapy with infiltration, trepanation, incision, drainage including tooth extraction, acute treatments for wounds, surgery with local anaesthesia, specific diagnostic procedures [for example ultrasound, model analysis, face bow, laser scanning], plate treatment for patients with cleft lips, jaws and palates, radiological diagnostics [panoramic X-ray and X-ray micro images] and dental laboratory work.

The main rooms needed are examination and treatment rooms with facilities for ultrasound and laser, X-ray room, a planning room [for X-ray assessment, model measurement, and operations planning], an operation unit with a laboratory for plaster of Paris and plastics processing. One of the ancillary rooms should be an appliance | instrument preparation room. It is helpful to be close to the X-ray diagnostic and laboratory medicine facilities. Short distances to the related specialist fields of ear, nose and throat medicine, plastic surgery and possible ophthalmology are also useful.

Other facilities

HOSPITAL OUTPATIENT DEPARTMENTS | Subject to the categories of DIN 13080, in Germany every hospital has the functional category 1.02 Medical Service, which covers all outpatient departments as part categories.[18] Since every hospital has a different medical service profile, the number and type of outpatient departments must be ascertained in each case. The content of the individual outpatient departments is determined by their main rooms. In addition, there are the first-line communication, ancillary and staff rooms, which are arranged centrally, part centrally or non-centrally, always depending on the need and organisation format.
– The first-line communication rooms include the registration point, medical records office, waiting rooms for ambulant or non-ambulant patients, toilets and the disposal room.
– The ancillary rooms include work rooms – clean and unclean, stores – medical goods and laundry, appliance room and cleaner's room
– The staff rooms include the medical director's study [head of staff – physician], administration department, senior physician's study, wash rooms for doctors and staff, meeting room plus library, standby emergency room, changing cubicles and lavatories for staff and if necessary a study room for managerial nursing staff in the specialist field and study room for technicians.

As hospital outpatient departments examine and treat ambulant inpatients, inpatients occupying beds and ambulant patients arriving from admissions, a good deal of thought needs to be devoted to the siting of facilities in the hospital. Compared with normal medical practices, there is a need here for wider corridors and doors, larger waiting rooms and changing cubicles for inpatients transported on stretchers or mobile beds. In particular, however, the main rooms for examinations and treatments must be of adequate dimensions to allow for appropriately sized usable floor spaces.

MEDICAL CARE CENTRE | In 2008, there were already over 1,100 medical care centres in Germany[7]. There is an increasing trend towards building more medical care centres because more and more doctors are looking to work in this sort of centre. No standard can be devised for this functional and spatial allocation plan because the individual facilities have quite different characteristics depending on their location, range of services, size and organisation. Each medical care centre contains numerous specialist medical practices and a range of more or less jointly used functional units for diagnosis and therapy – for example, a laboratory, X-ray diagnostics and endoscopy units, physiotherapy and a unit for outpatient operations. Medical care centres exclusively treat outpatients.

DESIGN FEATURES

External architecture

The appearance of many buildings indicates what purposes they serve. Since medical practices can be important in life, some sort of means of recognition would certainly be helpful. Designers should consider it part of their remit to devise suitable architectural features that would serve as a means of easy identification.

BODY OF THE BUILDING | Medical practices frequently occupy parts of multi-storey buildings which may be residential or commercial blocks or even medical centres. In such cases, the medical practice generally has to fit in with the existing external architecture. Often, however, medical practices are also designed as separate buildings. In such cases an independent architectural design is possible. The forms of such structures, generally to a maximum of three storeys high, can attract patients and indicate that their visit will be an interesting one. In general there are no limits to the imagination and formal language of the architects. It is a good tip, however, to ensure that, despite the desire to place a somewhat individual stamp on the building, the architecture of the medical practice is in keeping with the landscape and the surrounding buildings.

FRONT ELEVATION | Depending on whether the medical practice forms part of a larger building or has the building to itself, the front elevation can display a variety of different features. In the former case, it is often

Medical Practices

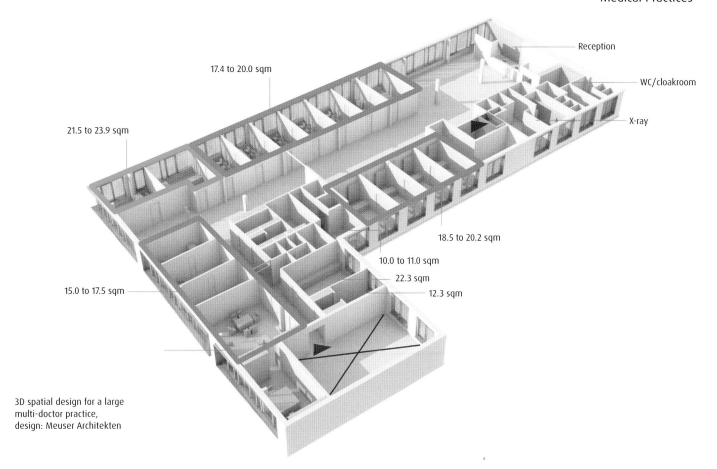

3D spatial design for a large multi-doctor practice, design: Meuser Architekten

only possible to attach some lighting features showing the name of the medical practice. In the latter case, the scope for design is considerably greater because of the open and closed walls, and through using expressive materials, a variety of colours and light and shade. Here, too, one should guard against producing an excess of sensory impressions which after all are incompatible with the serious character of a medical practice.

ENTRANCE | Patients approaching a medical practice are often not in the best of health and so the main entrance to the practice should be clearly recognisable as such from some distance – as it should also be for the handicapped and those unfamiliar with the locality. It is often useful to include one or two secondary entrances as well to shorten the distances patients need to cover. Other helpful features would be an invitingly designed entrance and lighting to identify the building during the hours of darkness. Automatic doors and a lack of barriers make access easier for patients with wheelchairs and prams. Easy-to-read opening hours information should be displayed in the entrance area. Subject to the Industrial Code, the name of the owner can be displayed as well.

EXTERNAL LIGHTING | External lighting should help patients find their way. It should emphasise the entrance in particular. The intensity of the lighting should be in keeping with the ambient lighting to avoid any possibility of dazzle.

Interior design

Any planner will be aware that spread out in front of him is the whole wide field of architectural design for the interiors of medical practices. Taking his experience as the starting point, he must approach his work with creativity, a solid awareness of quality and knowledge of the effect of hard and soft forms, bright and muted colours, light and shade. He must also use his knowledge of materials paying attention to every single detail.

MATERIALS AND SURFACES | The facilities can be designed with natural surfaces displaying their own natural colouring combined with artificial surfaces in a wide variety of colours. Care should be taken with the whole concept to ensure that hardwearing and easy-to-clean materials are selected. Because of the demand for hygiene, appliances should be made of durable and hard-wearing materials. Walls can be varied as follows: generally, plastering is sufficient – they can be painted white or in colours or they can be papered. Washable wall coverings should be used for operational and sanitary rooms. Ceilings are usually lowered to accommodate the technical fittings for lighting, ventilation and air conditioning. As such fittings need to be accessible for servicing, ceilings are generally made of panels. These are also often used because they have excellent acoustic properties. So far there have been only a few examples of ceilings used as a showground for highly imaginative designs. Any decision on flooring needs to be made after consideration of a range of aspects. These include the intended effect of the surface within the design concept as a whole, especially the dimensions of the existing substrate [in the case of conversions], the robustness, sound insulation properties, anti-slip properties, ease of cleaning – and crucially its cost-efficiency. Flooring to be laid in the entrance area should include a large enough foot mat, built in and flush with the floor surface.

FURNITURE AND FITTINGS | The ambitious aim of a medical practice planner should be to create an individual character for the practice, especially easy to achieve with well chosen features. Reliance solely on mass produced furniture is less apt to produce this effect, but individually produced designer pieces are often out of reach of the budget. This is an area where the planner must work with sensitivity to select a coherent blend. All the fitting and furnishing components should fuse together into an over arching design concept that expresses the image desired by the medical practice. Small but important details are places to deposit bags and fittings to hold the walking aids of patients so that they can discuss their medical histories unencumbered.

INTERIOR LIGHTING | Suitably adapted, the principles discussed for the exterior also apply to the interior. Light accompanies patients as soon as they enter and makes it easier for them to find their way. Accent lighting highlights pathways and room notices. It is important to ensure that the lighting brings out the true colours and is warm enough to create a welcoming atmosphere. There will be places where the angle of the lighting will be important – to avoid creating disturbing shadows. Choices for the work rooms should strive for sufficient lighting intensities, lack of dazzle, avoidance of reflections in screens, low energy needs and low rated heat input.[21]

COLOURS | Colours are crucial for the appearance of all surfaces and impact psychologically on patients and staff alike. According to Frieling, colours can have the following effects: calming, aggressive, oppressive, encouraging, alienating, soft, alarming, stimulating, warming, eye-directing, constricting, reassuring, pleasant, irritating, distracting, cold, awareness arousing, heightening, cooling, deepening, illuminating, communicative, activating, lightweight, covering, cherishing and recuperative.[31] The right amount of colour can therefore aid way finding in the space, alter proportions and raise moods. Too many bright colours are inadvisable because they generally only generate a short-term effect. The surfaces of facilities, appliances, walls, ceilings and floors, always visible and also generally accessible to the senses through touch, are the main objects that can have design roles and create challenges.

PLANTS | The competent hygienists hold different views on the use of plants in medical establishments. There are rooms in particular need of protection from infection and harmful substances where plants have no place [see the chapter entitled "Environmental Conditions"] and there are rooms where there are no objections for reasons of hygiene against plants. Most of the rooms in medical practices fall into this group. Magnificently coloured flowers and leafy plants exert a refreshing and revitalising effect on patients and staff alike. Particularly, patients coming to the practice for the first time will register the friendly atmosphere the plants help create.

ART | The way a medical practice or some other medical care facility is decorated with artworks can in fact make a clear and memorable statement about the philosophy and spirit of the establishment. There are countless possibilities – so there is no risk of being boring and repetitive. And no artwork once selected has to be on show for ever: variatio delectat. Depending on the means of the operator, modestly priced photos and prints [ideally in clip-in picture frames] can be used to decorate the rooms of the practice – as well as valuable paintings and sculptures. One idea would be to work with art students or famous artists who are generally happy to use the walls of such rooms as display areas for their work.
Careful attention should be paid to artwork selection because it is intended to appeal to a sensitive group of people. So particularly colourful and friendly subjects are definitely to be preferred to sombre scenarios. Schiller's words from the prologue to Wallenstein are worth repeating here: "Life is serious, art serene."

PLANNING AND BUILDING STAGES

Preparatory work
After the initial bright idea to expand, convert or renovate the existing medical practice or to build a new one, the client must assemble a number of basic facts with the main parties involved. No more than brief details are needed for the following tips because these tasks must, of necessity, be taken on by specialist planners with as much experience and expertise as possible.

LOCATION ANALYSIS | As already mentioned in the chapter entitled Location, the choice of location must never be left to chance, whether a new building is being planned, or an extension of an existing practice or even a move to new premises or the taking over of another practice. Instead, before any planning decision, the determining factors of the location must be analysed with great care. Other important factors are the distances to existing competing medical practices and medical centres and other medical facilities, the transportation situation and links to public transport and parking facilities, population

Medical Practices

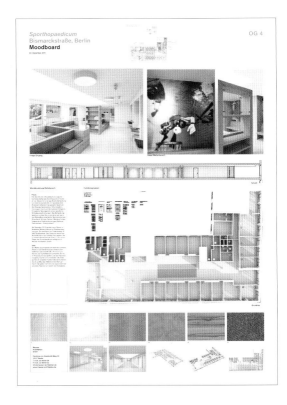
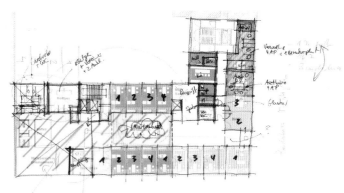

Moodboard and preliminary design for a large multi-doctor practice, design: Meuser Architekten

density, the probable incidence of passers-by, the size of possible target groups [families, singles, senior citizens], work and educational opportunities and also cultural and leisure amenities. Not least, the choice of the right location is also dependent on the rent and property prices in the region as well as on one's own financial resources. There may be a choice of locations. Much experience is needed to make the necessary evaluation and weigh up the many facts to be considered and then recommend a priority solution. Such matters are best left to an experienced specialist whose skills will pay for themselves.

OBJECTIVE PLANNING | Objective planning has been recommended for years by specialist groups because of the large number of flawed decisions made in construction. The principles of construction and operation can be worked out with these people [the location analysis discussed above can also form part of this process]. Before conversion work, a target-performance comparison is carried out and schematic variants are developed for achieving the objectives before the solution to be implemented can be proposed. This will be after the evaluative analysis made in the light of the probable investment and operating costs. Objective planning procedures should also reflect the future developments of the medical practice and possible options for any alterations and the associated extensions. Annex 4 to DIN 13080 "Concepts and Analysis for Objective Planning for General Hospitals"[18] contains helpful hints which naturally apply to medical practices.

SCHEDULING | As well as the ideas emerging from objective planning on the estimated planning process timeframe, a major element in planning in a specific building project is precise scheduling. All phases are subdivided into precise periods and dates for completion are set. An essential feature of effective scheduling is ongoing monitoring and, if necessary, updating. Under the motto "time is money", significant extra costs, which work overruns always entail, can be avoided if a building project is completed on time. This applies particularly to renovations and conversions carried out while business continues. Such situations generally cause disruption and inconvenience that customers and staff alike must tolerate, but they should not be allowed to continue a day longer than absolutely necessary.

FUNCTIONAL AND SPATIAL ALLOCATION PLANNING | Once the basic decisions have been taken on the medical practice being planned, the next essential step is the creation of a functional and spatial allocation plan. Here, too, specialised planners need to be called in, if possible, who will carry out this task right at the start of the whole project. They operate jointly with the manager[s] of the medical practice and designated colleagues. See the chapter on Construction Data for tips on the formal breakdown and the necessary tasks.
Care should be taken to ensure that each room with its required usable floor space is listed in the functional and spatial allocation plan. The best approach at this point is to prepare rough sketches for critical spaces to

show evidence of the given usable floor space values. In a good functional and spatial allocation plan no room is forgotten [not even the cleaner's room]. An appropriate area is reserved for the rooms to be developed and no unnecessary areas are required because these not only raise construction costs but also the running costs for the whole life of the practice.

COST PLANNING AND FUNDING | The investment costs incurred through building work are always of concern to the client, but especially at the beginning of the planning process. If at all possible, the client wants to have a clear understanding of the costs at an early stage so that he can make decisions with some confidence. Once he knows the cost frame he has the opportunity, for example, of opting for the right alternative among the drafts and a standard for the basic features.

A specific empirically established value for the costs per square metres of usable floor space, obtained by making costs analyses of comparable medical practices, will allow an estimate of the total costs to be made. This value needs to be understood as having limitations and must be modified by taking into account local and temporal conditions or different standards. For conversions and renovations, the process described can be applied only to a very limited degree because of the very different characteristics that are possible. This is precisely the point that needs the years of experience of specialist planners.

In Germany medical institutions administered by public authorities, which today are only rarely able to meet their statutory duty to promote investment, have been beset by problems for decades. This backlog of delayed investment has now reached the two-figure billion Euro mark. Since there is scarcely any hope of improvement in the short or long term, new investment models are being developed, for example in the form of public private partnerships [PPP]. These offer a variety of options for the public and private sectors to work together.

In the new Hospital Funding Reform Act[12] the legislation envisages a reform of public investment funding and in § 10 intends to develop standard investment assessment criteria to apply across Germany so as better to calculate the investment needs of the individual *Bundesländer*.

Planning services
In the fee code for architects [HOAI][32] § 15 sets out the services to be supplied by contractors. The work fields specified are: new buildings, new equipment, reconstructions, extensions, conversions, modernisations, enlargements to create more space, maintenance and repairs. The services break down into nine service engineering groups in terms of basic services and specialised services; most important of them are treated briefly here.

PREPLANNING | Essential services are analysis of basics, the production of a planning concept including examination of possible alternatives, clarification and illustration of the essential town planning, design, functional, technical, building, commercial and energy efficiency parameters, preliminary negotiations with authorities and a cost estimate subject to DIN 276[22]. Special services include, for example, the creation of a funding plan and a building work and operating cost benefit analysis.

DESIGN PLANNING | The basic services of design planning comprise first and foremost the design drawings to a scale of 1:100 and for the enlargements to create more space at a scale of 1:50 and 1:20, negotiations with authorities and others involved in planning as specialists as to the design's fitness for approval and a cost calculation subject to DIN 276[22].

EXECUTION PLANNING | This phase includes the execution, detail and structural drawings at scales of 1:50 to 1:1, including the materials specifications. These plans are of major importance for medical practices because they determine the organisation of the details.

SITE COORDINATION AND MONITORING | After delivery of the basic services outlined here – Preparation and Involvement in Contract Award, the job of site coordination and monitoring is to review the site to ascertain whether it conforms with the building consent and all relevant provisions regarding the details of the design. The basic services of this phase also encompass the creation of a schedule, the cost statement subject to DIN 276[22], the application for official inspection and approval and supervision of the elimination of any defects detected.

COMMISSIONING | Although this is not part of the HOAI service profile, it can be useful to plan the commissioning. This is especially desirable when building work is carried out while the business remains in operation to ensure a prudent, smooth-running transition to the new start. An opening ceremony, with a speech, drinks and small gifts is effective publicity. The author of this text would be delighted to be invited!

Sources

1. www.bundesaerztekammer.de
2. www.gbe-bund.de
3. Kuhrt, N.: Gut behandelt. In: ZEIT WISSEN 2009. Number 4. Pages 14–26.
4. Robert Koch-Institut [Ed.]. Böhm, K. | Müller, M.: Ausgaben und Finanzierung des Gesundheitswesens. Heft 45 der Gesundheitsberichterstattung. Berlin 2009. www.gbe-bund.de
5. Gemeinschaft fachärztlicher Berufsverbände. Menzel, H. [Ed.]: Viele Deutsche beklagen Verschlechterung der Gesundheitsversorgung. Facharzt Aktuell 03 | 2009.
6. Sozialgesetzbuch [SGB]. Fünftes Buch [V]. Gesetz of 20 December1988. BGBl. I. Page 2.477.
7. Kassenärztliche Bundesvereinigung: Aufgaben der Kassenärztlichen Vereinigungen. http://www.kbv.de/wir_ueber_uns/107.html
8. Institut für Qualität und Wirtschaftlichkeit im Gesundheitswesen: Über uns. www.iqwig.de/
9. GKV-Modernisierungsgesetz of 14 November 2003. BGBl. I. 2003. Number 55. Page 2.110 ff.
10. Vertragsarztrechtsänderungsgesetz of 22 December 2006. BGBl. Page 3.439.
11. Krankenhausfinanzierungsgesetz of 10 April 1991. BGBl. I. Page 886.
12. Krankenhausfinanzierungsreformgesetz [KHRG] of 17 March 2009. BGBl. I. Page 534.
13. Kassenärztliche Bundesvereinigung: Überwachungen und Begehungen von Arztpraxen durch Behörden. Berlin 2005.
14. Bundesärztekammer: [Muster-] Richtlinien über den Inhalt der Weiterbildung of 28 March 2008.
15. Schurr, M. | Kunhardt, H. | Dumont, M.: Unternehmen Arztpraxis – Ihr Erfolgsmanagement. Heidelberg 2008.
16. Ärztliches Zentrum für Qualität in der Medizin: Kompendium Q-M-A. Qualitätsmanagement in der ambulanten Versorgung. Cologne 20083.
17. DIN 277-1: 2005-02. Grundflächen und Rauminhalte von Bauwerken im Hochbau – Part 1: Begriffe, Ermittlungsgrundlagen | DIN 277-2: 2005-02. Grundflächen und Rauminhalte von Bauwerken im Hochbau – Part 2: Gliederung der Netto-Grundfläche [Nutzflächen, Technische Funktionsflächen und Verkehrsflächen] | DIN 277-3: 2005-04. Grundflächen und Rauminhalte von Bauwerken im Hochbau – Part 3: Mengen und Bezugseinheiten.
18. DIN 13080: 2016-06. Gliederung des Krankenhauses in Funktionsbereiche und Funktionsstellen | Beiblatt 1: 2016-06. Gliederung des Krankenhauses in Funktionsbereiche und Funktionsstellen – Hinweise zur Anwendung für Allgemeine Krankenhäuser | Beiblatt 2: 2016-06. Gliederung des Krankenhauses in Funktionsbereiche und Funktionsstellen – Hinweise zur Anwendung für Hochschul- und Universitätskliniken | Beiblatt 3: 1999-10. Gliederung des Krankenhauses in Funktionsbereiche und Funktionsstellen – Formblatt zur Ermittlung von Flächen im Krankenhaus | Beiblatt 4: 2004-07. Gliederung des Krankenhauses in Funktionsbereiche und Funktionsstellen – Begriffe und Gliederung der Zielplanung für Allgemeine Krankenhäuser.
19. Robert Koch-Institut: Anforderungen der Hygiene beim ambulanten Operieren in Krankenhaus und Praxis. Bundesgesundheitsblatt 1994. Number 5.
20. DIN 12924-4: 1994-01. Laboreinrichtungen. Abzüge, Abzüge in Apotheken, Hauptmaße.
21. Verordnung über Arbeitsstätten [Arbeitsstättenverordnung – ArbStättV]. Bundesgesetzblatt I. Page 1.595.
22. DIN 276-1: 2006-11. Kosten im Bauwesen – Part 1: Hochbau.
23. Müller, S. K. – Chemical Sensitivity Network [CSN]: Umweltbedingungen in der Arztpraxis des 21. Jahrhunderts – Gesünder für Arzt und Patient. www.csn-deutschland.de
24. Robert Koch-Institut: Anforderungen der Hygiene an die funktionelle und bauliche Gestaltung von Krankenhauseinrichtungen für die Versorgung ambulanter Patienten. Bundesgesundheitsblatt 1980. Number 11. Page 164–165.
25. DIN 4109: 1989-11. Schallschutz im Hochbau. Anforderungen und Nachweise [vorgesehener Ersatz durch DIN 4109-1: 2006-10].
26. DIN 4102-1: 1998-05. Brandverhalten von Baustoffen und Bauteilen – Part 1: Baustoffe, Begriffe, Anforderungen und Prüfungen.
27. Wiener Krankenanstaltenverbund [Ed.]. Aumayr, J. | Herbek, S. | Kastl, J. | Labryga, F. | Mejstrik, W. | Staudinger, Ch.: Planungshandbuch für Krankenhäuser und Pflegeheime. Vienna 2003.
28. Damaschke, S. | Scheffer, B. | Schossig, E.: Arztpraxen, Planungsgrundlagen und Architekturbeispiele. Leinfelden-Echterdingen 20032.
29. Teut, A. | Nedeljkov, G.: Die Gruppenpraxis. Düsseldorf 1973.
30. DIN 5035-8: 2007-07. Beleuchtung mit künstlichem Licht – Part 8: Arbeitsplatzleuchten.
31. Frieling, H.: Farbe am Arbeitsplatz. Munich 1992.
32. Honorarordnung für Architekten und Ingenieure in der ab 1.1.1996 gültigen Fassung. Stuttgart | Berlin | Cologne 1995. Revision on 17 July 2013.

Medical Practices

Orthopaedics at Rosenberg
Bhend.Klammer

Centre for Radio-Oncology
Brandherm + Krumrey

Dental Clinic
Klaus R. Bürger

ENT and Psychotherapy Clinic
Cossmann de Bruyn

Gastroenterology
Regina Dahmen-Ingenhoven

Dental Practice
Gruppe für Gestaltung

Children's Dental Practice
GRAFT Architekten

KU64 Dental Practice
GRAFT Architekten

MKG-Surgery Airport Clinic
Holzrausch

Radiology
Ippolito Fleitz Group

Orthodontic Clinic
Landau + Kindelbacher

SPORTHOPAEDICUM
Meuser Architekten

Orthopaedic Clinic
Mateja Mikulandra-Mackat

Matrei Health Centre
Gerhard Mitterberger

Medical + Dental Suite
pd raumplan

Centre d'Endodontie
pd raumplan

Municipal Clinic
Sander Hofrichter

Dental Practice
stengele+cie.

OMF Surgery
Wagenknecht Architekten

The unusual shape of the "Silver Tower", a former hotel dating back to the 1970s, with its striking oval façade and rounded windows provided the template for the practice conversion. Over one entire floor, the previous haphazard room structure was broken up and divided into several zones. In the centre is the new, transparent foyer, which extends between two treatment areas. Flush, amorphously shaped ceiling lights guide the patient from the entrance to reception and the waiting area. High-gloss white walls divide the open floor plan into separate areas, without hemming it in. The reflective surfaces visually extend the room – the flush inset pigeonholes, the visitor's coat-rack and the X-ray storage room do not detract from the overall impression of openness. All superfluous fittings have been removed, the bearing structure has been pared right down and the fire doors concealed. Dark furniture stands on the light grey rubber floor; the glossy varnished reception desk and the low waiting area tables are curved. The colour fluctuates between dark blue and petrol green depending on the angle of view. The twelfth floor of the "Silver Tower" affords stunning views over the city; light curtains can be drawn if necessary.

BHEND.KLAMMER

ORTHOPAEDICS AT ROSENBERG
ST. GALLEN

Medical Practices

Medical Practices

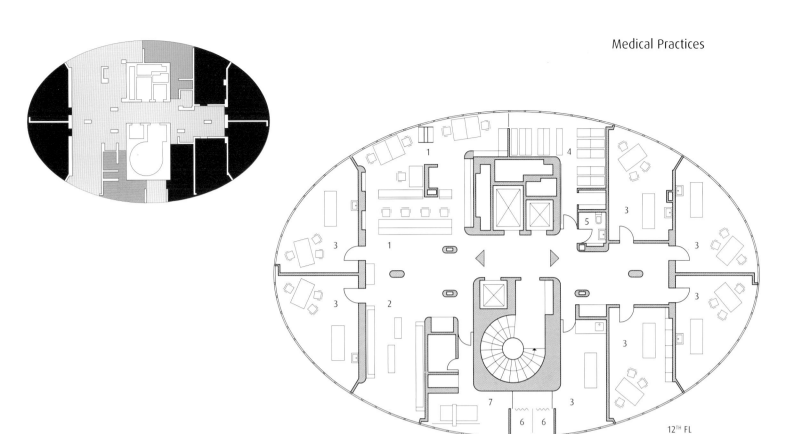

12TH FL

**Client	Operator**	Orthopaedics at Rosenberg
Planning time	2003–2004
Construction time	07 2004–08 2004
Usable floor space	291 sqm
Gross cubic capacity	1,460 cbm
Total cost	600,000 EUROS

a During conversion superfluous fittings were removed and the bearing structure pared right down.

b The open floor plan is broken up into individual areas by high-gloss white walls.

c The varnished reception desk, petrol green | dark blue depending on the angle of view, contrasts with the light grey rubber floor.

Diagrammatic plans, scale 1:400
Floor plans, scale 1:200

Usable floor spaces

Patient rooms	**yellow**	121 sqm	42 %
Examination	Treatment rooms	**red**	134 sqm	46 %
Specialist rooms	**pink**	17 sqm	6 %
Administrative rooms	**green**	19 sqm	6 %
Total		291 sqm	100 %

Floor plan layout
1 Reception
2 Waiting area
3 Treatment room
4 Filing room
5 Patient toilet
6 Changing cubicle
7 Examination X-ray

Performance data

Outpatients per year	10,000
Number and type of services per year	X-rays [6,500]
Clinic opening hours	5 days	50 hours
Waiting time [with	without appointment]	10–20 min.	45 min.
Number of staff	15
Clinic planning advice	Further expansion of reception and waiting area

Think positive! This uplifting motto provided the inspiration for the conversion of the existing premises of the Cologne Specialist Clinic. The reception and waiting area, which does not benefit from natural light, has been transformed into a "thoroughly feel-good environment". This Centre for Radio-Oncology is normally the preserve of seriously ill, mainly male patients. During their time at the clinic they need to feel that they are in good hands, well cared for and, if possible, should not be kept waiting for treatment. On this basis, the small waiting area located in a brightly lit corner is adequate. A light wall with a floral photo motif creates a relaxed atmosphere, while in the foyer a large backlit photo wall creates an impression of floodlit space. The materials are a study in contrasts: high gloss versus matt, translucent versus opaque colour. A distinctive technical feature of the practice is the ten-tonne X-ray protection room with its ultra-modern equipment. During radiation therapy patients have to spend several minutes alone in this room; its bright splash of colour is therefore intended to be especially attractive.

BRANDHERM + KRUMREY

CENTRE FOR RADIO-ONCOLOGY
COLOGNE

Medical Practices

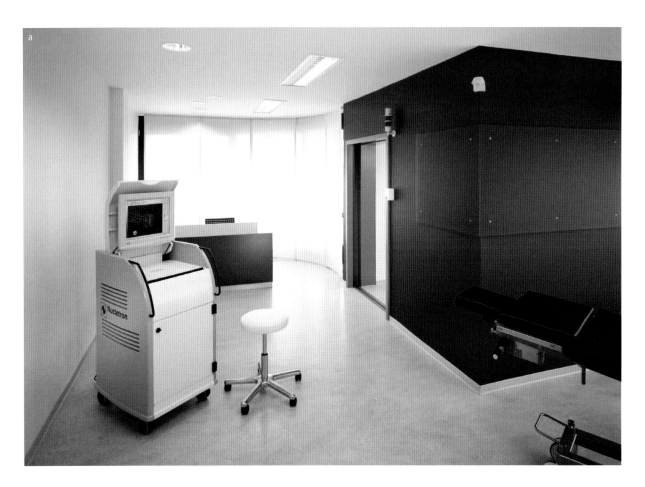

Medical Practices

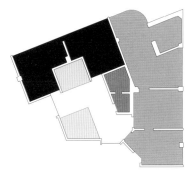

GF

Client	Dr. Gregor Spira
	Jürgen Metz
Operator	Dr. Gregor Spira
	Dr. Carsten Weise
Construction time	02 2003 – 04 2003
Usable floor space	165 sqm
	Redesign [in part]
	42 sqm

a The X-ray protection room behind the red wall is equipped with an ultra-modern system.
b The large backlit photo wall in the foyer creates the impression of a floodlit space.

Diagrammatic plans, scale 1:400
Floor plans, scale 1:200

Performance data

Outpatients per year	470 – 550
Inpatients per year	400
	[1,200 nursing days]
Number and type of services per year	LDR [170], HDR [500]
Clinic opening hours	5 days	40 hours
Waiting time	10 – 45 hours
Type and number of staff	2 specialists
	3 part-time staff

From the entrance to the triangular floor plan of the partnership practice, patients proceed into a large room, past the reception desk, with its unobstructed sight lines into the waiting area. On the other two sides of the triangle, the consultation, examination and recovery rooms, some of them interconnected, benefit from full natural light. A patient toilet, staff lounge, administrative office, storage and cleaning facilities would be useful additions.

Usable floor spaces

Patient rooms	**yellow**	19 sqm	12 %
Examination			
Treatment rooms	**red**	48 sqm	29 %
Specialist rooms	**pink**	89 sqm	54 %
Offices	Staff rooms	**orange**	9 sqm	5 %
Total		165 sqm	100 %

Floor plan layout

1 Reception
2 Waiting area
3 Consultation room
4 Staff toilet and shower
5 Small kitchen
6 Radiotherapy
7 Examination and recovery room

Close to the Allgäu mountains! The surrounding mountain scenery is reflected in the interior design of this new dental practice: backlit friezes provide decorative light sources in the waiting area and even the dominant colour, blue, references the nearby mountains. The colour palette – varying shades of blue contrasting with orange – was chosen specifically for its psychological effect: calming and relaxing, cooling and therapeutic. People often dread a visit to the dentist, and the interior concept is intended to allay their fears. Cool, soothing blue is counterbalanced by a warm cheerful orange, reinforcing the healing effect. Dark oak parquet flooring leads patients through the public areas. The upper floor of the gutted mansion in the centre of Marktoberdorf, which dates back to 1723, houses a cuboid sterilisation room, around which the treatment rooms are arranged. Glass doors backed by translucent film printed with poetry create a feeling of space and transparency, at the same time entertaining and relaxing the patients.

KLAUS R. BÜRGER

DENTAL CLINIC
MARKTOBERDORF

Medical Practices

Medical Practices

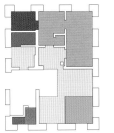 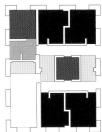 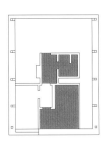

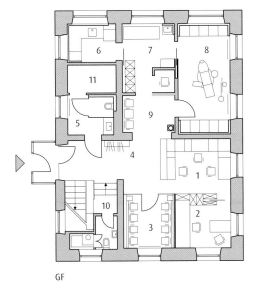
GF

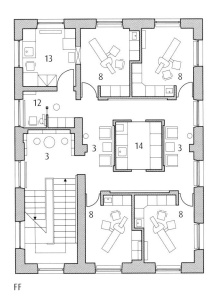
FF

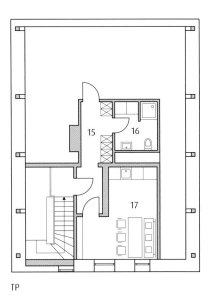
TP

**Client	Operator**	Dr. Johann Karg
	Dr. Volker Baumeister
Planning time	11 2004 – 03 2005
Construction time	01 2005 – 03 2005
Usable floor space	181 sqm

a Predominant cool blue is counterbalanced by a warm cheerful orange; dark oak parquet flooring leads the patient through the public areas.

b Glass doors create a sense of openness and transparency; poetry printed on translucent film helps to relax patients.

c The walls are decorated with backlit friezes of mountain landscapes and floral motifs.

Diagrammatic plans, scale 1:400
Floor plans, scale 1:200

Careful thought has gone into the layout of the rooms of this partnership clinic, which are spread over three floors of the rectangular floor plan. Every square metre has been put to good use, including the corridors and even the stair-heads. All the rooms needing natural light have windows. Reception, waiting area and office adjoin one another. Appropriately, only the staff rooms are located on the attic floor.

Usable floor spaces

Patient rooms	**yellow**	43 sqm	24 %
Examination	Treatment rooms	**red**	42 sqm	23 %
Specialist rooms	**pink**	33 sqm	18 %
Administrative rooms	**green**	9 sqm	5 %
Offices	Staff rooms	**orange**	28 sqm	15 %
Supply	Waste disposal	**brown**	21 sqm	12 %
Plant rooms	**blue**	5 sqm	3 %
Total		181 sqm	100 %

Floor plan layout

1 Reception
2 Office
3 Waiting area
4 Coat-rack
5 Patient toilet
6 Workroom
7 Recovery room
8 Treatment room
9 Information
10 Staff toilet
11 Boiler room
12 Examination X-ray
13 Store room
14 Sterilisation room
15 Staff changing cubicle
16 Staff toilet and shower
17 Staff lounge

The group practice of the ENT doctor and his wife, a qualified psychotherapist, extends over three floors in an open-plan studio house. The conversion has retained the openness of the rooms and the view through to the private garden, which the architects achieved by using ceiling height glass partition walls and an inspired lighting concept. On the garden side the extension has an impressive glass façade, while the street façade is broken only by small windows. Direct halogen lighting and indirect light from fluorescent lamps, combined with minimalist building materials – wood, aluminium and satinised acrylic glass – create a warm and light atmosphere in the practice rooms. In the basement, which receives minimal natural light, the smooth cast floor is a gleaming mood-lifting orange. The other rooms also play with colour – shades of green, yellow and orange give each one a different feel. The practice's three storeys are connected by a spiral staircase. On the top floor, an indirectly lit, translucent curtain over the round stairwell serves as a room divider.

COSSMANN_DE BRUYN

ENT AND PSYCHOTHERAPY CLINIC
DÜSSELDORF

Medical Practices

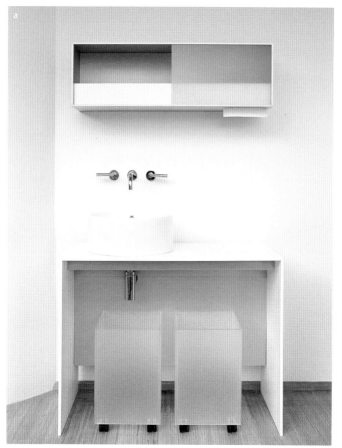

Medical Practices

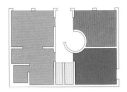 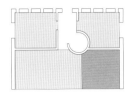 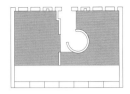

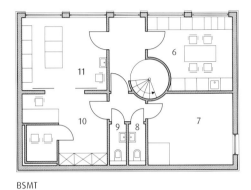
BSMT

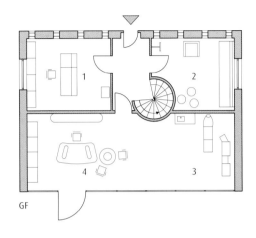
GF

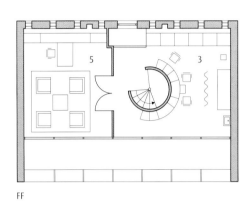
FF

Client	Dr. Claus Birken
Operator	Dr. Claus Birken
	Heike Kupka
Construction time	2003
Usable floor space	204 sqm

a The treatment room is defined by simple a functionality.
b Minimalist building materials – aluminium, wood and satinised acrylic glass – give the waiting area a warm and airy atmosphere.
c The translucent curtain over the round stairwell on the top floor serves as a room divider.

Diagrammatic plans, scale 1:400
Floor plans, scale 1:200

The resulting floor plan, for all three levels with only small corridor areas, is clearly divided. All rooms except one examination room benefit from natural light. The vertical connection in the shape of a spiral staircase may prove problematic for disabled and older patients and should perhaps be adapted.

Usable floor spaces

Patient rooms	**yellow**	65 sqm	32 %
Specialist rooms	**pink**	106 sqm	52 %
Offices | Staff rooms	**orange**	16 sqm	8 %
Plant rooms	**blue**	17 sqm	8 %
Total		204 sqm	100 %

Floor plan layout
1 Reception
2 Waiting area
3 Treatment room
4 Consultation room
5 Treatment room psychotherapy
6 Staff changing cubicle
7 Installations room
8 Toilet G
9 Toilet L
10 Examination audiometry
11 Organic resonance

"Diving in another world." Submersion in an underwater world, fluidity, glistening surfaces and a flowing continuum of rooms were the concepts that underpinned the design of this practice. Patients should feel that they have stepped through an invisible mirror, leaving their world behind and diving into a new one – strange and yet oddly familiar. "Go with the flow – don't stop!" This new practice, the creation of a young gastroenterologist, occupies part of a health centre. The layout of this spacious environment is designed to uplift and relax. From the moment they enter the open-plan entrance area – foyer, reception and waiting area flow seamlessly together – patients are made to feel at ease. The fibreboard and walnut seating elements are upholstered in silver-coated foam; walls and furniture are in various shades of blue and white, reinforcing the impression of a fantasy underwater world. Fish motifs play on the lines "Move like a jellyfish" from the song "Bubble Toes" by Jack Johnson, the inspiration behind the design. In the long corridor, on either side of which are treatment rooms and side-rooms, key words in yellow and orange stand out on the light walls. The floor covering is a smooth silk-matt, light blue epoxy resin, adding to the surreal, dreamlike effect produced by the practice rooms.

REGINA DAHMEN-INGENHOVEN

GASTROENTEROLOGY
REMSCHEID

Medical Practices

Medical Practices

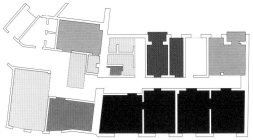

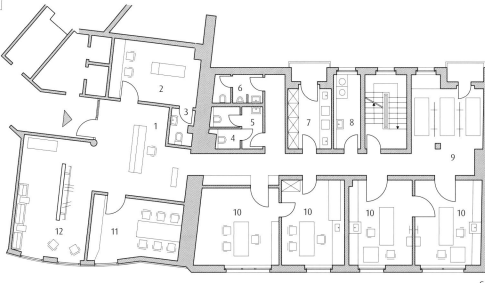

Client	Dr. Eric Jörgensen
Operator	Dr. Eric Jörgensen
	Ingomar Scheller
Planning time	07 2007–08 2007
Construction time	10 2007–12 2007
Usable floor space	172 sqm
Construction cost	50,000 EUROS

a In the long corridor, on either side of which are treatment rooms and side-rooms, key words in yellow and orange stand out on the light walls.

b Walls and furniture are in various shades of blue and white.

c The floor covering is a smooth, silk-matt, adding to the surreal, dream-like effect produced by the practice rooms.

Diagrammatic plans, scale 1:400
Floor plans, scale 1:200

Performance data

Outpatients per year	8,000–9,000
Number and type of services per year	Intestinoscopy
	Capsule endoscopy
	Colonoscopy
Clinic opening hours	5 days | 56 hours
Waiting time [with | without appointment]	0–30 min. | 0–60 min.
Type and number of staff	3 doctors
	8 clerical staff
	2 trainees
Clinic planning advice	Budget for potential extension

The rooms in this partnership practice, accessed by a spacious foyer and a straight corridor, have a functional layout which facilitates patient care. Reception and waiting area adjoin each other. Apart from one examination room, all rooms needing natural light have windows. The practice would benefit from the addition of an office for administrative tasks and dedicated space for storage and cleaning.

Usable floor spaces

Patient rooms	**yellow**	42 sqm	25 %
Examination | Treatment rooms	**red**	67 sqm	39 %
Specialist rooms	**pink**	31 sqm	18 %
Offices | Staff rooms	**orange**	16 sqm	9 %
Supply | Waste disposal	**brown**	16 sqm	9 %
Total		172 sqm	100 %

Floor plan layout

1	Reception	**5**	Patient toilet L	**9**	Recovery room
2	Exam. ultrasound	**6**	Patient toilet G	**10**	Examination
3	Patient toilet	**7**	Store room laundry	**11**	Staff lounge
4	Staff toilet	**8**	Workroom laundry	**12**	Waiting area

High-grade Canadian grey elm and coated glass – the exquisite fittings reflect the dentist's area of expertise, namely the preparation of top quality ceramic inlays. The transparent light-flooded architecture of the building is replicated in the interior design, with a glass wall running parallel to the slightly curved outer façade. This separates the treatment rooms and offices from the through corridor, which therefore also has the benefit of natural light. The field of vision on the glass wall is coated with a translucent film that varies in texture, ensuring patient privacy during treatment. The natural light illuminates the mainly orange film and gives what is actually a narrow passage the illusion of spaciousness. Ceiling height sliding doors punctuate the almost twelve metre long glass wall. The practice rooms on the other side have wood laminate flooring to the corridor, creating a design counterpoint to the glass wall opposite. High quality is also the hallmark of the reception area, with its cosy sofas and a small but stylish counter. This practice has a bespoke nameplate concept: pithy inscriptions such as "Aesthetics" and "Rivestident" [for the insertion of ceramic inlays] designate the individual rooms.

GRUPPE FÜR GESTALTUNG

DENTAL PRACTICE
BREMEN

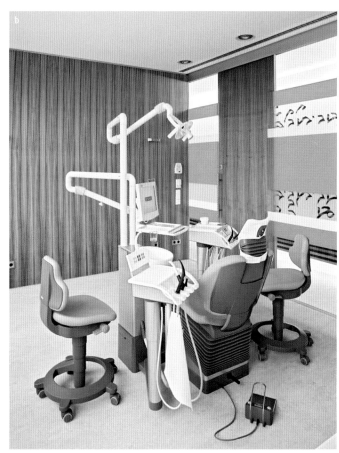

Medical Practices

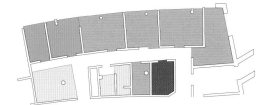

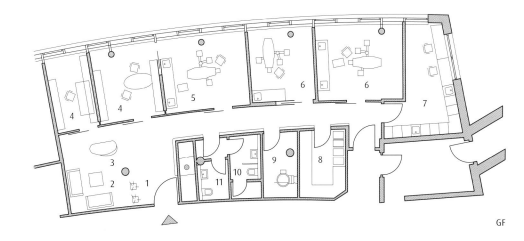

GF

Client	Dr. Inge Mittag
	BCB Building Consult Bremen GmbH
Operator	Dr. Inge Mittag
Planning time	05 2004 – 02 2005
Construction time	02 2005 – 03 2005
Gross floor area	170 sqm
Usable floor space	129 sqm
Gross cubic capacity	518 cbm

a The building's transparent, light-flooded exterior architecture mirrored by its interior design, with a glass wall running parallel to the slightly curved outer façade.

b The field of vision on the glass wall is coated in a translucent film that varies in texture, ensuring patient privacy during treatment.

c The luxurious fittings reflect the practice's area of expertise, the preparation of high quality ceramic inlays.

Diagrammatic plans, scale 1:400
Floor plans, scale 1:200

Performance data

Outpatients per year	1,000
Number and type of services per year	Root canal treatment [200]
Services to other clinics	CT [50]
Clinic opening hours	5 days | 50 hours
Waiting time [with | without appointment]	10 min. | 45 min.
Type and number of staff	2 doctors, 4 full-time staff,
	2 part-time staff, 1 dental technician
Clinic planning advice	Another office would be useful.

From the main entrance in the first third of the slightly curved, long rectangular building that houses the partnership practice, patients proceed to the treatment rooms, most of which are on the convex side. A second entrance in the rear section of the floor plan is used for deliveries. While the rooms on the convex side benefit from natural light, those on the concave side have no windows.

Usable floor spaces

Patient rooms	**yellow**	24 sqm	19 %
Specialist rooms	**pink**	75 sqm	58 %
Administrative rooms	**green**	22 sqm	17 %
Supply | Waste disposal	**brown**	8 sqm	6 %
Total		129 sqm	100 %

Floor plan layout

1 Coat-rack	5 Treatment room	8 Sterilisation room
2 Reception	6 Treatment room	and store room
3 Waiting area	prevention	9 Examination X-ray
4 Office	7 Ceramic room	10 Toilet G

At the entrance to this Berlin dental practice specialising in the treatment of children, young children and their parents stare in amazement at the curling blue wave that "flows" through the whole building and serves as ceiling, wall, reception and lounge. A structural space has been created in which the rooms are interwoven by the movements of the wave. The split-level design draws people entering the building "beneath the wave" into an underwater world. Children, their dread of a visit to the dentist quite forgotten, dive into a mysterious universe, where shoals of fish swim along the walls and light points adorn the ceiling like rising bubbles. The architects spurned the conventional white colour scheme and sterile, hygienic atmosphere normally found in dental practices, opting instead for soft, flowing shapes. An atmosphere has been created in which children can forget the world around them – and in particular the fact that they have come to see the dentist. The underwater theme permeates every area of the practice and is emphasised by targeted use of light and materials. Echoing the wave that "rolls" through the rooms, a sea-blue colour concept was devised for the practice in collaboration with STRAUSS & HILLEGAART, embracing ceilings, walls and floors and even the furniture and dental equipment. A visit to this practice is a special experience – and not just for the children!

GRAFT

CHILDREN'S DENTAL PRACTICE
BERLIN

Medical Practices

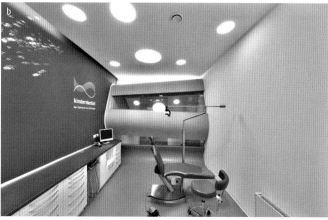

Medical Practices

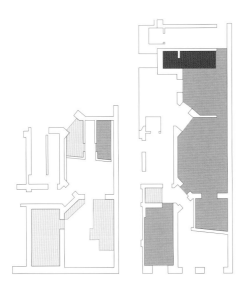
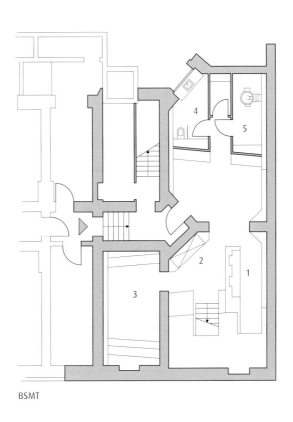
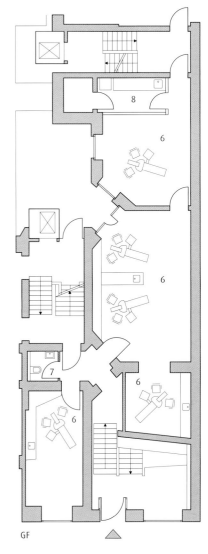

BSMT　　　　　　　　　GF

**Client	Operator**	Dr. A. Mokabberi
Construction time	02 2007–08 2007
Gross floor area	177 sqm
Usable floor space	157 sqm

a A curling blue wave "flows through" the foyer, interweaving the two floors of the practice.

b The sea-blue colour concept devised for the practice embraces ceilings, walls and floors and even the furniture and dental equipment.

c Shoals of fish on the walls and light points dotting the ceiling like some rising bubbles create a mysterious world.

d The basement waiting area, which creates the illusion of diving into the ocean depths.

Diagrammatic plans, scale 1:400
Floor plans, scale 1:200

Basement steps take patients into the spacious area housing reception, coat-rack and waiting area. The X-ray room is also located here, while all other treatment is provided at ground floor level. On the upper ground floor, natural light shines through doors into two of the four treatment rooms.

Usable floor spaces

Patient rooms	**yellow**	43 sqm	27 %
Specialist rooms	**pink**	104 sqm	66 %
Supply	Waste disposal	**brown**	10 sqm	7 %
Total		157 sqm	100 %

Floor plan layout

1 Reception
2 Coat-rack
3 Waiting area
4 Patient toilet L
5 Examination X-ray
6 Treatment room dentist
7 Patient toilet G
8 Sterilisation room

Few other environments have such negative connotations as a dental practice. Here the architects made a conscious decision to shun the normal practice setup, replacing it with a spectacular red and yellow interior themed on a landscape of sand-dune. Patients visiting this practice are made to feel like beach goers choosing a quiet spot in the sand, spreading their towels and settling down to enjoy the peace and quiet. The floor curves and waves billow on the ceiling. In the middle of this undulated landscape, flat built-in furniture – the unusual loungers and sofas are more than just seats – provide an idyllic beach setting, and a sun deck represents the sea. An open woodburner suspended from the ceiling conjures up a campfire idyll in Charlottenburg practice. The dental treatment room is fitted out as spa area: a geometric, subtly lit "spring grotto" with a large basin reflects gentle water movements on the ceiling; glass washbasins float like islands in this artistic water world. As a counterpoint to the colourful and vibrant design of the public areas, the high-tech treatment rooms are more restrained. The two practice levels are connected by an open staircase at the end of the main corridor; a canyon-like passageway offers stunning views over the surrounding roofscape.

GRAFT

KU64 DENTAL PRACTICE
BERLIN

Medical Practices

Client | Operator	Dr. Stefan Ziegler
Construction time	01 2005 – 07 2005
Gross floor area	940 sqm
Usable floor space	536 sqm
Gross cubic capacity	3,077 cbm
Total cost	519,680 EUROS

a   An open wood-burner suspended from the ceiling conjures up a campfire idyll in the waiting area above the roofs of Berlin.

b   A staircase at the end of the corridor connects the practice's two levels.

c   The wall motifs, which appear normal or distorted always depending on the angle of view, were designed in collaboration with STRAUSS & HILLEGAART.

d   The surprisingly colourful and vibrant corridor on the 5th floor.

Diagrammatic plans, scale 1:500
Floor plans, scale 1:200

Floor plan layout

The main patient entrance into this large, two-level multi-doctor practice is through the upper level which houses the reception and a spacious waiting area. On the lower level are two secondary entrances. All the rooms apart from a few side-rooms benefit from natural light.

Usable floor spaces			
Patient rooms	**yellow**	160 sqm	30 %
Examination | Treatment rooms	**red**	43 sqm	8 %
Specialist rooms	**pink**	203 sqm	38 %
Administrative rooms	**green**	41 sqm	8 %
Offices | Staff rooms	**orange**	57 sqm	11 %
Supply | Waste disposal	**brown**	19 sqm	3 %
Plant rooms	**blue**	13 sqm	2 %
Total		536 sqm	100 %

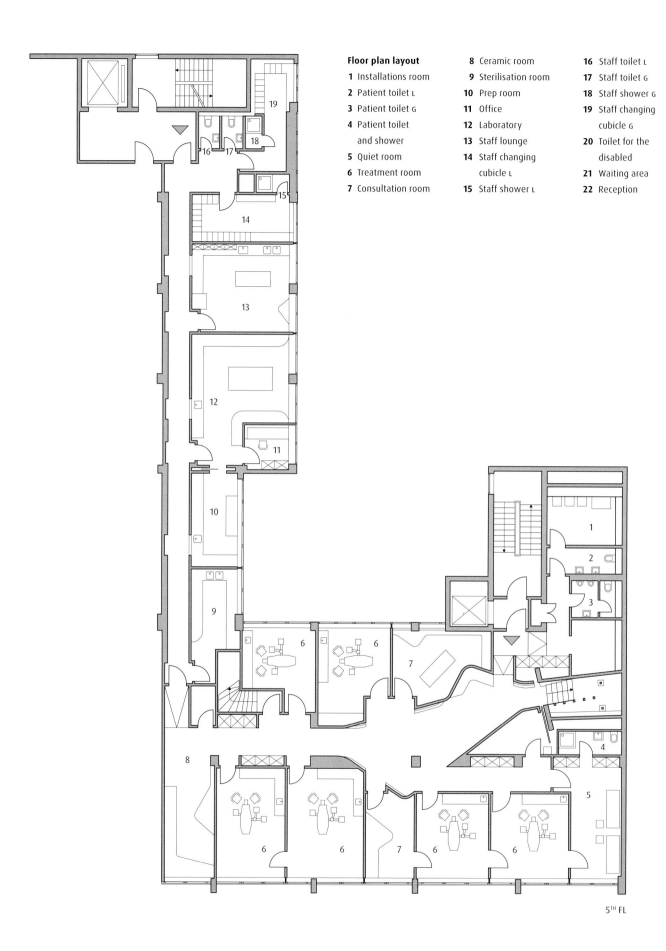

Floor plan layout
1 Installations room
2 Patient toilet L
3 Patient toilet G
4 Patient toilet and shower
5 Quiet room
6 Treatment room
7 Consultation room
8 Ceramic room
9 Sterilisation room
10 Prep room
11 Office
12 Laboratory
13 Staff lounge
14 Staff changing cubicle L
15 Staff shower L
16 Staff toilet L
17 Staff toilet G
18 Staff shower G
19 Staff changing cubicle G
20 Toilet for the disabled
21 Waiting area
22 Reception

Medical Practices

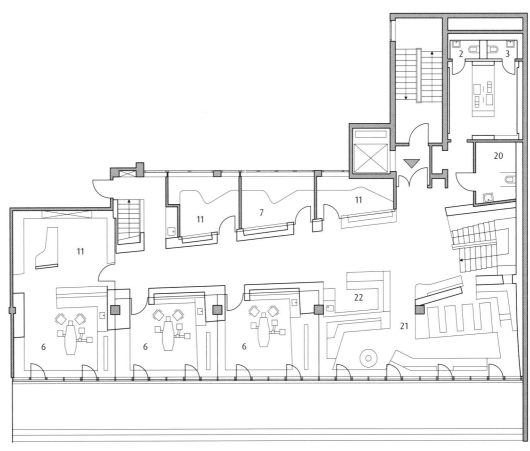

6TH FL

After climbing the long entrance staircase with its back-lit handrail and ball lights set at different heights, visitors to the Freising Airport Clinic reach a reception area that is impressive in its minimalist design and affords a wide view of the surrounding neighbourhood through frameless fixed glazing. The centrally located oak veneer wooden box, which houses the reception area and other functional rooms, forms the heart of the practice. The rounded corners of the corridor walls seem to emphasise the angularity of the platform-mounted wood-frame structure, while the surrounding light gap further accentuates the already imposing cube. The waiting area opposite, with its low ceiling, indirect lighting and felt covered rear walls, creates a cosy space for patients, that can relax during their wait by watching the programmes of their choice on the TV set into the wall. A smooth levelled grey-black concrete floor provides a strong counterpoint to the matt white walls. Both wooden box and walls indicate directions and emergency exits; these are slightly offset to avoid the impression of long monotonous passages. The light points of the recessed spots in the public area do not follow any pattern, but are randomly placed. The orange corporate identity colour is a repeating element throughout the practice rooms – handrail, taps and desk – adding an opulent bright touch to the deliberately simple colour scheme.

HOLZRAUSCH

MKG-SURGERY AIRPORT CLINIC
FREISING

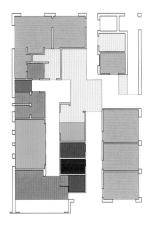

Medical Practices

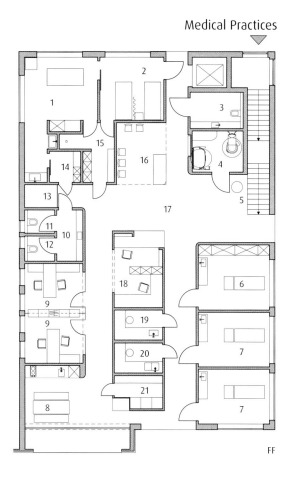

Client	Operator	Dr. Andreas Jauch
Gross floor area	250 sqm
Usable floor space	157 sqm

a A cosy corner for patients: the television set into the wall informs and entertains them as they wait.
b Entrance area with the high-quality wood reception desk.
c Simple, classical functionality in the treatment room.
d Large Sky-Frame sliding doors open outwards from the staff lounge and provide plenty of natural light.

Diagrammatic plans, scale 1:400
Floor plans, scale 1:200

Performance data

Outpatients per year	1,000
Number and type of services per year	Implantology [200]
	Cosmetic surgery [100]
	Orthodontics [200]
	Dentistry [300]
Services to other clinics	Tumour surgery
Clinic opening hours	5 days	50 hours
Waiting time [with	without appointment]	5 min.	15 min.
Type and number of staff	1–3 doctors
	4 specialists

Patients ascend a staircase or use a lift in one corner of the rectangular floor plan to reach the upper floor with its central reception and waiting area, from which they can then access all the rooms in the multiple practice via two corridors. The operating area is favourably located in a corner section and most of the rooms benefit from natural light.

Usable floor spaces

Patient rooms	**yellow**	24 sqm	15 %
Examination	Treatment rooms	**red**	39 sqm	25 %
Specialist rooms	**pink**	37 sqm	24 %
Administrative rooms	**green**	25 sqm	16 %
Offices	Staff rooms	**orange**	24 sqm	15 %
Supply	Waste disposal	**brown**	6 sqm	4 %
Plant rooms	**blue**	2 sqm	1 %
Total		157 sqm	100 %

Floor plan layout

1 Operating theatre
2 Recovery room
3 Patient toilet
4 Examination X-ray
5 Coat-rack
6 Treatment room prevention
7 Treatment room
8 Staff lounge and small kitchen
9 Office
10 Toilet anteroom
11 Staff toilet L
12 Staff toilet G
13 Installations room
14 Staff sluice
15 Patient changing cubicle
16 Waiting area
17 Reception
18 Office
19 Sterilisation room
20 Laboratory
21 Staff changing cubicle and store room

Following its move, the group radiology practice in the basement of the Schorndorf Health Centre now occupies an area of around 600 square metres. At the centre of this new practice is the spacious waiting area, decorated in light colours. Ceiling height, upholstered side walls keep waiting patients safe and cosy in keeping with the brief to the designers to make the patient experience as pleasant as possible. A referral for radiological examination often has serious medical implications. Many of the patients arriving for their appointment are aware of this and feel particularly vulnerable. They need to be given a sense of security and a simple layout makes it easier for them to find their way around. In contrast, the technical installations are almost invisible. On entering, patients make their way along the tapering corridor to the central waiting area. Black and white backlit walls with cloud motifs – the unusual colour scheme references an X-ray image – frame the two reception desks in the waiting area. The various functional units for nuclear medicine, MRI and CT, X-ray, mammography and ultrasound and the doctor's offices are grouped around the central waiting area and reached via the U-shaped access hallway. As a deliberate counterpoint to the calming colours in the waiting area, a distinctive orange has been chosen here to assist with orientation and create a suitable backdrop to this busy area.

IPPOLITO FLEITZ GROUP

RADIOLOGY
SCHORNDORF

Medical Practices

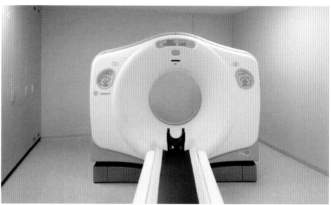

Medical Practices

Client | Operator	Dr. Gebhard Wittlinger	
	Dr. Christoph Hahn	
	Dr. Wolfgang Stern	
Planning time	06 2006	
Construction time	04 2007	
Gross floor area	800 sqm	
Usable floor space	577 sqm	
Gross cubic capacity	2,400 cbm	

a Coat-rack and reception office in the centrally located entrance area.
b As a deliberate counterpoint to the calming colours in the waiting area, the U-shaped access hallway is painted bright orange.
c View into the examination room CT, where the latest technology takes centre stage.
d Patients go past the reception desk straight to the central waiting area.
e The waiting area is decorated in light colours with ceiling height upholstered side walls.
f The waiting area for private patients.

Diagrammatic plans, scale 1:500
Floor plans, scale 1:200

Usable floor spaces

Patient rooms	yellow	225 sqm	39 %
Examination | Treatment rooms	red	20 sqm	3 %
Specialist rooms	pink	206 sqm	36 %
Administrative rooms	green	34 sqm	6 %
Plant rooms	blue	92 sqm	16 %
Total		577 sqm	100 %

Performance data

Outpatients per year	25,000
Number and type of services per year	Diagnostic radiology
Clinic opening hours	5 days | 50 hours
Waiting time [with | without appointment]	15 min. | 30 min.
Number of staff	25

Medical Practices

Floor plan layout

1. Coat-rack
2. Reception
3. Waiting area
4. Nuclear medicine switch room
5. Doctor's consultation room
6. Nuclear medicine treatment room
7. Laboratory
8. Treatment room spray booth
9. Sluice
10. Patient toilet
11. Staff toilet
12. Examination magnetic resonance tomography [MRI]
13. Installations room MRI
14. Patient changing cubicle
15. MRI switch room
16. Treatment room
17. CT switch room
18. Changing cubicle for the disabled
19. Examination CT
20. Examination X-ray
21. X-ray switch room
22. Examination ultrasound
23. Examination mammography
24. Private patients' waiting area
25. Office
26. Telephonist's office
27. Toilet for the disabled
28. Patient toilet L
29. Patient toilet G

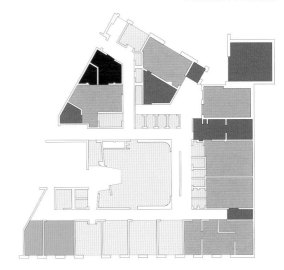

BSMT

In the listed "Schranne", the former grain storehouse of the city of Mindelheim, an extensive, open-plan dental practice has been created which spans the whole storey. The loft-like room, not supported by any pillars, is structured with the aid of free-standing room sculptures placed in its different functional areas – reception, waiting area and treatment unit – accentuating the room's effect. The deliberate absence of colour blurs the transitions between furniture, walls and ceiling and condenses them to the point where they overlap. The unbroken unity of the room is effectively supported by the jointless, dark grey Pandomo stone flooring. The organic design concept of the furniture is continued in the illuminated ceiling which spans the whole room. An innovative lighting concept with changeable light colours for different moods hides the ceiling panels hanging behind. The rooms intended for patients are almost all open-plan, while office and staff rooms and the laboratory are designed to be self-contained. The idea behind the unconventional interior design of the large treatment room is to minimise the distances for the attending physicians and to make it easy for patients to move around.

LANDAU + KINDELBACHER

ORTHODONTIC CLINIC
MINDELHEIM

Medical Practices

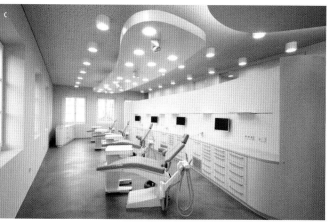

Medical Practices

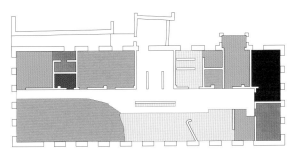

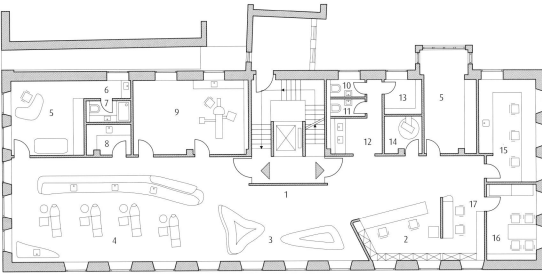

Client	Operator	Dr. Eleni Stylianidou
Planning time	09 2005 – 04 2007
Construction time	10 2006 – 04 2007
Usable floor space	197 sqm
Total cost	600,000 EUROS

a The absence of colour – apart from the red office chair which acts as an eye-catcher – blurs the transitions between furniture, walls and ceiling and condenses them to the point where they overlap.

b The loft-like reception and waiting area is structured with the aid of freestanding room sculptures, which accentuate the room's effect.

c The four side-by-side treatment units mean shorter journeys for the doctors, but the open layout may seem rather unusual for the patients.

d The unbroken unity of the room is impressively supported by the jointless Pandomo stone flooring.

Diagrammatic plans, scale 1:400
Floor plans, scale 1:200

Performance data

Outpatients per year	800
Services to other clinics	Dental treatments
	Orthodontic interventions
Clinic opening hours	5 days	35 hours
Waiting time	10 – 30 min.
Number of staff	11

Patients access a large area, occupying almost the entire length of the floor plan, via the entrance situated in the centre of the long rectangular floor plan of this single-doctor practice. All the rooms in the medical practice can be accessed from this room. The layout means that almost all the rooms receive natural light, but it does make it necessary to walk long distances and might bring a certain turmoil to the open-plan treatment rooms.

Usable floor spaces

Patient rooms	yellow	50 sqm	25 %
Examination	Treatment rooms	red	16 sqm	8 %
Specialist rooms	pink	80 sqm	41 %
Administrative rooms	green	31 sqm	16 %
Offices	Staff rooms	orange	16 sqm	8 %
Supply	Waste disposal	brown	4 sqm	2 %
Total		197 sqm	100 %

Floor plan layout

1. Coat-rack
2. Reception
3. Waiting area
4. Treatment room
5. Office
6. Small kitchen
7. Staff toilet and shower
8. Sterilisation room
9. Consultation and treatment room
10. Patient toilet G
11. Patient toilet L
12. Hygiene
13. Filing room
14. Examination X-ray
15. Laboratory
16. Staff lounge and small kitchen
17. Office

Providing a functional and creative image to a highly frequented medical practice – with up to 250 patients a day, over almost 700 square meters of usable space – poses a great challenge of spatial organisation. This was particularly true of the floor extension of the Sporthopaedicum – the orthopedic-surgical joint practice – in Berlin-Charlottenburg, conveniently situated near the two underground lines U2 and U7. The patient enters the lofty space from a central point. The functional combination of reception, waiting room, and circulation has created a spacious area that serves as a kind of public plaza, lined with the sixteen individual examination and treatment rooms, therefore reducing the amount of movement for the injured patients. The equipment of this central waiting area consists of wooden seating niches, and a sitting element, designed as a room sculpture.
As an optical element, the niches are equipped with large photos by the Frankfurt photographer Eberhard Hoch, showing competitive athletes in action – to illustrate the treating physicians' specialisation in sports accidents and joint disorders. Indirectly-lit ceiling panels locate the private examination rooms.
The custom-developed signage system, mediates the design branding of the Sporthopaedicum and the sleek interior design. Circular ceiling-mounted luminaires guarantee bright lighting throughout the entire practice.

MEUSER ARCHITEKTEN

SPORTHOPAEDICUM
BERLIN

Medical Practices

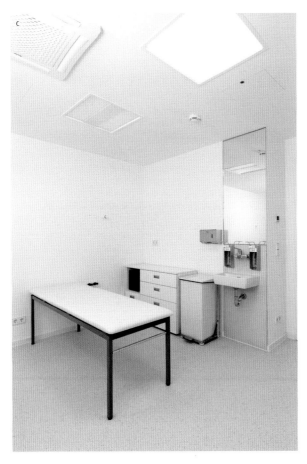

**Client	Operator**	Sporthopaedicum Berlin	
Planning time	08 2012–12 2012	
Construction time	01 2013–08 2013	
Gross floor space	1,170 sqm	
Usable floor space	698 sqm	
Gross volume	3,276 cbm	

a   The central piece of furniture is made of white finished oak and looks like a room sculpture.

b   The wall niches are designed with pictures by the photographer Eberhard Hoch.

c-d   The clarity of the architectural posture is also used in the examination rooms.

e   The reception greets the patient with a long counter. The oak bench next to it can accommodate patients with reduced mobility.

f   Upon entering, the visitor directly comes to a wall-high photograph of a triathlon swim; the location of the reception is suggested by the swimmers' direction of movement.

g   The treatment rooms are white; the chairs set the colour scheme.

Diagrammatic plans, scale 1:500
Floor plans, scale 1:200

Floor plan layout

This L-shaped building is used for a multiple practice with a compact layout. The longer wing presents itself as a corridor lined with examination rooms and a longitudinal waiting area in the middle, therefore shortening the distance between waiting and examination rooms. With the exception of the operation room, all the rooms are lit naturally.

Usable floor spaces

Patient rooms	**yellow**	212 sqm	30 %
Specialist rooms	**pink**	370 sqm	53 %
Administrative rooms	**green**	59 sqm	8 %
Offices	Staff rooms	**orange**	57 sqm	8 %
Plant rooms	**blue**	5 sqm	1 %
Total		703 sqm	100 %

Performance data

Ambulatory patients per year	11,632 [2014]
Inpatients per year	ca. 1,500 [2014]
Type of service per year	Ambulatory and operative treatment, sports traumatology, arthroscopy, special joint surgery, joint replacement.
Facility opening hours	5 days	50 hours
Waiting time	Depending on appointment or urgency
Number and type of staff	5–6 doctors, ca. 35 employees
Notes on planning the facility	Each doctor can move between four treatment rooms without having to use the publicly accessible routes.

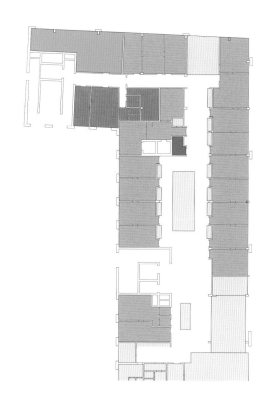

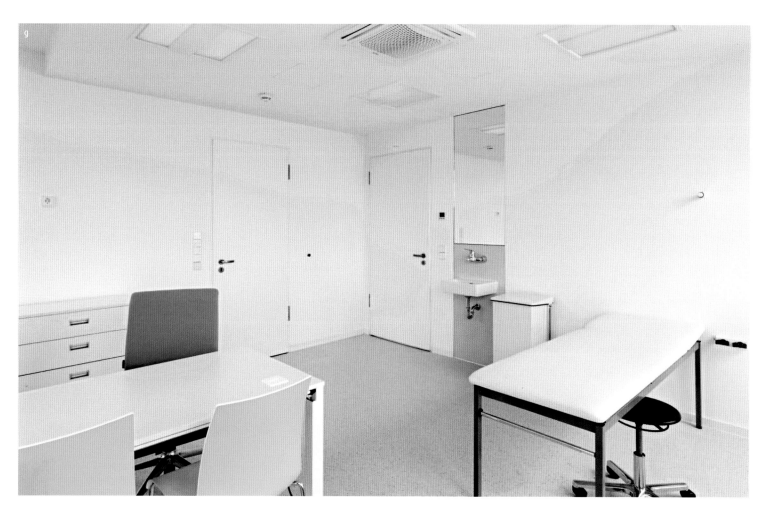

Floor plan layout

1 Reception
2 Waiting area
3 Patient toilet L
4 Patient toilet G
5 Patient toilet for disabled
6 X-ray room
7 Examination and treatment room
8 Procedure room
9 Procedure preparation room
10 Patient changing cubicle
11 Server room
12 Staff changing cubicle
13 Staff toilet L
14 Staff toilet G
15 Staff lounge with kitchenette
16 Doctor's office
17 Administration + office
18 Coat-rack

307

The conceptual idea for the newly-designed rooms at the orthopaedic practice is based on the essence of orthopaedics. The area of interaction, inherent in the treatment of the human musculoskeletal system, between movement and posture, modern and traditional, is conveyed by the architecture of the rooms. The curved reception desk with the portrayal of an athletic body is shaped like a spinal column. A colour scheme underlies the design of the practice, which concentrates on three areas. The largest area with corridor, reception, waiting and therapy area is orange. This colour has a warming, bright, communicative effect; it can also be found on letterheads and visiting cards. In order to make it easier for patients to find their way around the practice, the intensity of the orange increases from the reception up to the centre of the practice and decreases again in the same way from the centre out. The four treatment cubicles are also in orange shades which vary in intensity, so that the cubicle can be selected according to the patient's treatment duration. In the two consulting rooms, however, a predominant pastel-green combined with a cherry blossom motif creates a rather more cheerful, carefree and calming atmosphere. An airy, cool, light blue tone prevails in the sanitary rooms. One element connecting the three areas is a recurrent dark brown shade, both in the furniture and the wall elements and in the parquet flooring.

MATEJA MIKULANDRA-MACKAT **ORTHOPAEDIC CLINIC IN ADLERSHOF HEALTH CENTRE** BERLIN

Medical Practices

Medical Practices

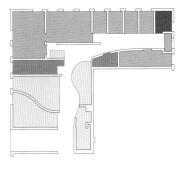

Client	Dr. Matthias Finkelstein
Operator	Dr. Matthias Finkelstein
	Dr. Carl Neisser
Planning time	11 2003 – 05 2004
Construction time	06 2004 – 01 2005
Gross floor area	179 sqm
Usable floor space	122 sqm
Gross cubic capacity	492 cbm
Construction cost	190,000 EUROS

a In order to make it easier for patients to find their way around, the intensity of the orange increases from the reception towards the centre of the practice and decreases again in the same way from the centre out.

b Private patient waiting area: vibrant warm orange combined with a brown colour, recurrent in furniture and wall elements.

c The curved reception desk is shaped like a spinal column.

Diagrammatic plans, scale 1:400
Floor plans, scale 1:200

The angular compact floor plan of this partnership practice with a similarly-shaped corridor means that the rooms are accessible without having to walk far. From an entrance situated in the centre of one branch of the building, patients directly access the reception with the two separate waiting areas for national health and private patients.

Usable floor spaces

Patient rooms	**yellow**	44 sqm	36 %
Examination | Specialist rooms	**pink**	60 sqm	49 %
Administrative rooms	**green**	2 sqm	2 %
Offices | Staff rooms	**orange**	12 sqm	10 %
Supply | Waste disposal	**brown**	4 sqm	3 %
Total		122 sqm	100 %

Floor plan layout

1 Panel patient' waiting area
2 Private patients' waiting area
3 Coat-rack
4 Staff lounge
5 Examination and treatment room
6 Lounge
7 Treatment room
8 Store room
9 Office
10 Multipurpose room
11 Staff toilet and shower
12 Reception
13 Patient toilet
14 Examination X-ray
15 Laboratory
16 Staff lounge and small kitchen
17 Office

When upgrading and converting the former rural medical practice to a group practice for four doctors, the defining themes were views of the interior of the practice and of the outside. The grey, exposed concrete cube lined with window strips and apertures has been built around the existing structure and an extension from the 1970s and, with its defiant architecture, deliberately contrasts with the rural type of construction. The architect designates his construction positively as "stone" or "rock", which represents the "traction" and "heaviness" of the old-established practice. The large reception and waiting area is used jointly for the entrance, the office and the toilet facilities. Ample window strips afford a fantastic panoramic view over the wooded Tyrolean mountains, thus integrating the modern building with its idyllic surroundings. The clear interior design concept forms a delightful contrast to traditional customary local construction. Green benches, which are combined to form a seating unit on a low platform, are more reminiscent of a waiting area than a rural medical practice. The basement in the three-storey extension houses the dental treatment rooms and the ground floor accommodates the premises for general medicine, internal medicine, X-ray and laboratory. On the upper floor there is even space for a small flat with a roof terrace.

GERHARD MITTERBERGER

MATREI HEALTH CENTRE
MATREI

Medical Practices

**Client	Operator**	Dr. Gerhard Gamper	
	Dr. Cornelia Gamper	
	Dr. Isabella Troyer	
	Dr. Johann Trojer	
	Nikolaus Trojer	
Planning time	2003–2004	
Construction time	2004–2005	
Usable floor space	735 sqm	
Gross cubic capacity	1,587 cbm	
Construction cost	917,500 EUROS	
Total cost	1,000,000 EUROS	

a Rough materiality of the exposed concrete cube is also present everywhere.

b Superior transparency: this "glass" examination room is additionally illuminated by an ample skylight.

c The red linoleum, laid almost everywhere, makes the practice rooms seem cosy.

d Generous window strips in the joint waiting and reception area offer a panoramic view.

e The ENT examination room is simple and unpretentious.

f Defining design principles, when converting this group practice, were views within the practice and outside.

Diagrammatic plans, scale 1:400
Floor plans, scale 1:200

Floor plan layout

This multiple practice is spread over two levels; most examination and treatment rooms are on the ground floor due to the requirement for natural light. In the basement, the rooms are easy to reach from the waiting area which is situated at the stairwell. On the ground floor, with spacious reception and waiting area, patients are called directly by the doctor.

Usable floor spaces

Patient rooms	**yellow**	129 sqm	17 %
Examination			
Treatment rooms	**red**	184 sqm	25 %
Specialist rooms	**pink**	162 sqm	22 %
Administrative rooms	**green**	11 sqm	1 %
Offices	Staff rooms	**orange**	34 sqm	5 %
Supply	Waste disposal	**brown**	189 sqm	26 %
Plant rooms	**blue**	26 sqm	4 %
Total		735 sqm	100 %

Performance data

Outpatients per year	12,000
Clinic opening hours	5 days	40 hours
Waiting time [with	without appointment]	10 min.	up to 60 min.
Type and number of staff	5 doctors
	10 clerical staff

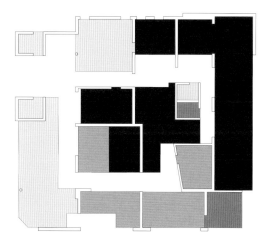
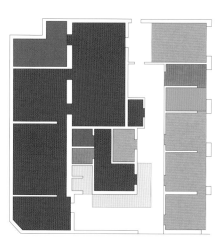

Medical Practices

Floor plan layout
1. Patient toilet L
2. Reception
3. Examination
4. Patient toilet
5. Staff shower
6. Examination X-ray
7. Staff lounge
8. Treatment room infusion
9. Laboratory
10. Waiting area
11. Patient toilet G
12. Store room
13. Installations room
14. Prosthetic dentistry
15. Small kitchen
16. Office
17. Dental treatment room
18. Sterilisation room
19. Staff toilet
20. Toilet anteroom
21. Examination sonography
22. Examination ENT

"Travel healthily" – the newly-established "medical + dental suite" at Cologne Bonn Airport has succeeded in promoting itself with this slogan. In the centre of the terminal, at the transition to the underground long-distance railway station, a joint general medicine and dentistry practice has been created. The centrepiece of the interior design is a 25-metres long wall, extending across the practice, whose curved shape is reminiscent of the wing of an aircraft. Many other details also evoke associations with aviation: the apertures in the waiting area, round-shaped like aircraft doors, the spotlights on the walls, resembling the marker lights of a runway, or the nose-shaped reception desk, which is clearly reminiscent of a cockpit on the floor plan. Except for the waiting area, which is laid with dark parquet flooring of Wenge wood, a brownish-coloured concrete floor has been chosen for the entire practice. One fine detail is a 25-centimeters wide stainless steel strip, which clads the aperture to the waiting area, matches the greyish paint of the curved wall. The absolute highlight of this exceptional, extremely dynamic practice, however, is the breathtaking view of what's going on at the airport. Behind Terminal D, which you view from the waiting and treatment area, you can watch aircrafts taking off and landing and forget that you are actually at a medical facility.

PD RAUMPLAN

MEDICAL + DENTAL SUITE AT COLOGNE BONN AIRPORT
COLOGNE

Medical Practices

a

b

Medical Practices

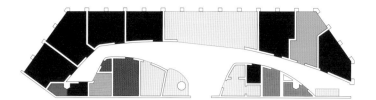

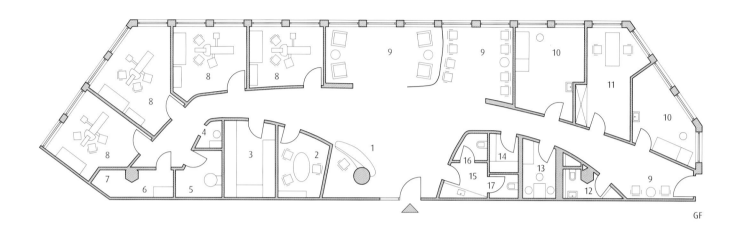

Client	Jochem Heibach
Operator	Jochem Heibach
	Dr. Jens Knitter
Planning time	11 2006 – 03 2007
Construction time	04 2007 – 07 2007
Gross floor area	240 sqm
Usable floor space	202 sqm
Construction cost	300,000 EUROS

a   Small details evoke the world of aviation: the apertures in the waiting area, round-shaped like aircraft doors, the spotlights on the walls, which look like marker lights on a runway ...

b   ... or the nose-shaped reception desk, which is reminiscent of a cockpit on the floor plan.

Diagrammatic plans, scale 1:400
Floor plans, scale 1:200

Performance data

Outpatients per year	18,000
Services to other clinics	Dental laboratory [2,500]
Clinic opening hours	7 days	70 hours
Waiting time [with	without appointment]	5–15 min.	15–45 min.
Type and number of staff	4 doctors
	4 full-time staff
	1 trainee

In the centre of the long polygonal floor plan is the entrance to this multiple practice, and from there the visible reception and waiting areas can be accessed directly. The suite of rooms opposite the entrance is fully equipped with natural light. The rooms at the side of the entrance, including the staff lounge, have to make do without natural light.

Usable floor spaces

Patient rooms	**yellow**	63 sqm	31 %
Examination	Treatment rooms	**red**	94 sqm	47 %
Specialist rooms	**pink**	5 sqm	3 %
Administrative rooms	**green**	16 sqm	8 %
Offices	Staff rooms	**orange**	9 sqm	4 %
Supply	Waste disposal	**brown**	13 sqm	6 %
Plant rooms	**blue**	2 sqm	1 %
Total		202 sqm	100 %

Floor plan layout

1 Reception
2 Consultation room
3 Sterilisation room
4 Cleaning room and store room
5 Examination X-ray
6 Laboratory
7 Installations room
8 Treatment room
9 Waiting area
10 Examination
11 Office
12 Staff toilet
13 Staff lounge
14 Laboratory
15 Toilet anteroom
16 Patient toilet L
17 Patient toilet G

This small dental practice, not far from the Arc de Triomphe in the centre of Paris, is characterised by aluminium panelling and an extensive use of glass. Before the conversion, the two-storey shop was occupied by a lighting manufacturer's showroom. In a relatively small space a group practice with four treatment rooms has been created, in total harmony with the magnificent façade of the art nouveau building. In order to make best use of the natural light, which only comes in from the street frontage, the partition wall between the two adjacent treatment rooms and the wall between the treatment room on the left of the entrance and the corridor was built in frosted glass. The other wall surfaces are covered in aluminium panels. Only the wall to the staircase is furnished with Wenge-veneered panels to add some warmth to the dominant aluminium look. In the treatment rooms, the contemporary atmosphere in the foyer is toned down and warmth is added with panelling and maple-veneer fittings. After the extremely straight-line design of the entrance, softer lines and curves are introduced to lead patients into the rear part of the practice. With its trapezoidal shape, the backlit reception desk of aluminium laminate combines straight and round lines and is modestly integrated into the foyer. A water wall is mounted behind the desk: It serves as an object of art and at the same time contributes to create a pleasant atmosphere.

PD RAUMPLAN

CENTRE D'ENDODONTIE
PARIS

Medical Practices

Medical Practices

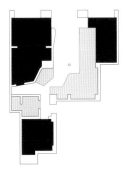

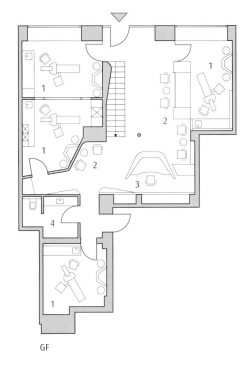
GF
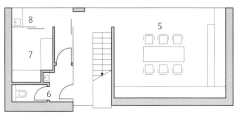
BSMT

Client	Operator	Dr. Jacob Amor
	Dr. David Bensoussan
	Dr. Herve Uzan
Planning time	05 2005 – 08 2005
Construction time	09 2005 – 12 2005
Gross floor area	160 sqm
Usable floor space	121 sqm
Construction cost	290,000 EUROS

a The waiting area is furnished with Wenge-veneered panels, which add a touch of warmth to the cold aluminium look.

b The straight-line design is later replaced by curves, which escort patients into the rear section of the practice with a softer line.

c The patient toilet is elaborately designed.

d This trapezoid-shaped reception desk of aluminium laminate combines straight and curved lines and is modestly integrated into the foyer.

Diagrammatic plans, scale 1:400
Floor plans, scale 1:200

Via the multiple practice entrance in the centre of the street frontage, the patients go straight past two treatment rooms and waiting area to the reception, whence two short connecting corridors lead to the remaining two treatment rooms. Rooms not found on the ground floor are accessed via a central, single flight of stairs. The large room there designated as an office may also be used as staff lounge.

Usable floor spaces

Patient rooms	**yellow**	29 sqm	24 %
Examination | Treatment rooms	**red**	51 sqm	42 %
Administrative rooms	**green**	29 sqm	24 %
Offices | Staff rooms	**orange**	6 sqm	5 %
Supply | Waste disposal	**brown**	6 sqm	5 %
Total		121 sqm	100 %

Floor plan layout

1 Treatment room
2 Waiting area
3 Reception
4 Patient toilet
5 Office
6 Staff toilet
7 Sterilisation room
8 Small kitchen

The new outpatient department at the Frankenthal Municipal Clinic, a building dating from the 1970s, combines emergency treatment and elective care. Situated close to one another, all the individual outpatient departments are easily accessible. All consultations check-ups take place in this part of the clinic. Emergency admissions are brought straight into the emergency room, where they receive the necessary care, via a new approach for non-ambulant patients. A new bed-passenger lift goes directly up to the intensive care unit and to the operating theatres. Those patients able to walk are processed at a wooden desk and transferred for medical admission to examination rooms, which are designed to be used by physicians from all disciplines. Special examination rooms, functional diagnostics and radiology are nearby. Thus, each patient can be thoroughly examined without the need to cover long distances. The examination will determine whether the patient needs to receive further treatment as an inpatient, a part-inpatient or even as an outpatient. Patients whose diagnosis has not been fully determined are kept in the newly developed ten-bed admissions unit until final clarification. This unit also functions as a day clinic for chemotherapy patients and as a rest facility for those who have had operations. The logical interior design concept helps the rooms exude freshness and tranquillity. Thanks to the elaborate lighting system, even inner areas are bright.

SANDER.HOFRICHTER

MUNICIPAL CLINIC
OUTPATIENT DEPARTMENT
FRANKENTHAL

Medical Practices

Client	Frankenthal
Operator	Municipal Clinic Frankenthal
Planning time	05 2005–05 2006
Construction time	05 2006–04 2009
Gross floor area	4,750 sqm
Usable floor space	2,330 sqm
Gross cubic capacity	18,335 cbm
Construction cost	2.1 million EUROS
Total cost	2.9 million EUROS

a The use of a continuous colour and material concept creates a pleasant atmosphere in the bright, light rooms.
b The sedate reception area for elective patients is situated right next to the bright yellow emergency room.
c Behind the reception desk for elective patients are rooms that can be used interdisciplinarily for medical | administrative admission.
d The patients' attention is drawn to the signalling colour at the emergency patient admission at the entrance of the new right of way for patients unable to walk.

Diagrammatic plans, scale 1:450
Floor plans, scale 1:450

Floor plan layout

The spatial concentration of numerous hospital outpatient departments offers much potential for thorough examination and for treatment. In addition, with the convenient layout on one level, further extension of the floor plan is inevitably associated with very long distances to walk and in particular with many rooms that do not have natural light. The main corridors, accessible from both sides of the central hub area, and the secondary corridor running in parallel with the longer one, from which connecting corridors exit, prove beneficial to the orientation system.

Usable floor spaces			
Patient rooms	**yellow**	605 sqm	26 %
Examination	Treatment rooms	**red**	370 sqm	16 %
Specialist rooms	**pink**	563 sqm	24 %
Administrative rooms	**green**	170 sqm	7 %
Offices	Staff rooms	**orange**	312 sqm	13 %
Supply	Waste disposal	**brown**	157 sqm	7 %
Training rooms	**violet**	38 sqm	2 %
Plant rooms	**blue**	115 sqm	5 %
Total		2.330 sqm	100 %

Performance data

Outpatients per year	13,000 [entire hospital]
Inpatients per year	8,200
Clinic opening hours	7 days	168 hours
Waiting time [with	without appointment]	0–100 min.	0–160 min.
Type and number of staff	2 doctors, 2 nursing staff, 1 trainee, 2–3 administrative staff [daytime]
Clinic planning advice	Too far from the care support point, to the procedure rooms or to the plaster room.

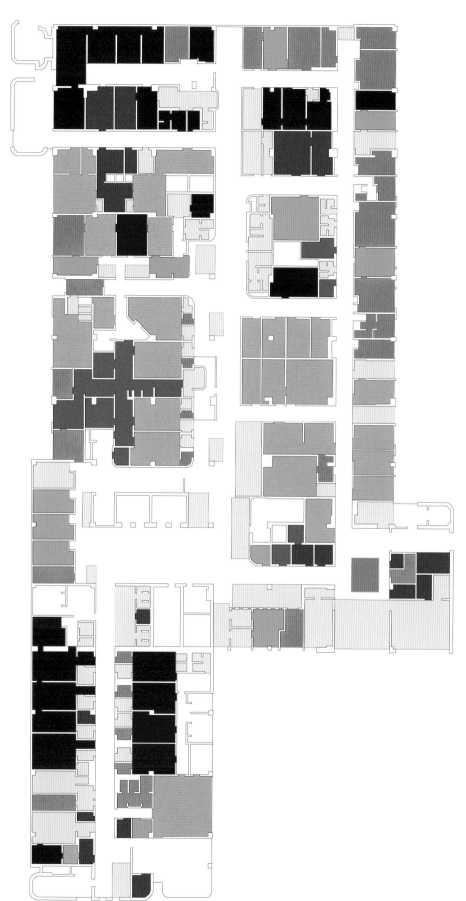

1	Waiting area	54	Library
2	Patient toilet G	55	Examination room psychiatry
3	Store room	56	Blood analysis laboratory
4	Patient toilet L	57	Washroom
5	Staff changing cubicle	58	Urine laboratory
6	Patient changing cubicle	59	Laboratory
7	Treatment room	60	Staff lounge
8	Patient shower	61	Laboratory monitoring
9	Patient toilet	62	Laboratory reception
10	Training room	63	Specimen collection room
11	Staff toilet	64	Gynaecology administrative office
12	Staff shower	65	Doctor's office [gynaecology]
13	Washroom	66	Examination
14	Equipment room	67	Administrative office
15	Office	68	Admissions consulting room
16	Fitness room	69	On duty office
17	Store room	70	Doctor's office [surgery]
18	Inhalation room	71	Doctor's administrative office
19	Consultation room	72	Patient toilet and shower
20	Clean linen store room	73	IT room
21	Anteroom	74	AV Elektro
22	Office	75	Consultation room
23	Reception	76	Staff toilet and shower
24	Dirty linen store room	77	Emergency doctor on duty office
25	Examination and treatment room	78	Functional diagnostics reception
26	Admissions manager's office	79	Examination ECG
27	Clerical services office	80	Examination pulmonary function
28	Coding office	81	Procedure room
29	Relatives' lounge	82	Plaster room
30	Assistant dietician's office	83	Examination admissions
31	Film processing	84	Treatment room surgery
32	Examination HRT	85	Examination ultrasound
33	Patient transfer room	86	Examination ergometry
34	Examination mammography	87	Examination EEG
35	Installations room	88	Patient room
36	Examination X-ray	89	Washroom
37	Admissions room	90	Patient sluice
38	Examination CT	91	Treatment room shock therapy
39	Computer room	92	Examination gynaecology
40	Switch room	93	Waste disposal room
41	Darkroom	94	Waste disposal room
42	Doctor's office	95	Supply room
43	Psychiatry office	96	Filing room
44	Doctor's office	97	Emergency room
45	Examination endoscopy	98	Clean store room
46	Prep room endoscopy	99	Small kitchen
47	Quiet room	100	Bed store area
48	Sterile goods store room	101	Workroom unsterile
49	Equipment preparation	102	Admissions room
50	Samples laboratory	103	Admissions room for short-term inpatients
51	Outpatient lounge	104	Admissions room for emergencies
52	Administrative office	105	Reception control centre
53	Examination psychiatric outpatient	106	Accounts office

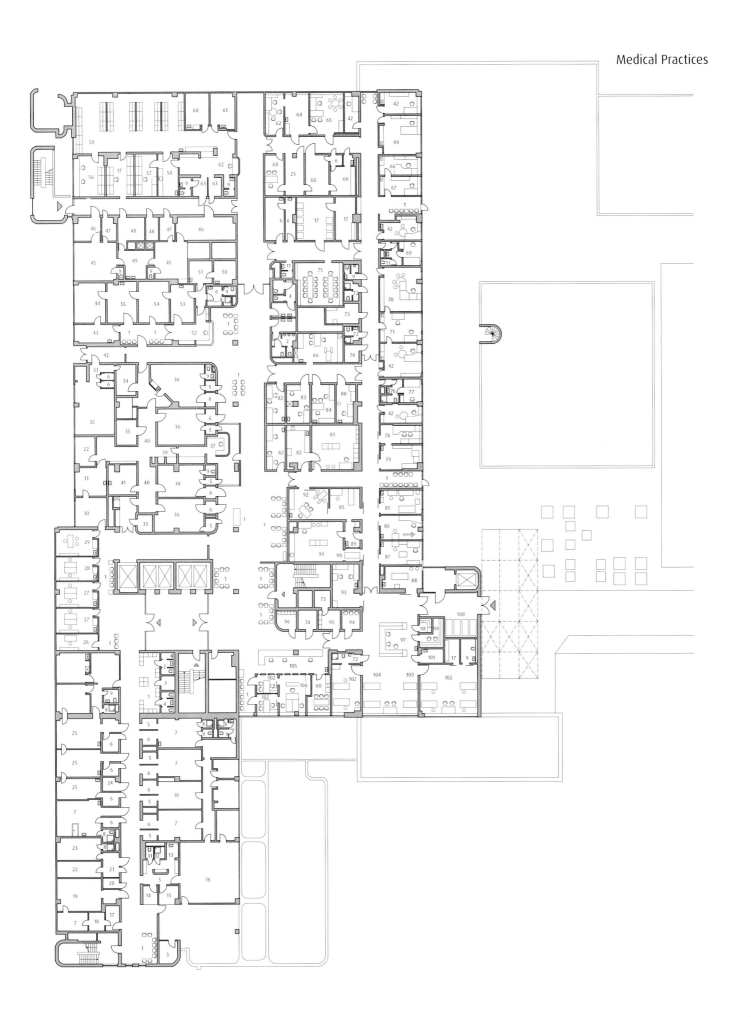

Medical Practices

A room on the ground floor, which was previously used as a hair salon, has been refashioned to extend the dental practice, based in a house in Frankfurt's West End dating from the turn of the 19th century. The required adjoining spaces, such as changing and store room, are located in the basement. The contained current premises have benefited from a smooth, flattering and calming design concept. The three main design elements – an adjusted box intended for consultation purposes, a non-structural wall and a suspended ceiling – playfully break away from the axial alignment and the right-angled geometry of the existing walls and create a new perception of space. The bright, lime-yellow flooring stands out in the simple white rooms, adding to the general elated feeling created by the high-gloss surfaces. Light and shadows of the white silhouettes reflect on the floor, creating interesting spatial impressions. In the evenings, this vibrant room looks like a stage set in special lighting, the leading actor being the modern treatment chair, which is actually a small work of art in itself.

STENGELE + CIE.

DENTAL PRACTICE
FRANKFURT AM MAIN

Medical Practices

a b

c

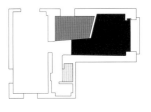

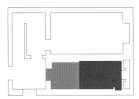

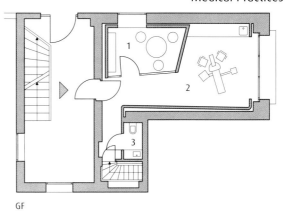

GF

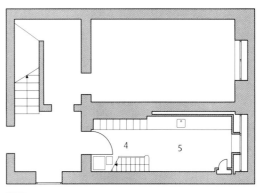

BSMT

Medical Practices

**Client	Operator**	Dr. Dr. Frank Sanner
	Dr. Andreas Sanner
Planning time	04 2004 –12 2004
Construction time	01 2005 –03 2005
Gross floor area	72 sqm
Usable floor space	55 sqm
Gross cubic capacity	259 cbm

a An adjusted box, a non-structural wall and a suspended ceiling playfully break away from the alignments and right-angled geometry of the existing walls creating a new perception of space.

b The treatment room is furnished with functional facilities.

c The white silhouette of the consultation box is reflected in the high-gloss surface of the bright yellow flooring, which leads to interesting spatial impressions.

Diagrammatic plans, scale 1:400
Floor plans, scale 1:200

From the staircase on the ground floor, patients access the extension of a partnership practice, which has its main rooms on the first floor. The extension consists of a larger treatment room with a positioned consultation box and a toilet. A single-flight staircase leads to another room in the basement, which houses the staff changing cubicle and a store room.

Usable floor spaces

Patient rooms	**yellow**	2 sqm	4 %
Examination				
Treatment rooms	**red**	21 sqm	38 %
Offices	Staff rooms	**orange**	19 sqm	34 %
Supply	Waste disposal	**brown**	13 sqm	24 %
Total		55 sqm	100 %

Floor plan layout

1 Consultation room
2 Treatment room
3 Patient toilet
4 Staff changing cubicle
5 Store room

A blue glass wall greets visitors entering the spacious Lübeck multiple practice. It typifies the central aspect of the spatial floor plan design, which, at the request of the operator is based on the principles of Feng Shui. Opposite the "water element" in the form of the glass wall is the "earth element" in the shape of the desk and the back office in maple. The reception desk and the blue glass wall are arranged diagonally to form an inviting trapezoid. The guide wall, also clad with maple and horizontally-extending aluminium profiles, connects the reception desk to the surgery and treatment wing, whose units are accessible via a central distribution corridor. Light is a defining element in the design; the cove lighting running parallel to the corridor aids the guiding effect of the maple wall. Using illuminated ceiling panels and selectively-placed built-in lights, individual areas are accentuated. Ceiling-high glass elements, partly frosted, make the transitions appear as if they are floating and enable light to penetrate deep into the rooms. Clear shapes, a few choice quality materials and clear-cut accessibility all have a calming effect on the patients and convey – in accordance with the ambitions of the practice – feelings of high standards and quality.

WAGENKNECHT ARCHITEKTEN

**OMF SURGERY
LINDENARCADEN**
LÜBECK

Medical Practices

Medical Practices

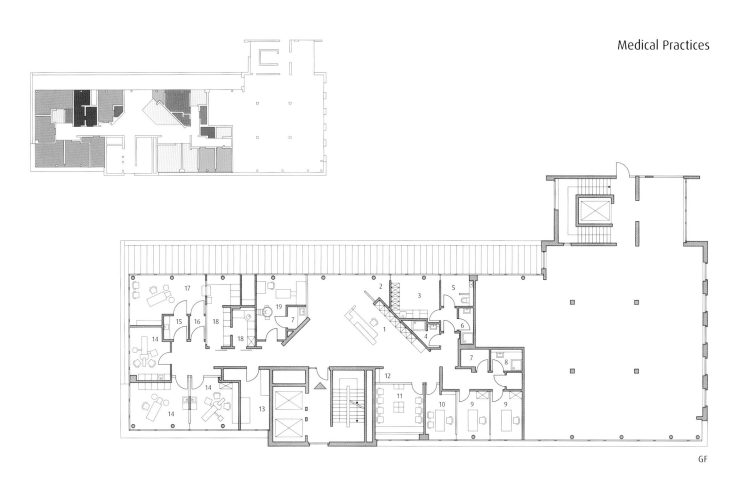

GF

Client	Dr. Dr. Hans-Peter Ulrich
Operator	Dr. Dr. Hans-Peter Ulrich
	Dr. Wilma Poeschel
	Dr. Dr. Stephan Otten
Planning time	2003
Construction time	2004
Gross floor area	378 sqm
Usable floor space	179 sqm
Gross cubic capacity	1,137 cbm
Construction cost	350,000 EUROS
Total cost	490,000 EUROS

a Illuminated ceiling panels and selectively-placed built-in lights accentuate individual areas in this practice.
b Reception desk and blue glass wall opposite are arranged diagonally to one another to form an inviting trapezoid.
c The guide wall clad with maple and horizontally-extending aluminium profiles connects the reception desk to the treatment wing.

Diagrammatic plans, scale 1:500
Floor plans, scale 1:250

Performance data

Outpatients per year	4,000
Clinic opening hours	7 days	70 hours
Waiting time	according to urgency
Type and number of staff	3 doctors
	7 clerical staff

From the centre of this long rectangular floor plan, patients of the multiple practice access the reception and the visible waiting area situated to the right of the entrance. From there they have to find the examination and treatment rooms to the left of the entrance and the operating unit via a narrow central corridor. It is encouraging that all the important rooms receive natural light.

Usable floor spaces

Patient rooms	**yellow**	43 sqm	24 %
Specialist rooms	**pink**	76 sqm	42 %
Administrative rooms	**green**	19 sqm	11 %
Offices	Staff rooms	**orange**	19 sqm	11 %
Supply	Waste disposal	**brown**	17 sqm	9 %
Plant rooms	**blue**	5 sqm	3 %
Total		179 sqm	100 %

Floor plan layout
1 Reception
2 Office
3 Staff lounge
4 Staff shower L
5 Toilet for the disabled
6 Staff shower G
7 Installations room EDV
8 Patient shower
9 Office
10 Consultation room
11 Waiting area
12 Coat-rack
13 Recovery room
14 Treatment room
15 Patient sluice
16 Staff sluice
17 Operating theatre
18 Sterilisation
19 Examination X-ray

Pharmacies

Philipp Meuser
Pharmacies, Between Tradition and the Modern Market

Klaus Bergdolt
From Herb Garden to Mail-Order Pharmacy

Klaus R. Bürger
Corporate Identity and Design

Franz Labryga
Planning Pharmacies: the German Principle

Philipp Meuser

Pharmacies, Between Tradition and the Modern Market

The colours flow fluidly from a soft blue into orange, and then into red and green. As many glances as the shop window attracts during the day, in the dark it is an absolute eye-catcher, even at a distance. One could almost believe that it is a light installation produced by art students trying to draw attention to grievances at their university.

Not all that long ago, the local pharmacy was still solidly fitted out with wooden furniture. The pharmacy still had the appearance of a workshop for health, where, in keeping with centuries-old tradition, the pharmacist was a craftsman who continued to produce some medicines himself. This practical experience in handling the materia medica produced an in-house manufactured insecticide, which was excellent in every respect. So everything was as it should be: The pharmacy employees were competent and friendly; the atmosphere was simple and pleasant – just like being in the corner shop; trust in the pharmacist's expertise existed on the basis of precisely such alchemical brews as the insecticide.

For me, at least, the "apothecary's shop" had proved itself to be more than just a storehouse (Greek: apo = away and tithenai = lay) for a range of goods and a hatch for passing through prescribed drugs. Suddenly, however, the old, familiar pharmacy was closed down. Subsequently this new pharmacy, designed in white and enhanced by warm light colours, was opened, as if out of thin air. There was no longer a clear physical separation between the pharmacist, as an expert adviser, and the customer on the other side – only a mental division. After all, the pharmacist knows that his position of privilege has been accorded to him step by step under law since the 12th/13th century, and that his practical, craftsman-like profession has been recognised as an academic discipline since the 19th century. According to representative surveys undertaken by various opinion research institutes, alongside doctors, pharmacists enjoy an undiminished level of trust amongst the German public, with a rate of 84 to 93 per cent. "Ask your doctor or pharmacist" is not only a slogan to warn against the side effects of certain medicines, but is also a formula aimed at arousing trust based on specialist expertise. The ladies and gentlemen in white coats as an outpost of the medical practice, who serve as its extended arm, are well aware of the respectful distance which the patients as customers accord them by virtue of their office. This means that the line separating the customer from the pharmacy's specialist, private area, once represented by the sales counter, can now be relaxed in optical and constructional terms. Only the window at the night service counter remains as it was.

Under law, pharmacies are obliged to ensure that in the public interest, the population is adequately supplied with medicines. In order to practice their profession, pharmacists are required to obtain a licence. The ever-present trust in the art of the pharmacist has been legally guaranteed under law since the 13th century by controls imposed by the authorities. In this way, the pharmacist was protected from the competition of hawkers, miracle healers, and other charlatans. Has any pharmacist ever gone bust under the described system?

One of the last guild privileges to be preserved in today's modern market and consumer society also entails monitoring. Names such as "Ratsapotheke", "Hofapotheke", "Klosterapotheke", and "Spitalapotheke" (city, court, monastery, and hospital pharmacy) demonstrate the prestige of pharmacies for the cities and territorial authorities, be they spiritual or secular. Frequently, outstanding architects were commissioned with the construction and interior decoration of these pharmacies. That many of them possessed luxurious furnishings was simply a matter of good taste. It is therefore nothing new that pharmacies were lent an identity, or a confirmed reputation, through location, furnishings, or quality of craftsmanship. The court doctor and apothecary formed an essential double act which was responsible for the well-being and health of His Grace and his family – precisely because of the already proven professional

expertise. One could not train to become a court or city apothecary, one achieved this status by appointment. The pharmacy as a location therefore only reflected the respect which the pharmacist's skills had earned.

The pharmacy as designer shop

The difference to today lies in the egalitarianism of pharmacies. The aristocracy has fallen, the royal households have been abolished, and city councils no longer have their own pharmacist with corresponding premises. Pharmacists and their business premises are part of a great whole beneath the pharmacy sign reserved for guild members, and as such they are already privileged. How, therefore, can an individual member of a free profession amidst a network of 21,500 pharmacists make himself stand out and – above all – in comparison with whom? As made clear by the Federal Union of German Associations of Pharmacists, competition and danger threaten not from amongst their own ranks, but rather from outside – and in several respects, as well.

In a sophisticated culture such as ours where, in the public perception, medicine has practically banished illness – thus also its worst case scenario: death – and it is now a question of remaining healthy, that means young and full of vitality, for as long as possible, the pharmacist's role has changed. Nobody goes to the pharmacy anymore because of questions of life and death – or at least if they do, they don't let it show. What people go to the pharmacy for is to register for a yoga course, or holistic healing, or to buy a dietary supplement. The trend which is changing and will continue to change pharmacies is known as "wellness". This is the word, written in large, clear letters that one sees first in the window of the local pharmacy on the corner. And it is this sense of well-being that determines the fit-out. The trend is towards a health system which spends a fortune on the luxuries of "forever young" and "living forever" and has allowed these to penetrate the system at every level. Even the statutory health insurance funds have erased illness from their self-image, and have exchanged it for the fresh scent of flowers and the happy faces of naturally healthy people. It's a question of health. "Health fund" or "the enterprise of life" are the new advertising slogans. So the change lies in the system.

The other change is the market which results out of this. Thanks to the trend described above, the pharmacies' product ranges are competing on this level with health food stores and drugstores. This is the result and is down to the fact that the classical pharmacies are suddenly competing with health food stores and drugstores, as well as completely new types of pharmacy such as branch outlets of certain companies like DocMorris, the international pharmaceutical mail-order company, or online sales with their incredible aggressive pricing policies. However, it is the citizen himself who decides. Green awareness has not only given many people to understand – at least theoretically – that man is part of a greater natural diversity, but has also led to a new physical awareness. In contrast to orthodox medicine, the so-called alternative healing methods and practices which treat physical ailments based on the idea of the person as a body-mind-spirit being have discredited the pharmacist as a public supplier of pharmaceuticals, which are prescribed by doctors and produced in the laboratories of chemical companies.

The orthodox medical troika of doctor, pharmaceutical industry, and pharmacist as the pharma-vendor no longer functions. The cause lies with the citizen whose awareness has been heightened, thus offering him a choice with regard to the concept he wants to use to become or remain healthy. Surveys notwithstanding, the orthodox school of medicine is tarnished and the pharmacist who is right in the middle has a problem. The visible and undeniable strength of nature, its healing plants, and the substances extracted from them are just as much in demand as is knowledge about the body's powers of

Pharmacies

Leibniz-Pharmacy, Berlin
Illuminated display window of a pharmacy

self-healing. In the midst of all this, the pharmacist's expertise as an advisor and person of trust, traditionally developed through practical experience and sanctified with academic titles, is asserting itself more than ever before. There is also the cash crisis within the health system. Medicines from the pharmaceutical companies are becoming more expensive and the health funds are nearing financial collapse. The orthodox exploitation chain of pharmaceutical industry, doctor, pharmacist, and customer is also financially creaky, compared with the affordable competition from mail-order and online purchasing which is aggressive and highly successful. This is why the pharmacy has diversified from being a legally privileged dispensary of prescription medicines, into a store for natural healing remedies and health consultancy. The binding force between all of these elements is expertise, that means training and certification.

Architecture as a creator of identity

In a market where, on the one hand, borders are blurring and, on the other, its privileges are being removed, just like any other business pharmacies need to demonstrate quality to the outside world and make their own role – apart from certified privileges – visible. The magic formula is "corporate identity", which must be visible both from a distance and on entering the pharmacy with regard to attitude, behaviour, and the communication of specialist skills, all the way through to the internal organisation. It is a question of corporate design. "Who am I, where am I, what am I, what do I have to offer?"

All of these questions concerning the pharmacy should be answered clearly for all the senses by architecture. For these are precisely the questions to which the customer expects an answer, which he subconsciously registers and which help him to decide whether to go into the pharmacy or not. The range of designed identity extends from the business's own logo and all communication material, to the design of the business and the corresponding internet platform. The architect has to bear all of these elements in mind because, as with any other provider of branded goods, this is an issue which involves nothing less than creating instant brand recognition, a question of ensuring and intensifying the customer's identification with "his pharmacy". It is the coherent unity of all these elements which allow a pharmacy to create a good external impression.

Inside, the colours, materials, and structure need to be coordinated. A restrainedly elegant and harmonious use of all three elements is essential here. Designing a pharmacy is, first and foremost, the job of the interior designer. It is evident that it is the offices who are equally at home with the disciplines of design, interior design, and architecture, who predominantly prove best at designing pharmacies. Or the designers may also be architects who have adequately demonstrated skills of the highest order in other areas such as living, gastronomy, offices, commercial premises or in different selected sectors.

The underlying design rule is: "Form follows content and function". In other words, a clear shape vocabulary and superior materials are required, and the two must be combined to create an atmosphere of integrity, as well as a spatial separation between the customer area and the pharmacy's working areas. A pharmacy is not a drugstore, and its design is required to meet the demands of a designer store – for example an optician's shop. This is because, in keeping with their professional claim to provide expert guidance and specialist knowledge as the customer's person of trust, the pharmacy ranks in the premium league of retail trade. Based on their products, pharmacies are superior sales outlets. They have to be – and that's the way their furnishings have to look as well. Design details are oriented to the size, location, and self-image of the pharmacy. Apart from having friendly personnel who are highly skilled at what they do, a good pharmacy primarily distinguishes itself externally through the use of colour. At a time when even delivery rooms and

Leibniz-Pharmacy, Berlin
Illuminated display window of a pharmacy

hospitals or medical practices are decorated in cheerful colours, nobody wants to go into a pharmacy that is chalk-white. As far back as 1791 Johann Wolfgang von Goethe was already examining the effect of colours on people's behaviour and mood. We have the Brazilian theologian Dom Hélder Pessoa Câmara to thank for the following finding, formulated in simple yet moving words: "The light which falls on things transforms them." Therefore, for a professionally designed pharmacy, the choice of colour for floor, ceiling, walls, and furnishings is elemental for the design of the spatial atmosphere, whereby the effect of external light and the reproduction of colours by means of artificial light should also be considered. It is a question of atmosphere. If this is right, and goes hand in hand with competent advice that creates trust, and friendly, skilled service, the pharmacy has won: both the customer and the turnover. If the first is not right, all the expertise in the world is of little or no use. Dark shades for the floor ensure that the customer feels looked after and secure.

A change from dark to light flooring makes it possible to show the border between the public area and that of the personnel. The colours used should be calming and solid, and this should extend to the furniture, as a relaxing atmosphere is also an aid to well-being. If restrained background and material colours are also used, what is truly central will automatically become the focus of attention: the products.

It is important that these materials are correctly chosen. Because the pharmacist as a profession has a long-standing tradition, materials should be chosen which will age gracefully. Nothing is worse than a dispensary which looks like a DIY store. A corporate culture is a thing of lasting value which, like wine or love, matures with age, and should never be allowed to become tarnished. Fashion trends come and go, but a good pharmacy is like a mature individual, who is serene in himself, regardless of trends. Change may, at most, emerge out of the pharmacy's own corporate self-image.

Further reading

Becker, Helmut: Zur Geschihcte der Krankenhausapotheke im Königreich Bayern. Münster 1977

Bedürftig, Friedemann: Geschichte der Apotheke. Von der magischen Heilkunst zur modernen Pharmazie. Cologne 2005

Gutmann, Siegfried: Alte deutsche Apotheken. Ausschnitte aus 700 Jahren deutscher Apothekengeschichte. Ettlingen 1972

Kallinich, Günter: Schöne alte Apotheken. Munich 1975

Leidecker, Kurt | Papsch, Walburga: Die Geschichte der Pharmazie. Eine Chronik. Florstadt 2002

Meuser, Philipp [ed.]: Pharmacies. Construction and Design Manual. Berlin 2008

Klaus Bergdolt

From Herb Garden to Mail-Order Pharmacy

Old pharmacies exercise a huge fascination over many people. In many places they have been lovingly refurbished and their original appearance restored, so that one can sense how people in the Middle Ages and the early Modern era must have felt when entering such places. Hidden away behind artistically numbered drawers, beneath tables, in boxes, tins, mortars, retorts, leather pouches, distilling apparatus and bottles, in secret cupboards or poison cabinets were those substances with which the apothecary [as the pharmacist was once known] mixed his medicines. It was not merely a question of assisting the individual in his everyday, often fruitless attempt to cure serious illnesses – the apothecary also sought cures for plagues which threatened both man and beast, for diseases which brought war in their wake, and for risks to the life and limb of the rulers. For centuries, the apothecary's shop was a "workshop for health", mysterious and with a magical aura. In the view of the simple people, alchemy and magic were tangibly close there; and where else would one expect the manufacture of gold, or the creation of the "homunculus" than somewhere like a apothecary's shop whose technical equipment was more or less predestined for such uses? Following doctor's prescriptions or based on the apothecary's own imagination and expertise, pills, creams, drops, infusions, emetics, comfits, suppositories, embrocations, cosmetics, and herbal mixtures for both internal and external use were expertly prepared. The apothecary's expert knowledge was equal to the doctor's when it came to medication ["materia medica"]; there is documentary evidence that rivalry already existed between the two disciplines in the Middle Ages. The gathering, drying, crushing, distilling, and colouring of medical plants and substances said to have curative properties, provided first practical experience. What they tended to lack was the theory, that means physicians trained with regard to plants and healing herbs using the works of antique and medieval authors. Until into the 19th century, there was no standard training for pharmacists, so that their number included respected natural scientists, as well as harmless hucksters and respectable artisans.

In general, however, access to the apothecary's shop appeared almost hermetic, something that was further reinforced by Latin inscriptions and enigmatic shorthand. That they were often housed beneath ancient vaults or in cellars strengthened this impression still further. The alchemical tradition, that of searching for the "quintessence" of things, the barely disputed existence of miracle pills, the collecting of gemstones, unicorn horn powder, bezoar stone, or the mysterious, human-shaped mandragora – all of these aroused astonishment and, indeed, fear. "Dosis facit venenum" – danger is simultaneously implicit in the old principle of healing: What was useful or did good, what could preserve life could also – if wrongly administered – lead to death. Poisons and their antidotes have frighteningly common properties, and every visitor sensed this. That "Composita" had strength greater than the sum of their parts, and that "Simplicia" possessed an additional "Vis occulta" had already been stressed by Galen and Avicenna. Life and death, salvation and danger rubbed shoulders on the shelves, so that weighing scales and measuring unit took on a special significance. A beautiful apothecary's shop demonstrated nothing less than the cultural standing of the given municipality. Indeed, without the dispensing of medicines, doctors' efforts to preserve health were condemned to failure. Apothecaries, irrespective of the diversity of their training, enjoyed a very special level of trust amongst the people; an important quality which also reaped great rewards. Since time immemorial, apothecaries have been crucial advisors, as were those drug and root traders who took on the role of apothecary where none was available.

In early Modern times, from 1600 to 1800, the various types of apothecary [city council apothecary, court apothecary, monastery apothecary, or infirmary apothecary] were objects of prestige belonging to the municipalities, princes or clerics. Not infrequently, outstanding architects were commissioned with their design and construction. Many possessed noble, almost luxurious furnishings, which fulfilled all possible aesthetic demands. Magnificent examples have been preserved, predominantly from the 18th century – from Portugal to Russia. The court apothecary in particular enjoyed high status for centuries and, just like the personal physician, accompanied the prince or king with his "travelling apothecary", a beautifully crafted portable cabinet, which contained space for small bottles, cinnamon canisters, tins of unguent, and even books. The idea of the apothecary's shop as the central place for the production and sale of medication, separate from a medical practice, had already been born in the high Middle Ages. It is likely that the example was set by the Arabs, although such establishments also existed in Byzantium.

Pharmacological literature which not only legitimised the profession of apothecary, but also elevated its authority, can be traced back to ancient times. Works ascribed to Theophrastus, Pliny, Galen, and Dioscurides already had a legendary, almost axiomatic status. No less influential at the time were Avicenna's "Canon Medicinae", the "Antidotarium Nicolai", and a work known as "Circa instans", ascribed to the Italian physician Matthaeus Platearius. It was above all in the monasteries of the early Middle Ages that healing with plants and pharmaceuticals enjoyed great popularity. The floor plan of St. Gallen monastery [around 800] already foresaw a building for physicians and apothecaries ["armarium pigmentorum"]. Almost every monastery had a herb garden, and this is now once again in fashion. An important element here was "signature reading": It was believed that the therapeutic effectiveness of a plant could be recognised from its external shape and colour. It was above all in the 13th century that early forms of the pharmacy, that means shops for medicines, were established in many European cities. The expanding international trade in drugs and herbs [Byzantium, Venice, Pisa, Nuremberg] had a crucial influence on this development. Whether the professional pharmacist ultimately developed out of the monasteries or from the guild of travelling drug traders ["huckster theory"] is still debated today. It were precisely the exotic, imported healing substances, whose origins possessed an aura of the legendary, which were considered to be highly effective, at latest since the Carolingian period.

From the 14th century, apothecaries were sworn in by urban or state authorities; firstly in Italy and southern Europe and then later, north of the Alps, state concessions for establishing apothecaries were granted. The "Assises" of the Norman king, Roger II [1140], ratified at the assembly of Ariano, contained the first [preserved] European Apothecary's Code. This was followed – firstly for the Italian-ruled area – by Frederick II of Hohenstaufen's "Constitutiones medicales" [1231]. In the West, pharmacy was now considered to be an independent science, separate from medicine, but equally necessary. This did not prevent that, in fact, it was practised more as a skilled trade. Regular inspections of apothecaries, as were prescribed in the Venetian "Capitulare de specialibus" [1258], were soon obligatory, but the prices were only partially fixed by the authorities. Apothecaries were generally considered to be free traders. However, stringent restrictions [oath, fixed prices, punishment for violations] were foreseen in the "Breslau Medical Code" of 1352. In central Europe, too, those wishing to open an apothecary's shop generally required a licence. Until the 19th century, most apothecary's shops were astonishingly small, their core elements formed by folding tables, a fireplace, a few cupboards, and a small storeroom.

Important substances are extracted from healing plants

From left to right:
Sage (Salvia officinalis)
Peppermint (Mentha x piperita)
Stinging nettle (Urtica dioica)

The term "aphothecarius" is only documented in central Europe since the middle of the 14th century. This was then the master of the apothecary's shop or his deputy. Prior to 1300 there is documentary evidence of apothecaries – in the later, "pharmacist" sense of the word – in only a few German cities for example [1190 in Cologne, 1241 in Trier, 1275 in Mainz]. One special form was the northern German-Hanseatic "city council apothecary" [Ratsapotheke], a "wholesale business" for the sale of drugs and medication of all types, owned by the city itself. In Italy, for example in Florence, apothecaries ["speziali"] were for a long time members of the same guild as doctors and painters – the mixing of herbs, medicines and colours being seen as a common feature. Since the income of many apothecaries was not very high, they also traded in candles, culinary herbs, cosmetics, amulets, and even relics. There were high penalties for forgeries. In most southern European countries, the training of apothecaries corresponded with the guild rules for skilled tradesmen in the late Middle Ages. Except in the monastery apothecaries, apprentices, journeymen, and masters all worked side by side in Italian apothecaries. Irrespective of this artisanal structure, a certain knowledge of Latin was required.

In Germany too, the nature of the profession remained typical. In the late Middle Ages, the master examination was taken in Cologne and was attended by a member of the university's medical faculty. The "materia medica" [term dates back to Dioscurides] prepared in the apothecary comprised a huge range of raw materials which could be processed into medicine. Paracelsus [1493–1541] radically updated the theory of "materia medica". Illnesses were no longer considered to be the result of the wrong mixture of humours, but rather as local changes, which disturbed the organism as a whole, whose chemical processes were controlled by a life force ["Archaeus"]. Illnesses were to be fought with chemical medication, so that tinctures, extracts, essences – produced on the basis of chemical processes – took on a crucial role. Naturally, Paracelsus's theory met with resistance. The Leiden medical professor Franciscus Sylvius [1614–1672] proposed a compromise which, although it assumed that chemical fermentation processes took place in the body, equated health with the correct balance of acid and alkali, which was influenced by these fermentations. This "clinical chemistry", whose founder is said to be Sylvius, was combined with the old teaching of "euchrasy" and/or the theory of the humours. Pharmacy became ever more complicated–and today it after all remains a chemistry-oriented profession.

In the 16th and 17th centuries, the number of healing plants known in Europe grew rapidly. In the newly founded botanical gardens [such as in Padua, Pisa, Florence and Leiden], exotic growths were planted which had been imported from America and other colonies, and these were then integrated into the established canon of healing plants. However, the professionals often argued about their effects; exotic forgeries were common. Conservative doctors and medical faculties rejected the new plants out of hand. Both doctors and pharmacists long debated whether the traditional mercury cure or rather guajak wood imported from America should be the preferred choice for treating syphilis, a disease which was brought by Charles VIII's army from France to central Europe [hence "the French disease"].

Around 1500, the invention of printing encouraged the production of detailed herbal books, which popularised botanical and pharmaceutical knowledge to a certain extent. The works of Jacobus Tabernaemontanus, Otto Brunfels, Hieronymus Bock, Leonhart Fuchs or Pier Andrea Mattioli appeared with beautiful illustrations by famous artists. The word "Herbarium" – originally relating to collections of dried plants – now also referred to these printed illustrations of plants. Another new type of publication were the so-called pharmacopoeia, authoritative works in which pharmaceutical formulas

Sales area of a modern pharmacy, Munich

Pharmacists are trading on the access to the Internet, as well

Counseling interview in a modern pharmacy

were fixed and standardised by means of drug and formula lists [for the first time in Nuremberg in 1546]. There were also apothecary manuals, which contained all known "Simplicia" and "Composita" and usually also included a specialist dictionary. They were indispensable sources of information for apothecaries and doctors alike. In the 17th and 18th centuries, these works were constantly revised and adapted to correspond with the latest scientific and medical paradigm shifts.

From 1530, the universities at Padua, Bologna, and Pisa had their own chairs of medical botany. This allows one to conclude that the typical cooperation with doctors, which exists today, developed early, particularly in Italy. In the cities of the German imperial cities, most apothecaries were under the control of the authorities who – in keeping with Hohenstaufen tradition – granted them a "privilegium" for a specific city or territory, thus protecting them against the uncontrollable competition of hawkers, miracle doctors, herb gatherers, theriac sellers, and charlatans. Simple people often turned to the monastery apothecaries, where highly valued specific medications with promising names were, to some extent, being developed [Jesuit powder, Capuchin powder, Pulvis Carthusianorum]; not to mention the various different liqueurs.

In the 18th century it was quite common that pharmacist's apprentices were offered lectures in general botany. The academisation of the profession, which complemented medical activities like no other, advanced rapidly. In keeping with the medical edict passed by Frederick Wilhelm I of Prussia [1725], on completion of his apprenticeship and at least seven years as a journeyman, the would-be pharmacist was required to pass a "Processus pharmaceutico-chymicus" at the "Collegium medico-chirurgicum" in Berlin and take a semi-academic examination. Those who wished to set up as pharmacists in small towns could be exempted from the academic examination. The pharmacist's scientific and social standing depended not least on his chosen location. The 19th century saw a further scientification of pharmacy. Many pharmacists saw themselves as natural scientists – to a far greater extent than did many physicians at the beginning of the century. Talented and ambitious students of the "materia medica", whether their training was "artisanal" or "scientific", now had the opportunity to isolate and artificially produce the active ingredients in traditionally-used plants. Thus in 1817 Friedrich Wilhelm Sertürner presented morphine in a famous work. Justus von Liebig [1803–1873] rose from being a pharmacist's apprentice in Heppenheim to the position of pharmacy inspector in Hesse, in addition to becoming one of the leading chemists of his age. Thanks to his dedication, pharmacy in Germany was finally academised. It was above all the discovery of alkaloids and glycosides which now influenced the daily lives of pharmacists. While orthodox medicine was in crisis and the Krakow professor Josef Dietl loudly proclaimed its incompetence in 1840 ["therapeutic nihilism"], pharmacology and biochemistry enjoyed a boom. During the second half of the 19th century natural scientific positivism asserted itself within the faculties of medicine.

Countless germs were discovered within a relatively short space of time. In close cooperation with physicians, and using the identical methodology, pharmacy and pharmacology were finally promoted to a scientific university subject. The western governments supported this process. The Société de Pharmacie de Paris, founded in 1803, saw itself as an elite scientific society, as did the American Pharmaceutical Association which was formed in 1852. In many European countries, standardised pharmacopoeia were published on the basis of the latest research, for example, in Germany, the "Pharmacopoea Germanica", the first comprehensive work of this kind in that country. In the old, respected pharmacies, industrially manufactured pharmaceuticals

Pharmacist in the storage of goods (source: DocMorris)

began to dominate, whereby some pharmacists, such as Heinrich Merck of Darmstadt, even managed the leap to become major industrialists. The subsequent great wars put the new system to a very special kind of test: The organised storage of huge amounts of pharmaceuticals, vaccines, and various prophylactics made the pharmaceutical industry interesting with regard to military medicine. Mass-produced medication was significantly cheaper than that produced manually, although naturally the pharmacists could no longer vouch for the quality or effectiveness of such products; their job was largely restricted to explaining their effects. In the USA, cheap drugstores were set up, while in Europe "druggists" and herb traders competed with the pharmacists. Interestingly, France differentiated between first and second class pharmacists. In an anti-scientific backlash, both advocates of alternative medicine and charlatans flexed their muscles in the area of pharmacy. Around the middle of the 20th century, the market became clearer to both patients and physicians: Leave was finally taken from "custom-made" medication. The West German authorities now tried to gain control of the situation by introducing stringent legislation and monitoring the training of pharmacists. The "Obercollegium Medicum et Sanitatis" in Berlin drew up highly rigorous standards in order to regulate the training of pharmacists. The external appearance of the pharmacies developed – as critics had already bemoaned at the end of the 19th century – into elegant stores. The pharmacist was elevated to become an esteemed member of polite society. Much that was once associated with the pharmacy still is: drugs, medication, herbs, prescriptions, scents, and, above all, well-informed individuals who, in complement to or competition with doctors, offer advice and assistance in case of illness. Such advice is highly valued as medication is still regarded as treacherous: Warnings against side effects are still, indeed increasingly, being made. "Ask your doctor or pharmacist" is a well known slogan. One particular challenge facing pharmacists is the boom in alternative therapies and the astonishingly high level of trust that many people place in non-orthodox medicine. The pharmacist's workplace has undergone a radical change.

Modern pharmacies boast a practical design, and the market situation forces the owner to sell products which have only the slenderest connection with health or illness. The trade in industrial pharma products stands to the fore, while the manufacture of remedies and tinctures takes a back seat. Many pharmacies meanwhile resemble health food stores. In terms of quantity, however, the exchange of prescriptions and pharmaceuticals still plays a crucial role. Customers expect pharmacists, who mutated into "academics" in the 19th century, to possess sound scientific knowledge of the various pharma products, their chemical composition, and their side effects.

Pharmacies are also, now more than ever, integrated into the state "health system" all the way through to its mandate as local provider, which cannot be interrupted. Today, as independent businesspeople, many pharmacists feel that they are being led by the nose. Whilst in the golden years of West Germany's economic miracle new pharmacies shot out of the ground like mushrooms and payment for medication from the health insurance funds whose financial scope seemed inexhaustible, presented no problem at all, the current crisis in the health system has also been affecting pharmacies for some time. Their obligations remain but their privileges are dwindling.

The international competition from aggressively advertising mail-order pharmacies, online orders, as well as patients and/or customers informing themselves, and the public's increasing disdain for expert advice, are additional sources of pressure. Just like medical practices, hospitals and out-patient clinics, the pharmacy too is facing drastic changes.

Further reading
Cowen, David L./Helfand, William H.: Die Geschichte der Pharmazie in Kunst und Kultur. Cologne 1991.

Gaude, Werner: Die alte Apotheke. Eine tausendjährige Kulturgeschichte. Stuttgart 1986.

Richter, Thomas: Apothekerwesen. In: Gerabek, Werner et al. [ed.]: Enzyklopädie Medizingeschichte. Berlin/New York 2005. Page 80–86

Schmitz, Rudolf: Geschichte der Pharmazie I. Von den Anfängen bis zum Ausgang des Mittelalters. Eschborn 1998.

Schmitz, Rudolf: Geschichte der Pharmazie II. Von der Frühen Neuzeit bis zur Gegenwart. Eschborn 2005.

Klaus R. Bürger

Corporate Identity and Design

Every company or organisation has a clear goal: to achieve success in its most diverse dimensions and forms. This success is only achieved, however, if, alongside purely economic success such as increased turnover or profit maximisation, another factor is added: the people who work for a company or organisation need to have a sense of belonging in order to stand up for and work towards a common goal.

The more clearly the goal is formulated, the more unambivalently an institution's purpose is defined, the easier it is to translate this corporate identity into the language of architecture and support it in a targeted way. This applies just as much to pharmacies as to globally operating companies, to religions and churches, to cities and countries. In major companies, the subject of corporate identity has long been a matter of course and, just as is the case for large companies, pharmacies also need to create a corporate identity, a mission and an awareness of purpose, which extends beyond increasing sales figures. An identity which places people's health and health promotion centre-stage and which takes not only monetary success but also the spirit and the soul of pharmacy into account. Corporate identity doesn't just happen – it has to be developed – to cover every aspect which relates to the company, the organisation, and the pharmacy: corporate behaviour, corporate communications, and corporate design. The market too, with its demands and competitors, together with the products and services, plays a major role in corporate identity.

All of these elements mutually influence one another and stand in an integral relationship with regard to perception and effect. Once defined, however, a corporate identity is not valid for all eternity but requires regular examination and, where necessary, adjustment to changes in framework conditions, attitudes, and values. Of course, one should not lose sight of the company's individuality, its real purpose, and its strengths, simply because a few market parametres have changed. Hasty or premature changes to mission and strategy made in order to achieve short-term success, or in the hope of long-term sustainability, usually have precisely the opposite effect. The corporate identity becomes blurred, the overall image is no longer consistent or authentic.

But it is precisely an authentic, truly lived, and clearly visible corporate identity with which the pharmacy needs to establish itself in the market. Similarly to a brand article, a business such as a pharmacy can likewise anchor itself in the consciousness of patients and customers by means of specific characteristics and features. Consumers are quite willing to accept a higher price for the quality of branded goods because the brand communicates trust and security, and reduces the wish to try out something different or new: This should be the aim of a pharmacy's corporate identity. The identity of the pharmacy as a corporate entity must be so clear that it serves as a benchmark for all transactions and activities. It is not only the products that are sold in the pharmacy, but also the people who work there, who need to be able to communicate these standards and values. It is equally important that the building, its location, and its interior design all express and communicate this identity.

The creation and development of a corporate identity is a process in which all of the said sub-aspects should be analysed in detail and with a long-term perspective in mind – from the standpoint of the pharmacy manager, as well as from the standpoints of the employees, the customers, and the patients.

Furthermore, the location, competition situation, and demographic factors of the surrounding area need to be integrated into the analysis, because economic success is also essential in securing the pharmacy's long-term existence.

As such, the range of over-the-counter products and advisory services being offered must correspond with the needs and demands of the immediate area. The identification of existing and potential target groups gives rise to a range with which the pharmacy can position itself with good prospects of success – provided that there

1 Alpin Pharmacy am Klinikum, Kempten Location as the central factor of corporate philosophy – applied to interior design: the Alps are reflected in the colours and materials of the sales area. The panorama picture of the Alps creates a clear cohesion between the interior and the surroundings.
2 Allmann'sche Pharmacy, Biberach Immediately at the entrance, the logo illustrates the corporate design.
3 Allmann'sche Pharmacy, Biberach Colours and logo are evident in details throughout the entire dispensary.

not already some competitor wooing the same target groups. In consequence, the search for a corporate identity is influenced by internal factors – the strengths and weaknesses upon which the pharmacy and its personnel have a direct impact, as well as those external aspects – opportunities and risks inherent in the local area – which cannot be controlled directly.

The elements of a corporate identity are also its instruments: behaviour, communication, and design. All three should convey the identity, the entrepreneurial personality, both internally and outwards, creating a sense of belonging amongst the employees and developing a sense of identification amongst clients. It is the corporate design which is generally first perceived by the outside world. All communication material, from logo to the design of the business stationery, and every individual print or online measure, must have a uniform quality and visual. Visual communication is visible and these visual impressions and judged, consciously and subconsciously, giving rise to either acceptance or rejection. A standardised presence can however also do more: It can enhance the recognition value and influence affective behaviour by increasing the customer's identification with "his pharmacy". A consistent visual appearance also includes a dress code and, not least, the visual design of the pharmacy's own product range. And, finally, one should not forget the interior and external architecture of the dispensary.

An important, but not directly quantifiable component of corporate identity is individual behaviour. On the one hand, the behaviour of the pharmacy manager to his staff with regard to management style, personnel development, and further training opportunities and in terms of how he deals with criticism, are important. On the other hand, the behaviour of all employees in contact with customers and interested parties, suppliers and service providers, competitors and the public at large, plays an important role – for, after all, the business's inner attitude is being communicated to other people. That applies all the more to health amenities which stand in the service of the people. Visual appearance and design are components of communication which are in a dialogue with the outside world. When taken together all elements of corporate identity form a harmonious whole, which is positively perceived and accepted by customers and patients: this finally is the business's image.

The architecture and, in particular, the interior design plays an important role in terms of corporate identity. It must not only create spaces, but should and must also reflect the pharmacy's identity – in other words, these spaces too must encourage positive and open behaviour and the willingness of employees and customers to communicate. At the same time, a genius loci must also be shaped, in which customers and patients feel safe, content, and well looked after. Good design has the aim of inspiring people, of making them stay, of arousing wishes, and of stimulating the imagination and ultimately, in terms of shop design's functional aspects, the aim of enhancing people's willingness to buy. This cannot be achieved in functional rooms in which the product-filled shelves lack emotional character.

But where interior design serves its true purpose, it can convey aura and atmosphere, personality and a personal feel to people, thus rendering the corporate identity visible and making it tangible. As such, the design of pharmacies and other health amenities should, in particular, employ creative and socio-psychological expertise to complement the purely technical aspects of interior design.

The elements which interior design can employ are both simple and complex: colour, shape, and material. Each element makes its very own statement and, in its individuality, has an effect on all those who come into contact with it. However, as soon as two or more colours, two or more shapes, two or more materials come together, the statement made by the individual element may no longer be valid, and the elements either mutually strengthen or foil each other: they enter into a dialogue.

Pharmacies

4

5

6

This interplay is potentialised where all three elements are brought together – just as several different medications taken together may cause pharmaceutical interaction. The interaction profile of colour, shape and material should therefore be professionally coordinated with the corporate identity, so that it corresponds with the corporate philosophy. There, where people feel at home, heal or recover, the interplay between the interior design elements should be designed with the greatest possible care and circumspection. Depending on the given functional area of the pharmacy – whether over-the-counter sales and cash desk, self-service or over-the-counter display, whether prescriptions or advice – different goals are being pursued with regard to the stimulation of emotions and actions. The potential psychological effects of colours, shapes, and materials must be deployed and combined here in a carefully targeted manner.

The suggestive strength and unconscious effect of colours were known long before Goethe's colour circle. Depending on the intensity of the shade and degree to which they are used, yellow and red, blue and green, or black and white in all their tones and shades can spark off a huge variety of responses. In the healthcare area, and thus also in pharmacies, no general statement can be made on whether there are colours which are more likely to encourage recovery and well-being than others. Thus white was traditionally seen as the [non] colour of the healing professions, but has lost ground continually since colour psychology became a fixed feature of the interior design of pharmacies, medical practices, and clinics. The situation is similar with regard to the selection of shapes and materials.

Curves always appear softer than sharp edges and are also employed at a superordinate level in pharmacies: Wherever the space of the dispensary permits, bends and curves can be used to steer the paths of customers and patients. The psychology of paths can be applied, people's automatic tendency to veer to the right can be taken into account, and customers can be unconsciously steered to the self-service section if there is sufficient space in the sales area. The choice of suitable materials for a pharmacy is made in light of functionality and aesthetics. Where robustness and durability are required, they must also be guaranteed. For, just as a pharmacy's corporate identity is built to last, equally, the interior design of the pharmacy should not be subject to frequent change or, even worse, display signs of wear and tear.

A corporate identity is not a fashion statement and its visual must also communicate constant sustainability and tangible authenticity. If the personality of the pharmacy is best expressed in a glowing shade of orange then the communication of this personality can outlive any trend using precisely this colour, or possibly only using it as a highlight, in the pharmacy.

Fashion consciousness however is no contradiction to a corporate identity geared to the long term. A good corporate identity can also be implemented by means of effective, topical advertising strategies and measures. This is a question of nuances, details – it is not about the pharmacy as a whole. It should be made clear here, that "measures" refers not just to the design of eye-catching posters and advertisements.

Rather, it is the choice of product range, the type and way in which goods are placed, and the question of whether customers are addressed in a loud or quiet manner – all elements which are just as important as the consistent use of the corporate design. Even the training of employees with regard to providing expert advice during special, target group-oriented campaigns or in the areas of starting conversation, question techniques or handling objections, is an element of customer-oriented corporate identity. The interior design of a pharmacy is more than the harmonious composition of various functional areas to create a pretty dispensary, because an "attractive" pharmacy is either the expression of a corporate culture or it is an empty façade – and today's customers and patients are well able to differentiate between exactly the two.

4 Schiller Pharmacy, Heidenheim
White is no longer the rule of thumb when designing healthcare amenities:
deep red,

5 Michaels Pharmacy, Winterbach
... dark blue,

6 Adler Pharmacy, Göppingen
... delicate green

Design: Klaus R. Bürger
Photos: Uwe Spöring

Further reading
Antonoff, Roman: CI-Report 13. Darmstadt 2004.

Bürger, Klaus R.: Apothekenarchitektur – funktionsgerecht & ästhetisch. In: Offprint Anzag Magazin. Frankfurt/Main 1997.

Damaschke, Sabine/Scheffer, Bernadette: Apotheken. Planen, Gestalten und Einrichten. Leinfelden-Echterdingen 2000.

Kreft, Wilhelm: Ladenplanung. Merchandising – Architektur. Strategie für Verkaufsräume: Gestaltungs-Grundlagen, Erlebnis-Inszenierungen, Kundenleitweg, Planungen. Leinfelden-Echterdingen 1993.

Franz Labryga

Planning Pharmacies: the German Principle

The potential owners or leaseholders of the pharmacy and all of those involved in the planning, that means architects and interior designers, and possibly artists, engineers, and suppliers, must engage very closely with the operational and constructional principles of planning, whether new construction, expansion or redesign, if they wish to create what is by today's standards a high-performing, economically viable, customer-friendly and well-designed pharmacy. The following sections provide information on planning, construction, and furnishing.

TYPES OF PHARMACY

According to the Federal Union of German Associations of Pharmacists (ABDA), there are 19,880 pharmacies[1] in Germany. Depending on how one looks at them, they may be divided into the following categories:

Classification by size | A rough observation leads to classification into small, medium-sized and large pharmacies. Precise definition of the respective sizes is provided in the "Usable areas" section.

Classification by location | Because the services provided often vary, a differentiation is made between urban pharmacies and rural pharmacies.

Classification by neighbourhood | For both customers and pharmacy it is economically important to be close to other businesses or, better still, to be in the direct vicinity of other businesses in the form of a pharmacy within a department store. The greatest benefits accrue to the pharmacy in a healthcare facility, because the definition of a need and its satisfaction are in the same premises.

Classification by type of sale | A differentiation should be made between physical pharmacies that sell their products to customers locally and mail-order pharmacies which, while retaining the functions of a physical pharmacy have a special licence and are primarily focused on dispatching goods to fulfil customer orders.

Classification in accordance with ABDA | The vast majority are main or standalone pharmacies [with operating licence] – just under eleven percent being chain pharmacies.

Classification pursuant to the "Ordinance governing the operation of pharmacies" [Apothekenbetriebsordnung: ApBetrO[2]] | Here a differentiation is made between public pharmacies, hospital-supplying pharmacies, branch and emergency pharmacies, and hospital pharmacies.

Classification based on development of marketing | a very recent differentiation, whose importance will certainly increase in future, is that between premium pharmacies which feature high product quality and offer advice and care to customers, and discount pharmacies, who primarily advertise with bargain prices, aim for high product turnover, and largely dispense with service and customer advice. In the present work, only the first three categories are applied.

LEGISLATION, ORDINANCES, AND OTHER REGULATIONS

Special regulations apply to the construction of pharmacies, the observance of which is of crucial importance, particularly with regard to averting danger from customers and employees.

ApBetrO
This ordinance passed on 26 September 1995 [latest revision on 18 July 2017] regulates the operation of pharmacies and the supply of medication to the public. It is rooted in article 21 of the "Pharmacy Act" [Gesetz über das Apothekenwesen[3]], which forms the basis for the establishment of all pharmacies in Germany. In both main sections of

the ApBetrO employee prerequisites, requirements with regard to the manufacture, storage, and designation of medications, and the size and furnishing of the business premises are regulated. Individual details are specified in the relevant sections of the planning principles.

Other requirements
The text of the ApBetrO also draws attention to other legal requirements to be fulfilled for the operation of a pharmacy:
- German Pharmacy Act [Gesetz über das Apothekenwesen/Apothekengesetz]
- German Medicines Act [Gesetz über den Verkehr mit Arzneimitteln/Arzneimittelgesetz]
- Medical Products User Regulations [Medizinprodukte-Betreiberverordnung]
- Ordinance on Medical Devices Vigilance [Medizinprodukte-Sicherheitsplanverordnung]
- Business Trading Hours Act [Ladenschlussgesetz]

The guideline published by the Robert Koch Institute "Hygiene requirements for the functional and constructional design, and operation of in-hospital and hospital-supplying pharmacies"[4], also applies to other pharmacies with similar duties. The following details are taken from a list by Damaschke and Scheffer[5] of further regulations to be observed when setting up a pharmacy:

Laws
- Narcotics Act [Betäubungsmittelgesetz]
- Medicinal Advertising Act [Heilmittelwerberecht]
- Working Hours Act [Arbeitszeitgesetz]
- Young Persons' Employment Act [Jugendarbeitsschutzgesetz]
- Maternity Protection Act [Mutterschutzgesetz]
- Employee Protection Act [Beschäftigtenschutzgesetz]

Ordinances, regulations, and guidelines
- General Administrative Regulation on the Implementation of the Medicines Act [Allgemeine Verwaltungsvorschrift zur Durchführung des Arzneimittelgesetzes]
- Guideline for Awarding Supplier Licences [Richtlinie für die Erteilung der Genehmigung von Versorgungsverträgen]
- Official Approval and Inspection of Pharmacies [Amtliche Abnahme und Besichtigung von Apotheken]
- Granting of Operating Licence [Erteilung der Betriebserlaubnis]
- Hazardous Materials Ordinance [Gefahrstoffverordnung]
- Flammable Liquids Ordinance [Verordnung über brennbare Flüssigkeiten]
- Chemicals Ban Ordinance [Chemikalienverbotsverordnung]
- Narcotics Prescription Ordinance [Betäubungsmittel-Verschreibungsverordnung]
- Guidelines | Good Manufacturing Practices
- Federal Pharmacists' Code – BApO [Bundesapothekerordnung]
- Administrative and Professional Guidelines [Verwaltungs- und Berufsgenossenschaftliche Vorschriften]
- Accident Prevention Regulations of the Statutory Insurance Association for Non-State Health Services and Welfare Work [Unfallverhütungsvorschriften der Berufsgenossenschaft für Gesundheitsdienst und Wohlfahrtspflege]
- Workplace Ordinance [Arbeitsstättenverordnung]
- German Standards Regulations [DIN-Vorschriften]
- Federal State Construction Ordinances [Landesbauordnungen]
- Trade, Commerce, and Industry Regulation Act [Gewerbeordnung]
- Labelling Regulations [Kennzeichenvorschriften]
- Professional Code of the Pharmacists' Chamber [Berufsordnung der Apothekenkammer]
- Competition Law [Wettbewerbsrecht]

In case of need, reference will be made in the following explanations to the regulations mentioned above.

REQUIRED OPERATION DATA

The implementation of a planning idea starts with the consideration of the establishment's future operations. On this basis construction is later undertaken.

The following is an explanation of the operating requirements which are important for pharmacies:

Catchment area
One indicator of a pharmacy's importance and profile is the size of the catchment area into which the majority of its customers come. Some pharmacies also have a large catchment area because the population density, for example in rural areas, is relatively low. Where customers have to travel a long way to the pharmacy, whether on foot, with public transport, or by car, the pharmacist should consider what amenities he would like to offer them. He should consider providing seating and rest areas, particularly for old or disabled people, as well as refreshment in the shape of a mineral water dispenser, and a WC. The impact of demographic structure and type of customer within the catchment area on the pharmacy's economic development should not be underestimated. Successful pharmacies integrate this data when planning their sales strategies.

Architect in planning phase examining compliance with regulations

"Special regulations apply to the construction of pharmacies, the observance of which is of crucial importance, particularly with regard to averting danger from customers and employees."
Sketches: Hartmann

Location

The considerations which apply when choosing a good home also apply to pharmacies. The top three criteria are location, location, location. It is therefore urgently recommended that before any detailed planning is made, a thorough location analysis be undertaken, in order to find out the best location for the future pharmacy. The following locations are advantageous:
- locations with a high number of people passing by, whether on main roads, in pedestrian zones or in houses with highly varied frequency of passers-by, such as department stores or shopping centres, railway stations or airports
- locations in or close to healthcare facilities where medications are prescribed: group medical practices or health centres, the practices of GPs or specialist physicians, gymnastic and massage practices, care centres, podiatry clinics, beauty parlours or spas.

The following locations are a disadvantage:
- locations which are too close to existing pharmacies, except where level of footfall permits such competition and that this competitive situation even proves to be healthy
- locations with no connection to the local transport network: in general, locations which are more than 200 metres from a public transport stop are unsuitable.

Arrangement of products

The original meaning of the word "apothecary" [storehouse] still remains valid to some extent today. As in times past, storage still plays an essential role, although a quite different one to before.

DELIVERY OF GOODS | Goods for the pharmacy usually arrive at regular intervals, conveniently delivered in transport containers or on pallets via a dedicated route to the stockroom, which is practically located at ground floor level or in a basement accessible by elevator. In larger pharmacies, a room for receiving the goods comes before the stockroom itself; where the pharmacy itself consigns a lot of goods, it makes sense to have a space or room for dispatching goods.

NUMBER OF ARTICLES IN STOCK | To be precise, this refers to the number of different types of article and the number of units of the individual types which are constantly in stock at the pharmacy. A large number of medicines are ordered as required. Depending on the distance from the distributing warehouse, items can mostly be handed over to the customer within a few hours or on the following day.

STOCK TURNOVER | The size of a stockroom depends not only on the type and number of medicines constantly in stock, but also on how fast the stock is turned over. Once a pharmacist has agreed fast and reliable delivery with suppliers, storage areas can be reduced in size accordingly. The constantly increasing differentiation in the choice of goods available in recent years often forces existing pharmacies to enter into such agreements. However, in newly established pharmacies, this method can be useful. For pharmacies in rural areas, frequent deliveries may be more difficult to arrange because of the greater distances involved.

SIZE AND TYPE OF STOCKROOM | The determining factors for the size of the stockroom are the number of packaged goods kept in stock, the height of the stockrooms, and the type of storage. There are two different types of stockroom:
- ABC storage [alphabetical storage] | The items are arranged in alphabetical order. This method of storage makes it easier to find medicines and, because there is barely any technical dependency, permits a high level of supply security.
- chaotic storage | The items are arranged according to how fast they sell and the available storage capacity. This method of storage requires a good data processing system and the entry and dispensing of medicines needs to be automated. Alongside only a low rate of error in ordering, the space saved is a major advantage. Increasingly, pharmacies are using automated stockrooms, which are available in semi-automatic or fully automatic form. A thorough cost-benefit analysis should be carried out before deciding on which type of storage to use. Under special circumstances, pharmacies also have an emergency stockroom specially for those medicines which are most urgently required in emergencies.

GOODS TRANSPORT | Items may be transported by trolley or using automated systems in the shape of pneumatic delivery, conveyor belts, spindle-shaped chutes or, in large pharmacies such as those in hospitals, box-type conveyor systems which transport the ordered items across longer distances to their various destinations. Deciding which method to choose largely depends on local conditions and the performance level required. Owing to the not insignificant investment and operating costs for automated transport systems, a thorough analysis should be carried out before an automated stockroom is set up.

Waste disposal

Waste disposal is subject to the requirements laid down by the Federal Ministry of the Environment. Waste occurring in pharmacies in usually collected at regular intervals by waste management companies. Special attention is given to toxins, including medicines as well as organic and inorganic chemicals, and expired narcotics.

Sale of pharmacy products in a modern pharmacy

The disposal of them is regulated by the Narcotics Act. Special rules apply to medicines which are out of date: They are either collected and disposed of by wholesalers, or the pharmacies are required to organise their disposal themselves.

In-house manufacture

A total of six clauses of the ApBetrO concern themselves with the in-house manufacture and examination of medicines in pharmacies. Apart from general regulations on management and examination [§ 6], requirements with regard to formulation [§ 7], small scale production [§ 8], large-scale production [§ 9], examination and approval of medicines [§ 10], and the constituent ingredients [§ 11] must also be met. In light of the ever growing spread of industrially produced finished products, the present level of in-house manufacture will probably decrease considerably in future.

Specialist advice

The main duty of the pharmacy sales staff is to provide customers with specialist advice. This is provided in a discreet way. It may also include the query as to whether the customer has understood everything he has been told. The customers' experiences in this regard form the core of their judgment, which may be crucial to the pharmacy's survival. Advisory services have expanded significantly in recent years; today they are regularly assisted by the latest reference material and the internet.

Special services

The market situation has worsened in recent times due to increased competition and reduced purchasing power, pharmacists have continually expanded their range of services. Using imagination and creativity, numerous new services have been created which help to increase attractiveness. There are five areas in which pharmacies have become active:

Expansion of the product range
- expansion of the range in the area of "typical pharmacy products"
- preparation of individual formulations for teas, creams, capsules and cosmetic products
- inclusion of products typically sold in medical supplies stores
- manufacture of cytostatics

Expansion of advisory services
- allergy advice
- medicinal tolerance check
- skincare
- nutrition and dietary advice
- endocrinological advice
- homoeopathic advice
- babycare
- incontinency advice
- vaccination advice
- travel advice and mediation, particularly for long-haul trips
- medical and scientific research

Performing healthcare tasks
- measuring blood pressure, uric acid, blood sugar, and cholesterol
- measuring weight
- measuring vein pressure
- incontinence and stoma care
- body structure analysis
- skin analysis
- facial analysis
- podiatry
- massage
- pregnancy testing
- fitting for compression tights
- cosmetic treatments
- spa/fitness treatments

Pharmacies

DocMorris Pharmacy, Munich: arrangement of products

Expansion of services
- information by e-mail
- collection of prescriptions
- introduction of a customer card
- home deliveries of medicines
- obtaining quotations from the health insurance funds
- taking orders by telephone, telefax, and e-mail
- supplying health facilities
- automatic fee calculations
- import and export of medicines
- lending baby scales
- organising self-help groups
- setting up a health café

Organisation of courses and seminars
- courses for healthy eating
- walking courses
- yoga courses
- fasting weeks
- lectures on clinical pictures
- readings on subjects which promote health
- seminars on dietetics and healthy eating

The planner should be aware of the type and scope of such special services, as these activities will have an impact on the type and number of rooms required and their floor areas.

Sale of pharmacy products

In economic terms, the pharmacy's main goal is to sell goods in an efficient way, achieving high turnover, and increasing profit. Ultimately, all efforts are made to achieve this goal. A lastingly effective corporate philosophy [corporate identity] means that while doing this, great attention is paid to the customer's well-being.

Supplying third parties

It is beneficial to a pharmacy's balance sheet if it can manage to constantly deliver pharmacy products to one or more healthcare facilities. These include, for example, sheltered housing, old people's homes, homes for the disabled, medical practices and group practices, medical supply centres, and hospitals. Where such deliveries are constant, because of the increased turnover of goods, the pharmacy requires both a larger stockroom, but also areas for receiving goods and, depending on the type of consignment, for packaging, inventory management, and dispatch. In any event, adequate space for vehicles with incoming and outbound goods is required in front of the pharmacy.

Personnel deployment

Clauses 2, 3, and 28 of the ApBetrO regulate pharmacy personnel. A pharmacy is required to have a manager as well as the necessary pharmaceutical and non-pharmaceutical staff. The pharmaceutical personnel includes the pharmacist, pharmaceutical technical assistants, pharmacist's assistants, pharmacy technicians, pharmaceutical assistants, and the respective trainees; the non-pharmaceutical personnel includes pharmacy assistants, skilled pharmacy staff, and commercial staff with pharmaceutical training. In a pharmacy that is integrated in a hospital, this very clearly defined range of personnel corresponds with the type and performance structure of the hospital. For other pharmacies, the ordinance specifies that the criterion for personnel requirements is that a properly operating business structure be guaranteed. The type and, above all, number of staff are crucial factors in determining the size of usable areas in service rooms, staff rooms, and also in changing rooms.

Laboratory

Office

Archive

Operating costs

Ultimately all operational considerations with regard to organising a customer-friendly and well-performing pharmacy have the aim of keeping whole the operating costs in reasonable check.

Firstly, there are the personnel costs which comprise around 70 percent of total costs. Careful thought should therefore be given to the deployment of pharmaceutical and non-pharmaceutical personnel. A basic prerequisite for economic staffing is the organisation of optimised workflows for the pharmacy's main work processes. With regard to material costs too, which are largely made up of energy costs, costs for consumables, the cost of the premises, and advertising costs, efficient deployment is necessary in order to achieve a favourable operating result.

DATA REQUIRED FOR CONSTRUCTION

Taking the specified operational data into account, the data required for construction can now be established and collated. The two data packages form the basis for further activities and decisions with regard to the layout and design of the pharmacy.

Urban development requirements

BUILDING PLOT | In the case of a new construction being planned, when choosing a building plot, both the square metre price as well as the size, and the existing building regulation must be taken into account. A farsighted purchaser will ensure that the size of the plot allows for later expansion, and thus for the potential development of the pharmacy itself.

TRANSPORT CONNECTIONS | Good accessibility has a crucial influence on whether or not a pharmacy can survive. This applies in particular for clients who come to the pharmacy on foot, as well as to those who are also reliant on public transport with nearby stops.

Preservation orders | An old pharmacy which is of outstanding architectural design and therefore listed is a gem, for both the owner and the customers, who enjoy visiting for this quality alone. Usually, however, the authorities impose strict conditions on any construction work, so that it is very difficult to make alterations and the pharmacy must often manage with functional shortcomings.

Measures to preserve the building's substance can also be very expensive. Consolation prises exist in the shape of occasional public funding, or financial support granted by sponsors to preserve such a historic legacy.

Usable areas

Apart from the public area, the construction area and the technical functional area, the usable area is the most important type of area in terms of a building's breakdown of gross area as laid down in DIN 277[6]. While the first three types of area are the same in every type of building, the usable area – as the name suggests – gives an insight into its project-specific use. It is therefore the most powerful key indicator in a floor plan.

Since size is one of the main differentiating features of pharmacies, the respective usable areas are used for differentiation. With regard to the examples presented in this book, and with regard to existing development trends, the following applies:
- pharmacies with usable areas of up to 175 sqm: small pharmacies
- pharmacies with usable areas of 176 to 350 sqm: medium-sized pharmacies
- pharmacies with usable areas of over 350 sqm: large pharmacies

Space allocation plan

In light of the previous explanations, it becomes clear that only the usable areas can be recorded in the space allocation plans required for planning pharmacies. All other areas, [public, technical functional, and construc-

Apotheke Zur Rose, Halle
Logistics in a shipment pharmacy

tion areas] are only defined during the drafting phase. The rooms should be specified by type and size in the space allocation plans. The following are useful indicators with regard to planning the illumination of rooms:
○ daylight required
◉ daylight desired
● daylight not required

DIN 13080 "Division of Hospitals into Functional Areas and Functional Sections"[7], a standard which has now been successfully applied in hospital planning for over 20 years, even beyond Germany's borders, contains a recommendation on the breakdown of space allocation plans. This states that rooms should be divided into four groups:
– main rooms
– auxiliary rooms
– connecting rooms
– staff rooms

The planner may decide whether he wishes to break down his space allocation plan in this way, or whether he prefers to use the breakdown by functional groups described in the following section. For the sake of consistency, this book uses functional groups throughout.

Breakdown into functional groups
In planning a pharmacy, a large number of different types of room need to be considered. In order to reduce this large number, some rooms with similar purposes are included under the same heading, for example stockrooms for infusions, drugs or tea are classified under "special storage". In planning terms it has proven practical to allocate the around 40 rooms which remain after the standardisation of designations, into specific room units. Since, compared with hospitals, the room units in a pharmacy are relatively small, the classification of "functional group" has been introduced. Similar to the method used in DIN 13080, colour markers are used, which considerably ease the tasks of collating the floor plan arrangements and of analysing and comparing plans. The eight functional groups briefly outlined below are not present in every pharmacy, but depend rather on the type of pharmacy and its performance profile.

CUSTOMER ROOMS | The rooms allocated to this functional group comprise the dispensary which forms the heart of the pharmacy, with facilities for self-service and counter sales and the cash desk. In general, it is only in medium-sized and large pharmacies that special purpose rooms, a reception area, a waiting room, a play area for children, a WC, one or more consulting rooms, an intensive consultancy room, a treatment room, and a beauty treatment room are foreseen. In any event, the customer rooms should include a night service counter and, where pharmacies are suitably located in transport terms, a drive-in counter. The customer room functional group is marked in yellow in floor plans.

BUSINESS ROOMS | It is frequently the case that business rooms, with the exception of the stockroom, are classified as all rooms which are not customer rooms. For better differentiation, however, the term is more closely defined here. Such rooms include those rooms which are essential to the fulfilment of core business: the alphabetically-organised stockroom, one or more workplaces, prescription room, and a lab. Cost-cutting solutions combine the prescription room and lab in a single room. This functional group is marked in red.

SPECIAL PURPOSE ROOMS | This functional group is usually only present in large pharmacies. Such rooms have special functions, for example a cytostatics room with sluice, a parenterals room, and a physical stockroom. The rarely encountered herb garden is also included in this group. This functional group is marked in pink.

ADMINISTRATIVE ROOMS | This relatively small functional group comprises the rooms required for the administration of the pharmacy. These primarily include one or more offices and, in large pharmacies, a separate room for purchasing and orders, a room for collections, and an archive. Administrative rooms are marked in floor plans in green.

SERVICE AND STAFF ROOMS | This functional group comprises those rooms which are required for the pharmacy manager's duties, for staff use, and for sanitary purposes, and include service rooms, the night service room, the staff rooms with pantry, and the sanitary rooms, that means changing rooms, showers, and WCs. In terms of making savings, it is possible to combine staff room with night service room or with the pantry. Very occasionally there are libraries, also used as consulting rooms, and a separate room for sales to staff. These service and staff rooms are marked in orange.

SUPPLY AND WASTE DISPOSAL ROOMS | The rooms required by the pharmacy for supply and waste disposal purposes largely comprise the storage areas. In larger establishments there are also rooms for incoming goods, special storage, an emergency depot, and an order room for the packaging and dispatch of goods. The cleaning room and waste disposal room also belong to this functional group. Supply and waste disposal rooms are marked in brown.

TRAINING ROOMS | Rooms for the manifold education of customers and staff are still rare. These are lecture, teaching, and seminar rooms, also functioning as multi-purpose rooms, which may also serve as reserve space for future expansion. Training rooms are subsequently marked in violet.

PLANT ROOMS | Dedicated rooms are required in larger pharmacies for the installation of technical systems. These are technical rooms and shafts for electrical and air conditioning systems, as well as for pneumatic delivery tubes and transport facilities in automated warehouses. This type of room is marked in floor plans in blue.

Individual rooms
Of the around 40 standard rooms in a pharmacy, one part largely comprises rooms for normal use. These include all rooms belonging to the functional groups of administrative rooms, service and social rooms, some of the supply and waste disposal rooms, as well as the training rooms. In the following text, indications of specific characteristics are sufficient. The remaining rooms characteristic of pharmacies should be considered a little more closely. Detailed descriptions of the requirements for fitting out and furnishing these rooms are contained in the publications "Apotheken" by Damaschke and Scheffer, and "Apothekenbesichtigung" by Spegg[8].

Generally it is the case that the following rooms do not necessarily need to be usable areas completely surrounded by walls. They may often be open or, in case, semi-open areas, so that as transparent an effect as possible is created overall.

DISPENSARY | The largest and most important room in the pharmacy is the dispensary. Today, this room is not only used to dispense medicines but to also fulfil the other duties specified in the ApBetrO, that means to provide information and advice.

In order to properly perform these duties, different areas have formed within the greater space, which break it down and give it its characteristic feel. Customers entering the pharmacy first see the self-service shelves; these contain what is actually the secondary range. In accordance with the ApBetrO, these are pharmacy-typical products. The customer may select these items himself and purchase them. Customers do not have access to the goods sold over-the-counter. These are shelves containing prescription-only medicines, or for which the pharmacist's advice is required. At the heart of the dispensary is the large sales counter [PoS], or as are increasingly being seen, smaller, individual PoS, which are described in more detail in the chapter "Design Elements". It is possible to create a lastingly pleasant impression on the customer by offering him services such as fresh water or other drinks, seating, a place to hang his coat, or the opportunity to check his precise weight.

SPECIAL PURPOSE ROOMS | Although the ApBetrO specifies that the spread of self-service products should not play a superordinate role to or obstruct the pharmacy's primary objective of dispensing medicine, the self-service range also expands for economic reasons. It is therefore not surprising that the customer area in new pharmacies is increasing in size, and that dispensaries are thus becoming larger. Special purpose areas can be counted as part of the dispensary or listed separately because of their special status. It is convenient for customers if product groups are clearly organised into main strategic selling areas, for example cosmetics, dental care, children, seniors, intimate hygiene, nutrition, sport, and holiday.

PLAY AREA FOR CHILDREN | A stroll through a pharmacy full of tempting products requires time and effort. It is pleasant for parents and those accompanying them if, when visiting the pharmacy, their children can be entertained, for example with painting, in a well-supervised place.

CUSTOMER WC | These are still rare but very customer-friendly, offering customers the opportunity of using the toilet. Separate toilets for men and women are only necessary where demand is high.

NIGHT DISPENSARY | The nocturnal sale of medicines calls for facilities which makes this already complicated procedure easier for customers. The customer should be able to check which pharmacy is on night duty, ideally while protected from rain and cold, and announce his presence by means of a well-lit bell. Prescriptions and medicines are exchanged via a hatch in the immediate vicinity which should not be too large, in order to avoid robberies. The more of the customer the pharmacist can see, the greater his sense of safety. Increasingly, pharmacists are picking up on the American example of drive-in counters, which allow drivers to have direct contact with the pharmacy.

CONSULTING ROOM | The provision of a room which is both optically and acoustically insulated meets patients' need for privacy. It is above all senior citizens who value confidentiality when consulting the pharmacist. Ideally, this should be a closed room with seating and adequate space for movement, as this room should also be available for the provision of special services. During the fit-out, thought should be given to soundproofing measures; practical features here include an intercom to the dispensary, a video recorder, a PC, and a monitor, preferably with touchscreen.

INTENSIVE CONSULTING ROOM | Large pharmacies in particular face increased demand for consultancies, so that a second consultancy room is required. This fulfils the same functions as the room mentioned above, although ideally it should have a somewhat larger usable area for a comfortable seating group, so that conversations can be held with more than two people.

TREATMENT AND REST ROOM | In order to provide certain special services, larger pharmacies require a treatment and rest room which should be equipped with a divan. This room also serves as a reserve space for the future expansion of the consultancy and service range.

BEAUTY TREATMENT ROOM | In order to protect the privacy of customers having facials, manicures or pedicures, a dedicated room is required which may also contain two or more treatment stations.

ALPHABETICAL STORAGE | This term is now being widely used, although often reference is simply made to the pharmacy stockroom, the main feature of which is the alphabetical arrangement of the pharmaceutical preparations. Ideally, the storage cupboards, usually drawer racks, or racks of refrigerated cabinets, should be directly accessible from the dispensary. The alphabetical system may be dispensed with once a decision has been made to install an automated stockroom with a goods transport system.

WORK ROOM | Ideally located next to the stockroom, the work room is for the pharmacy assistants who manage the pharmacy stockroom, that means who keep it stocked and ensure that expiry dates are checked. The number of workplaces depends on the performance profile of the individual pharmacy.

PRESCRIPTION ROOM | According to the ApBetrO, a compounding room is an essential room. It is used to directly manufacture certain delivery types of medicine, such as powders, creams, emulsions, tinctures, extracts, capsules, and mixtures. Special attention must be given here to the ApBetrO, as well as to hygiene and Good Manufacturing Practices. As such, the prescription room must be in a protected area, separate from the customer area. This does not necessarily mean that it is not possible to see into this room, which recalls the pharmacy's roots in the apothecary's shop. For practical reasons, eye contact with the dispensary is also recommended. In fitting-out and equipping the prescription room, the ApBetrO should be observed. Since industrially prepared delivery methods are increasingly pressing to the fore, prescription rooms are losing ground. However, individual pharmacies have recognised that there is potential to be activated here. Where space is limited, the prescription room can be combined with the laboratory. This jointly used room should have separated working areas and a minimum size of 15 square metres.

LAB | A well-performing laboratory requires Bunsen burners, heat sources, retorts, and annealing furnaces, and must also handle flammable liquids. It is therefore understandable that the construction and fit-out of a laboratory are governed by numerous regulations. The most important of these are the ApBetrO, the building regulations of the given federal state, fire safety regulations and DIN 12924, Part 4[9]. These regulations govern, for example, the minimum area [12 square metres], the fire-resistant construction of walls, ceilings, and floors, a second escape route and, where work is undertaken with flammable liquids, an extractor with suction device, and at least one fire extinguisher. Attention should also be paid to good lighting, colourfast light, natural or mechanical ventilation, and surfaces which are easy to clean.

WASHING ROOM | For extensive and specialised lab work, the installation of a washing room for cleaning containers and equipment actually is practical. Otherwise, as a standard feature, a washbasin and a special washing machine need to be built into the laboratory.

OFFICE | It is convenient to have a dedicated room for the undisturbed handling of office and administrative work which, despite computers, continues to increase. Under no circumstances should this room also serve as the night service room, as the personal feel is then lost. The office may also be used as a small in-house library for storing the scientific and other literature required under the ApBetrO, such as the German Pharmacopoeia [Deutsche Arzneibuch], the European Pharmacopoeia, the German Drug Code, the Synonym List [Synonymverzeichnis], the Register of Standard Licensed Medical Products [Verzeichnis der Standardzulassungen], and the Provisions of the Applicable Pharmacy, Medicine, Narcotics, and Chemical laws.

ARCHIVE | At least in larger pharmacies, a small archive for the orderly storage of files which inevitably accumulate over time is recommended, and this room can also be used for the storage of household items such as ladders or spare chairs.

CYTOSTATICS ROOM | If cytostatics [substances which inhibit cell growth and cell partition] are prepared, a dedicated room is required. For hygienic reasons, this room must contain a sluice for the personnel to change their clothes. In pharmacies which supply hospitals, depending on the medical performance profile, a room for parenteral nutrition [by subcutaneous injection] may be installed instead of or in addition to this room.

SERVICE ROOM – PHARMACIST | A personally organised service room generally meets the requirements of the manager of a larger pharmacy; this allows him to contribute to the pharmacy's performance. The room does not need to be large in size. Its well-designed interior represents the pharmacist's business philosophy.

NIGHT SERVICE ROOM | In accordance with the ApBetrO, a night service room is one of the essential rooms which, if well designed, has the appearance of a comfortable hotel room. In the event that it is also used as a staff room – not prohibited but undesirable – instead of the preferable bed, a couch should be installed. The employee on night duty must have access to the pantry.

STAFF ROOM | § 29 of the German Workplace Directive [Arbeitsstättenverordnung[10]] states that if an establishment has more than ten employees, an easily accessible break room is required. In pharmacies, even where there are fewer employees, such a room should be installed, because the consumption of food at the workplace does not correspond with hygiene requirements. It is practical to arrange a pantry which, in larger pharmacies, can also take the shape of an individual room. A view of greenery outside is recommended as this enhances the employees' sense of relaxation.

STAFF CHANGING ROOMS | According to the German Workplace Directive, only one clothing storage unit and one locker for the storage of valuables are required per person. However, in the interests of both hygiene and the personnel themselves, small, individual changing rooms, divided by gender, are more suitable.

SANITARY ROOMS | The ApBetrO requires that toilets for men and women, solely for the use of the staff, are required where an establishment has more than five employees. This means that a separate WC needs to be installed for customers. The installation of shower rooms depends on night service requirements. Shower rooms should contain a washbasin and may also be used by other employees. They should be beside the changing rooms and close to the night service room.

INCOMING GOODS ROOM | A dedicated room for incoming goods really is required because suppliers also deliver goods outside normal pharmacy opening hours, and these goods are initially deposited in this room. Packaging for example can be removed here and preliminary sorting undertaken, and goods can also be prepared here for dispatch.

STOCKROOM | The size and type of stockroom has already been outlined in the "Data required for operation" chapter. In this description of the individual rooms, the type of goods being stocked is explored. According to Damaschke and Scheffer, there are considerable differences in the types of stock:
– proprietary medicinal products | In accordance with the ApBetrO, the stockroom should contain at least one average week's supply of such products. There should be no negative impact on the quality of the medicines.
– vaccines | Defined temperatures subject to constant monitoring are prescribed.
– heat-sensitive medicines | Refrigerators or cooling units with three temperature levels are required.
– medicines which need to be stored very carefully | These are products which need to be kept under lock and key. Only specialised pharmaceutical personnel have access to them.
– highly toxic and toxic substances and toxic reagents | Here too, a lockable cabinet is required. The dispensing of these substances in regulated by the Chemicals Ban Ordinance.
– flammable liquids | The stockroom must be ventilated and should have no floor drain. Attention should be paid to escape routes.
– narcotics | The Narcotics Act requires that such products are separately stored and protected against unauthorised removal.
– medicines which require careful storage | Storage vessels must be marked in accordance with regulations.

Pharmacies

	Customer rooms	Business rooms	Special purpose rooms	Administrative rooms	Service and staff rooms	Supply and waste disposal	Training rooms	Plant rooms
Customer rooms		●	○		○	○	●	
Business rooms	●		●	●	●	●		●
Special purpose rooms	○	●			●	●		
Administrative rooms		●			●			
Service and staff rooms	○	●	●	●				
Supply and waste disposal	○	●	●					●
Training rooms	●							
Plant rooms		●				●		

● Various functional relationships
● Certain functional relationships
● A few functional relationships
○ No functional relationships

Matrix of functional relationships between the functional groups

– drugs | Containers are to be protected against light and damp. Special advice on how to store individual drugs is provided in the German Pharmacopoeia.
– self-service products | Here it is merely necessary to ensure that these items are not mixed up with prescription-only products and medicines.

SPECIAL STOCKROOM | It should be decided from case to case as to which of the ten mentioned types of medicine should be stored in a special stockroom.

AUTOMATED STOCKROOM | The automated stockroom, outlined in the "Data required for operation" chapter, is meanwhile a feature of many pharmacies. Much positive feedback has been heard. Since several companies offer modular elements, such storage can be adapted well to given local conditions. When combined with automated transport systems, there is no need to place the automated stockroom in the immediate vicinity of the dispensary. This means that it is also possible to locate it on the floor below or above the pharmacy's main rooms. The space thus gained in a functionally important place may be used for other purposes. The main advantage of the automated stockroom lies in the time saved by the pharmaceutical personnel, who no longer need to walk to and from the alphabetically-organised or other type of stockroom, and can therefore dedicate more time to advising customers. The use of new types of computer systems offers a further time saving to personnel: By improving ordering and supply capacity, stock is reduced and the administrative workload thus minimised.

ORDER ROOM | In pharmacies which constantly supply a significant percentage of their goods to wholesale buyers such as group medical practices or hospitals, it is recommended that an order room be installed. This is where goods are assembled, where necessary packaged, and held ready for collection.

WASTE DISPOSAL ROOM | In order to avoid that waste is left lying around unsorted, pharmacies, including small pharmacies, should have a room with individual bins for the different types of waste, from which collection can be easily effected.

PLANT ROOM | Although highly desirable, owing to a lack of space or to a lack of coordination on the part of technical service providers, a dedicated room containing all technical connections, known as the plant room, is often not included in draft plans. However, in terms of overview and servicing, such a room is most useful. In some of the presented projects it is regarded.

Key construction indicators
Alongside establishing qualitative features, some quantitative data is also important in evaluating a project. Gross floor area | DIN 277 stipulates that the GFA is the sum of floor areas of all floor plan levels of a building in square metres. In relation to the usable areas which are specified above, this value gives an indication of the compactness of the building solution.

GROSS ROOM VOLUME | DIN 277 stipulates that gross room area is the volume of the entire building contained within its outer boundary lines in cubic metres.

CONSTRUCTION COSTS | DIN 276 Building Construction Costs[11] serves as the basis for establishing construction costs. The cost groups – 300 Building-Construction and 400 Building-Technical Systems – are taken into consideration at this point. In order to achieve comparability, costs should always contain the statutory rate of VAT. The cost of construction per cubic metre of gross room volume is often used as a comparison value.

TOTAL COSTS | DIN 276 also applies here, with cost groups 200 to 700 being relevant. These are all costs including value added tax. The only exception are the costs of the plot, as these diverge widely depending on local conditions.

The total costs are also known as investment costs. Taken together with the operating costs, they are important in deciding on a new pharmacy, expansion or redesign. A particularly important indicator here is total costs per square metre of usable area.

Parties involved in planning
With regard to construction in the healthcare sector, the formation of a committee to accompany the planning process has proven useful. Its objective is to examine all issues around planning and construction, and it attempts to find practical solutions which are ideally acceptable to all.

The main person in this group is the developer – this being either the pharmacist himself, or an outside investor. The job of coordination and direction is undertaken by a pharmacy planner. In general, this will be an independent architect or interior designer although, increasingly, it may also be a representative from a design company who specialises in the construction of pharmacies. Where the developer has no expertise of pharmacies, the pharmacist in charge of the pharmacy or responsible for operations management, ideally together with a member of the pharmaceutical personnel, should be part of the planning team. Questions relating, in particular, to operational workflows, space allocation planning, and implementation planning are to be discussed in detail with these specialist individuals, who are involved in the pharmacy routine every day. The effort deployed here is worthwhile if constructional and organisational errors can thus be avoided. Since the pharmacy's main brief is to provide services to customers, it can only be of advantage if their opinions are integrated into planning decisions. It is therefore a good idea to find an interested customer, who can accompany the planning process, offering criticism and suggestions. Alternatively or in addition, a customer survey can be carried out and the results integrated into planning. It can be useful to invite representatives of the licensing authorities to some of the planning committee's meetings.

PROJECT ELEMENTS

In general, pharmacies do not belong to the type of building designed by standard architects' offices. Rather, this is a specialised area for individual architects and interior designers who often have years of experience. Frequently, however, the work may be undertaken by an architect for whom the job represents a major challenge. The following remarks should primarily be of assistance to the architects, who are interested in designing pharmacies. They are also suitable for potential developers and operators, in providing an overview of the planning process.

Functions
Observation of functional workflows in pharmacies often make it possible to identify shortcomings and optimisation potential. As such, before building work starts, it is worthwhile undertaking a thorough examination of functional workflows and the resultant functional relationships, in order to ensure the smoothest and most effective operation. The following considers a few aspects of project planning.

Functional relationships between the functional groups | One of the first areas to be considered for a project is the clarification of functional relationships between the functional groups. The strength of functional relationships is made clear in the corresponding matrix. There are three functional groups at the heart of the pharmacy: customer rooms, business rooms, and supply and waste disposal rooms. Depending on the size and performance profile of the pharmacy, the other functional groups cluster around this core needs to be taken into account.

Functionally practical arrangement of the function groups in pharmacies with two or three floors | The matrix also allows conclusions to be drawn with regard to the arrangement of the functional groups in pharmacies with two or three floors. With several different levels, a greater effort is inevitably required for vertical paths.

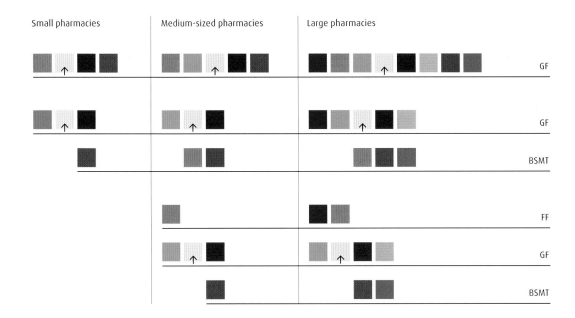

Arrangement of functional groups in one, two, and three-storey buildings

Whether these paths are via a well-located elevator, narrow, labyrinthine stairways or, possibly, even a tight spiral staircase is of major importance.

The schematic drawing shows the practical location of the functional groups for small, medium-sized, and large pharmacies arranged over two or three floors: The customer and business rooms are always immediately adjacent to one another. In multi-storey pharmacies, the basement is a suitable location for the stockroom, whereby it is recommended that an elevator be installed, connecting the stockroom with the business rooms. Where the basement rooms have adequate daylight, social rooms can also be installed here, while service rooms are better located close to the business rooms. The upper floor may be used for the entire functional group of service and staff rooms.

INDIVIDUAL FUNCTIONS | On principle, the rooms belonging to the same functional group should ideally be close together – after all, this aspect was one reason why functional groups as such were created. There are also functional operations which call for close relationships between the individual functional groups. Thus the dispensary, which belongs to the functional group "customer rooms", should have a very close relationship with the business rooms, in particular with the alphabetically-organised stockroom which, in some cases, even forms a single unit with the dispensary, separated from the customer area only by the sales counter. Furthermore, it should be easy to reach the stockroom from the incoming goods area, and there should be a short path from the stockroom to the "supply and waste disposal" functional group. Some priorities should also be considered within the functional groups: The work room should be close to the alphabetically-organised stockroom. The prescription room and lab should ideally be located adjacent to one another or, as part of more affordable solutions, form a single unit.

Formation of zones in the floor plan | The pharmacy's floor plan is largely dependent on available floor space in a given building. Where the floor area can be chosen, or where a freestanding pharmacy building is being planned, consideration should be given to zoning the floor plan. The type of zoning determines the constructional nature of the pharmacy. Here, it is a question of arranging the pharmacy rooms from the viewpoint of the entering customers. A differentiation is made between pharmacies with one, two or three zones. The individual room zones may be immediately adjacent to one another or separated by hallways. In the case of a single zone floor plan, all rooms are adjacent to each other; with two zones, the rooms are arranged behind one another while, with a three-zone floor plan, the rooms are staggered to an even greater depth. On principle, these arrangement variations apply in small, medium-sized, and large pharmacies in buildings with one, two or three floors, whereby three floors are not suitable for small pharmacies. The type of zoning and the number of floors are determining factors for the cost-effectiveness of a floor plan. It is generally held that because of the avoidance of vertical paths and the high degree of compactness, single-storey buildings with two or three zones offer the best solution in terms of building typology and energy consumption, irrespective of the size of the pharmacy. A measure of compactness is the circumference of the vertical outer surfaces [U]. This value is by far the lowest for the two recommended solutions.

Daylight

In recent decades, demand for rooms with daylight has risen considerably. On principle, it is considered that staff rooms and rooms in which work is performed for longer periods should receive daylight. There are different regulations with regard to the length of time in the various federal states. Solutions which provide daylight in as many work and staff rooms as possible are seen as the way forward. Bearing in mind the remarks about daylight mentioned in the "space allocation plan" section, the architects face the challenge of finding means, including technical ones, which provide the pharmacy's employees with daylight.

Orientation

That people and, in this case, senior citizens in particular have problems with orientation is surely connected with information overload. The entrance itself is very often a psychological threshold. Walking through the pharmacy past vast numbers of shelves packed with goods, can feel like wandering in a maze. Lighting and colour markings on ceilings and walls can be useful here, as can the choice of flooring to mark specific routes, the avoidance of excessive product choice and, especially, the clear and well-organised positioning of the built-in and mobile furniture.

Technical fit-out

The sources of requirements governing the technical fit-out were discussed above in "Legislation, ordinances, and other regulations". Because of their wide scope, only a brief outline of explanations with regard to technical fit-out can be provided here.

STATICS | In the case of refurbishment, the planner must work with the existing static conditions, largely for economic reasons. In the case of new constructions, however, he should use the statics in order to achieve as broad a span width as possible. This creates wide spaces which are not divided or interrupted by walls or supports, allowing greater scope for design and flexibility for changes.

VENTILATION | The pharmacy should convey as many impressions on the senses as possible, proffering the scent of medicines is a questionable undertaking, as individual sensitivities vary greatly. The installation of an easily-controllable, effective ventilation system is required to provide a neutral level of ventilation. Perhaps, after further experiments, scents with a health-promoting effect may be dispersed through ventilation systems.

HEATING, COOLING, AIR HUMIDITY, AND FILTERING | The German Workplace Directive requires a minimum temperature of 21º C and a maximum temperature of 25º C in work rooms. This also provides a comfortable temperature for customers. Comfort also means that on hot days, a soothing coolness and pleasant air humidity are provided, since dry air has a negative effect on the airways. With regard to allergy sufferers, filtering the incoming air is good. Conventional heating systems, cooling equipment, humidifiers, and filters can take care of these needs.

AIR CONDITIONING | The requirements mentioned above can all be resolved by a modern air conditioning system. In new constructions, the decision to install such equipment must be made before planning starts. Retrofitting is a major undertaking. A properly adjusted air conditioning system offers the advantage of controllable climate without draughts and irritating noise. Space is also saved as no heaters are required. It should be checked in good time whether the investment and operating costs can be kept within reasonable bounds, and an estimate made of how much this air – which might be the healthiest around – is worth.

ACOUSTICS | Reducing noise pollution is one of the main tasks of environmental protection today. The building regulations of the federal states already call for protection against the transmission of airborne sound within buildings. DIN 4109[12] contains requirements with regard to rooms and construction elements; the German Workplace Directive regulates noise levels in work and staff rooms. These requirements may be fulfilled by separating construction elements through the right choice of floors to reduce impact sound and airborne sound, the use of acoustic ceilings and sound-absorbent wall coverings, the type of fit-out and furnishing, and the use of sound-absorbing doors.

LIGHTING | Exterior and internal lighting requirements are discussed in "Design elements" above. The planning of lighting systems is a highly complex and interrelated issue and should therefore, in the interests of the pharmacy's overall appearance, be carried out by a qualified specialist.

One room zone | Two room zones | Three room zones

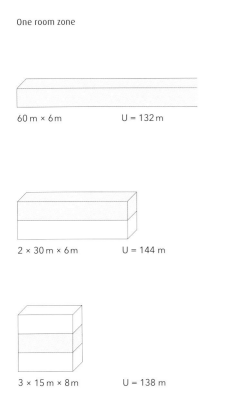
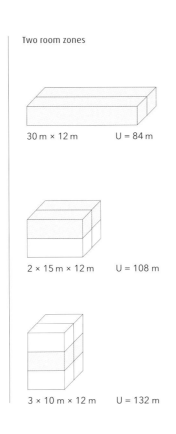
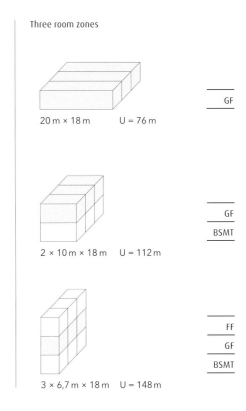

60 m × 6 m U = 132 m	30 m × 12 m U = 84 m	20 m × 18 m U = 76 m
2 × 30 m × 6 m U = 144 m	2 × 15 m × 12 m U = 108 m	2 × 10 m × 18 m U = 112 m
3 × 15 m × 8 m U = 138 m	3 × 10 m × 12 m U = 132 m	3 × 6,7 m × 18 m U = 148 m

GF

GF
BSMT

FF
GF
BSMT

Diagrams of one, two, and three-storey pharmacies with one to three room zones. For reasons of clarity, the verticals are shown in super-elevation. The enclosed space is the same size in all diagrams. "U" represents the circumference of the outer surfaces in metres.

BURGLARY PROTECTION | It is not merely the monetary value of the medicines, but also the attraction exercised by certain drugs which give rise to an increased risk of burglary. This cannot be prevented altogether but may at least be reduced with the large-scale use of technology. Fittings in this area are usually metal shutters and pull-down grilles as well as high-resistance safety glass and motion detectors with lights and sonic alarms. Video surveillance systems, in this case, have also proven useful, as these can help to identify burglars. A radio connection to the nearest police station finally can lead to an early arrest.

FIRE PROTECTION | Attention is dedicated to preventing fires in pharmacies, particularly in labs. The building regulations of the federal states, the German Workplace Directive, DIN 4102[13], the Health and Safety Authorities [Amt für Arbeitsschutz] and the Professional Accident Prevention and Insurance Associations [Berufsgenossenschaften] devote great attention to fire prevention in pharmacies, particularly in laboratory rooms. There are requirements regarding the fire-resistance class of supporting and non-supporting construction elements, and the fire behaviour of the building materials. The length and number of escape routes are also defined. Pharmacies must have a minimum of three fire extinguishers. In particularly dangerous areas, the installation of fire alarms and sprinkler systems is recommended.

Special features of refurbishment

In general, refurbishments are carried out to repair wear and tear in the fit-out and building substance. In addition, it is usually hoped to better exploit the area and space and to optimise workflows. In recent times, it is changes to the technical fit-out, such as the installation of a new lighting system, heating, ventilation, cooling or filtering systems, or the installation of complete air conditioning systems or, in conjunction with the establishment of an automated stockroom, the creation of a transport system, which are common reasons for minor or major construction work. While the construction of a new construction is subject only to the requirement to be completed by a given deadline, refurbishments, including expansions and redesigns, are associated with considerably greater difficulties. The temporary closure of the business premises usually ensures the fastest completion of the work but, for economic reasons, this solution is usually not chosen, so that construction work has to take place alongside normal business. Dustproof sheeting is required to protect customers and personnel and to preserve hygiene. Such sheets should also contribute to reducing the noise of the building work. It is usually necessary to relocate certain facilities or specialist rooms. In such cases, careful thought should be given to how the business can continue to operate during the individual building phases without too much additional effort.

PT	FUNCTIONAL GROUP FUNCTIONAL UNIT FUNCTIONAL ELEMENT	ILLUMINATION	IN SQM
1	**Customer rooms**		**56**
1.1	Sales area with self-service area, behind- the-counter shelves, cash-desk, and night dispensing area	○	50
1.2	Consultations	○	6
2	**Business rooms**		**46**
2.1	Laboratory	○	12
2.2	Prescription area	○	6
2.3	Workstation [and office]	○	12
2.4	Alphabetic storage	○	16
3	**Service and staff rooms**		**32**
3.1	Night shift room	○	10
3.2	Staff lounge and kitchen	○	8
3.3	Staff changing | Women		
3.3.1	Staff changing	◐	4
3.3.2	Shower	●	2
3.4	Staff changing | Men		
3.4.1	Staff changing	◐	2
3.4.2	Shower	●	2
3.5	Bathroom – Staff | Women		
3.5.1	Vestibule	●	1
3.5.2	Bathroom	●	1
3.6	Bathroom – Staff | Men		
3.6.1	Vestibule	●	1
3.6.2	Bathroom	●	1
4	**Supply and waste disposal**		**26**
4.1	Goods inwards	◐	4
4.2	Product storage	●	16
4.3	Cleaning equipment storage and disposal room	●	6
Usable area of a small pharmacy			**160**

Example of a functional and spatial allocation plan for a small pharmacy

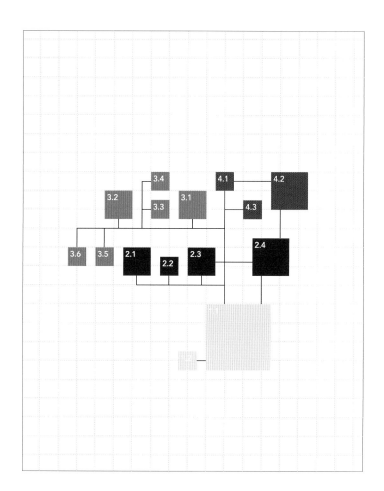

Function scheme of a small pharmacy

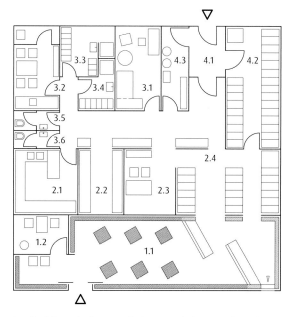

Standard floor plan for a small pharmacy, dark grey: self-service range, light grey: over-the-counter products, scale 1:200

STANDARD SPACE ALLOCATION PLANS, FUNCTIONAL DIAGRAMS, AND STANDARD FLOOR PLANS

The space allocation plans, functional diagrams, and floor plans shown here for small, medium-sized, and large pharmacies serve to explain the principles of operation and construction in one standard example each. Standard in this instance means a sample or model for the minimum effort required to achieve a functional and cost-effective solution.

For the space allocation plans, the explanation provided in "Breakdown into functional groups" above is used.

The space allocation plans also contain a comment on lighting [see "Space allocation plans" above]. A differentiation is made between individual rooms [functional elements] and rooms which belong directly together [functional units]. The suffix plan ID numbers designate the functional group and, consecutively in each case, the functional units and functional elements.

The functional diagrams show all rooms and their functional relationships. The different squares used as symbols for the rooms should give a rough indication of the size of the given usable area.

The standard floor plans are not intended to be a model for plans. In the deliberately chosen single-storey form shown here, they merely represent a visualisation of the space allocation plans and are thus confirmation of the latter's feasibility and an aid to orientation when commencing planning. Plans are the sole remit of architects, interior designers, and other pharmacy planners who, using their creativity and design skills, work together with pharmacists and their employees to come up with individual solutions.

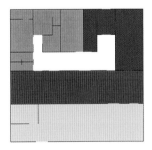

Diagrammatic plan of the floor plan, scale 1:400

Standards for a small pharmacy

For small pharmacies with a usable area of up to 175 square metres, it is particularly important to comply with the provisions of the ApBetrO with regard to minimum size. These state that a pharmacy must consist of a dispensary, a laboratory, a storage, and a night service room. These rooms must have a usable area of at least 110 square metres. These provisions do not make reference to the necessary service and staff rooms.

SPACE ALLOCATION PLAN AND FUNCTIONAL DIAGRAM FOR A SMALL PHARMACY | The minimum size required by the ApBetrO is met by the following space allocation plan, which has an area of 116 square metres [dispensary with consulting room 56 square metres, laboratory 12 square metres, prescription room 6 square metres, alphabetical storage and stockroom 32 square metres, and night service room 10 square metres]. Together with necessary expansions, this gives a usable area of 160 square metres, a value which should not be undercut. In selecting this size, it should be ensured that expansion is possible through the construction of an annexe, or by using rooms on a lower or upper floor in order to take future developments into account. The functional relationships between the rooms in the space allocation plan are relatively simply and easily implemented.

Because of the economical layout of the space allocation plan, dedicated rooms are not available for some functions. Thus, for example, there are no dedicated areas for special purposes [children's play area, washing room, special lab, archive, pharmacist's service room, special stockroom, and waste disposal room]. Furthermore, the "training" functional area is absent. If space-saving automatic systems are used, usage of the areas for alphabetical storage and the medical stockroom can be optimised.

FLOOR PLAN FOR A SMALL PHARMACY | The schematic floor plan is broken down into three zones with slight overlaps: the customer zone, the business room zone, and the storage and staff room zone. Incoming goods are transferred to the medical stockroom or to alphabetical storage, which has a direct link with the medical stockroom. The dispensary, which is parallel to the façade, has plenty of daylight and a curved customer path which leads to the two sales counters via numerous products displayed on wall shelving and in mobile containers. The dispensing counter for night service is conveniently located for the pharmacy personnel. All rooms in the central zone can be supplied with daylight if generous glazing is used.

PT	FUNCTIONAL GROUP FUNCTIONAL UNIT FUNCTIONAL ELEMENT	ILLUMINATION	USEABLE AREA IN SQM
1	**Customer rooms**		**124**
1.1	Sales area with self-service area, behind- the-counter shelves, and cash-desk	○	74
1.2	Special purpose rooms		
1.2.1	Beauty treatment	○	8
1.2.2	Spa treatment	○	8
1.2.3	Other areas	○	10
1.3	Night service dispensary	●	2
1.4	Consulting	○	8
1.5	Treatment and rest room	○	12
1.6	Bathroom – Customers	Unisex		
1.6.1	Vestibule	●	1
1.6.2	Bathroom	●	1
2	**Business rooms**		**66**
2.1	Laboratory	○	14
2.2	Prescription area	○	10
2.3	Workstation	○	10
2.4	Washing room	◉	8
2.5	Alphabetic storage	○	24
3	**Administrative rooms**		**20**
3.1	Office	○	14
3.2	Archive	●	6
4	**Service and staff rooms**		**54**
4.1	Service room – Pharmacist	○	8
4.2	Night shift room	○	12
4.3	Staff lounge and kitchen	○	16
4.4	Staff changing	Women		
4.4.1	Staff changing	◉	6
4.4.2	Shower	●	2
4.5	Staff changing	Men		
4.5.1	Staff changing	◉	4
4.5.2	Shower	●	2
4.6	Bathroom – Staff	Women		
4.6.1	Vestibule	●	1
4.6.2	Bathroom	●	1
4.7	Bathroom – Staff	Men		
4.7.1	Vestibule	●	1
4.7.2	Bathroom	●	1
5	**Supply and waste disposal**		**56**
5.1	Goods inwards	◉	8
5.2	Storage		
5.2.1	Product storage	●	30
5.2.2	Drug storage	●	8
5.3	Cleaning equipment storage	●	6
5.4	Waste disposal room	●	4
Usable area of a **medium-sized pharmacy**			**320**

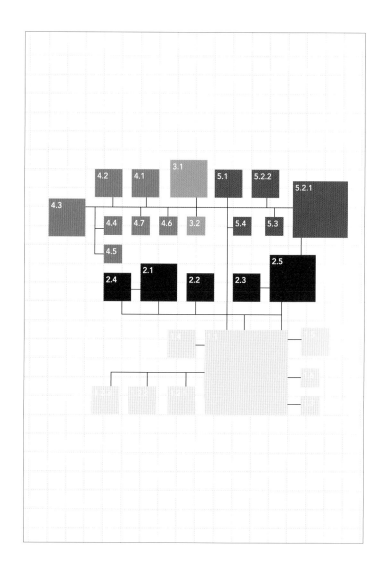

Function scheme of a medium-sized pharmacy

Example of a functional and spatial
allocation plan for a medium-sized pharmacy

Standards for a medium-sized pharmacy

This category's upper size limit of 320 square metres of usable area, as a basic principle, allows for a well-equipped pharmacy.

Space allocation plan and functional diagram for a medium-sized pharmacy | With the exception of the special purpose rooms and the rooms for training, all functional groups are in evidence. The special purpose room also provides space for expanding the product range, while the treatment and rest room can also be used for healthcare services. The pharmacist has his own service room, and there is a special stockroom.

Floor plan for a medium-sized pharmacy | Because of the short paths involved, a roughly square floor plan divided into three zones has been chosen: the customer zone, the business zone, the zone for the administrative, service and staff rooms, and the supply and waste disposal rooms. Where generous glazing is used in the rooms adjacent to the customer area, all rooms in which the staff spend longer periods of time have daylight. Practical transport is installed from the incoming goods area to the stockroom, and from there to alphabetical storage.

The business rooms are conveniently located adjacent to each other; the same goes for the service and staff rooms. From the office, incoming goods can be directly monitored. The dispensary, also arranged lengthwise here, makes it possible for passers-by to get a glimpse in through the shelves. There is also plenty of light in the room above the shelves. The customers reach the three sales counters after they have had the chance to look at the self-service range. Other products are displayed on the shelves and in the hanging elements of the over-the-counter section. In a medium-sized pharmacy, the dispensary should also have a consultancy room, a treatment and rest room, a waiting area, and a WC. Customers have access to the night dispensary immediately beside the slightly recessed entrance.

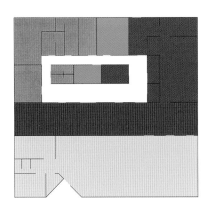

Diagrammatic plan of the floor plan, scale 1:400

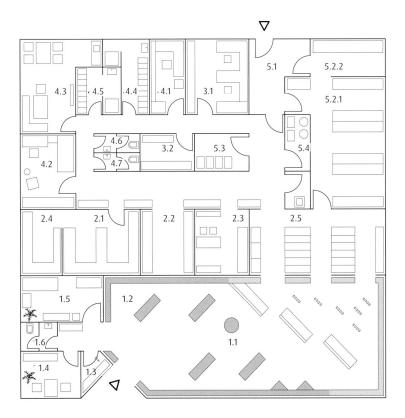

Standard floor plan for a medium-sized pharmacy,
dark grey: self-service range,
light grey: over-the-counter products,
scale 1:200

PT	FUNCTIONAL GROUP FUNCTIONAL UNIT FUNCTIONAL ELEMENT	ILLUMINATION	IN SQM
1	**Customer rooms**		**176**
1.1	Sales area	○	90
1.2	Special purpose rooms		
1.2.1	Beauty treatment	○	12
1.2.2	Spa treatment	○	12
1.2.3	Other areas	○	14
1.3	Children's play area	○	6
1.4	Night service dispensary	●	4
1.5	Consulting	○	6
1.6	Intensive consulting	○	10
1.7	Treatment		
1.7.1	Waiting room	◉	6
1.7.2	Treatment and rest room	○	12
1.8	Bathroom – Customers | Women	●	1
1.9	Bathroom – Customers | Men	●	1
2	**Business rooms**		**82**
2.1	Laboratory	○	16
2.2	Prescription area	○	12
2.3	Workstation	○	10
2.4	Washing room	◉	8
2.5	Alphabetic storage	○	36
3	**Special purpose rooms**		**10**
3.1	Special lab	○	10
4	**Administrative rooms**		**26**
4.1	Office	○	18
4.2	Archive	○	8
5	**Service and staff rooms**		**68**
5.1	Service room–Pharmacist	○	10
5.2	Night shift room	○	12
5.3	Staff rest area (lounge | kitchen)	○	22
5.4	Staff changing W incl. shower	○●	12
5.5	Staff changing M incl. shower	○●	12
5.6	Bathroom – Staff | Women		
5.6.1	Vestibule	●	1
5.6.2	Bathroom	●	1
5.7	Bathroom – Staff | Men		
5.7.1	Vestibule	●	1
5.7.2	Bathroom	●	1
6	**Supply and waste disposal**		**88**
6.1	Goods inwards	◉	10
6.2	Storage		
6.2.1	Product storage	●	36
6.2.2	Drug storage	●	14
6.2.3	Packaging material and window dressing	●	4
6.3	Order picker	○	12
6.4	Cleaning equipment storage	●	6
6.5	Waste disposal room	●	4
7	**Training rooms**		**30**
7.1	Training	●	
7.1.1	Cloakroom	◉	4
7.1.2	Lecture and seminar room	●	26
Usable area of a large pharmacy			**480**

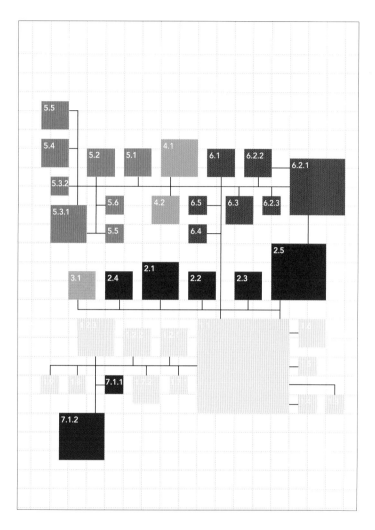

Function scheme of a large pharmacy

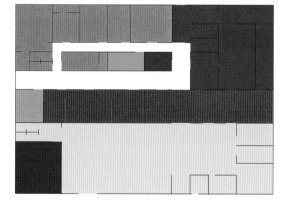

Diagrammatic plan of the floor plan, scale 1:400

Example of a functional and spatial allocation plan for a large pharmacy

Standards for a large pharmacy

SPACE ALLOCATION PLAN AND FUNCTIONAL DIAGRAM FOR A LARGE PHARMACY | All functional groups are represented here. In even larger pharmacies, for example those in hospitals, there are, in particular, several special purpose rooms and other usable areas for training. Using an area of 480 square metres, the space allocation plan represents the average size for large pharmacies.

FLOOR PLAN FOR A LARGE PHARMACY | A shape resembling a square is also chosen for the floor plan of a large pharmacy, as this shortens paths. Here too, three zones are foreseen which, because of the chosen floor plan shape, give rise to partial overlaps. Despite the necessary staggered depth of the floor plan, almost all rooms in which employees spend longer periods have daylight, whereby it is required that the walls oriented towards the customer area have generous expanses of glass. In the case of the centrally located order room, above the alphabetical storage and connected with the stockroom, a skylight can help to provide those working there with a glimpse of the sky. The pharmacist's room, located beside the office with integrated library, offers a good overview of incoming and outgoing goods. The staff rooms form an interconnected unit. Customers are offered rooms for discreet and intensive consultation, a treatment and rest room with waiting room, a lecture and seminar room with cloakroom, a children's play corner which they can see clearly, and toilets for men and women, all directly accessible from the dispensary. In large pharmacies too, the customer's path leads him to the sales counters at the rear of the room via a range of shelves and mobile containers displaying additional product offers. Customers requiring night service can avail of a small room, accessible from outside the pharmacy, which is also practical during harsh winter weather.

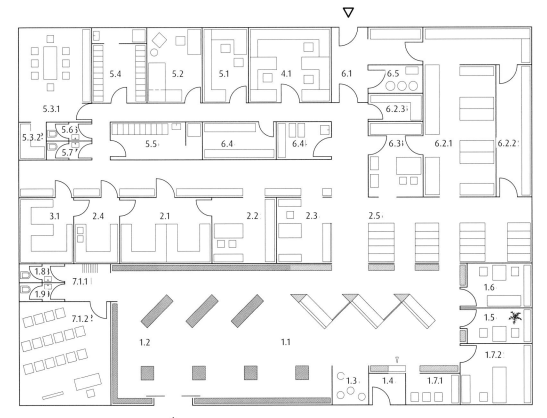

Standard floor plan for a large pharmacy,
dark grey: self-service range,
light grey: over-the-counter products,
scale 1:200

Signing the contract

Sketch design

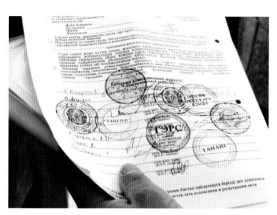
Building permit

DESIGN ELEMENTS

Exterior architecture
The appearance of some buildings reveals their purpose. In terms of the importance that pharmacies can have in daily life, this kind of identification would be most helpful. Designers should feel compelled to find suitable architectural means to create an unmistakable image for pharmacies.

BUILDING STRUCTURE | Pharmacies usually occupy parts of several-storey buildings, which may be residential or commercial properties, department stores, shopping centres, group medical practices, railway stations or airports. In such cases, the pharmacy is normally required to fit in with the existing exterior architecture. However, some pharmacies are also designed as individual buildings, and it is here that a distinctive pharmacy design is possible. The shaping of such buildings – mainly in a range from one storey to three storeys at most – can be used to make them attractive to customers and to suggest the appeal of a visit. In general, imagination and shape vocabulary have no boundaries. However, despite the desirable differentiation, it is recommended that the pharmacy architecture take the landscape and the surrounding buildings into consideration. As Mies van der Rohe famously exhorted – less is more.

FAÇADE | In terms of the façade too, very different looks are possible, depending on whether the pharmacy is part of a greater structure or an individual building. In the first instance, there is often only the possibility of affixing illuminated elements with the pharmacy "green cross" symbol ["A" in Germany] and the name of the pharmacy. In the second case, there is considerably more scope for design, using open and closed walls, expressive materials, a diverse range of colours, and by playing with light and shadow. Here too, information overload should be avoided, as this does not fit in with the essentially serious nature of the pharmacy.

ENTRANCE | For customers, who are often not in the best of health, the main entrance to the pharmacy should be clearly identifiable from a considerable distance, including for people who are unfamiliar with the area or those with disabilities. It is often wise to create one or more side entrances, in order to shorten customer paths. It is also helpful to design an entrance which acts as a magnet, and to install nocturnal illumination. Automatic doors and barrier freedom ease access for customers with wheelchairs and pushchairs/prams. The ApBetrO requires that the opening hours and the rota service list be displayed at the entrance. The Trade, Commerce, and Industry Regulation Act specifies that the proprietor's name must also be shown directly there.

SHOP WINDOWS | In shop design terms, the primary function of shop windows is to display the goods on offer inside. While old pharmacies often only had windows for illumination, newer pharmacies are usually fitted with large display windows, which advertise new products more or less aggressively.
Unfortunately, this often blocks the view of the usually interesting interior of the pharmacy. As such, many pharmacists today dispense with the presentation of products in the shop windows, and use the generous window space to create a sense of transparency and openness, in order to generate trust. In arranging shelving on glazed outer walls, clever design should ensure that the interior is still visible, thus arousing curiosity and the desire to enter the pharmacy.

EXTERIOR LIGHTING | The exterior lighting should attract customers, ease orientation, and it should as well create the right mood. Special attention should always be paid to the pharmacy's entrance. The strength of the illumination should be adjusted to the level of lighting in the surrounding area, in order to avoid glare.

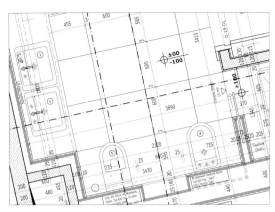
Planning and construction process I

Technical drawings

Interior design

The planner has a broad palette of architectural design of interior spaces in pharmacies from which to choose. Armed with his creative experience, a well developed sense of quality, and a knowledge of the effects of hard and soft shapes, harsh and soft colours, light and shadow, and the use of materials, he should approach his work with a loving eye to detail.

BUILT-IN AND FREESTANDING FURNITURE | One of the primary goals of the pharmacy planner should be to create an individual feel, and this can be best achieved with built-in furniture. The exclusive use of mass-produced furniture is less suitable for this purpose; however, the cost of custom-made designer furniture is often incompatible with the available budget, so the planner needs to find a carefully balanced solution. All of the furnishing elements should form an over-arching design concept, which expresses the desired image of the pharmacy. In order to preserve customer privacy, the once usually large single point of sale [PoS] is now commonly replaced by several individual PoS, which should be kept clear of goods. Such units should ideally include an integrated till, an integrated coin tray, and the necessary computer equipment [monitor, keyboard, price display, scanner, and photocopier]. Other small but important details include bag hooks or holders for customers' mobility aids, so that they can carry on a conversation with the pharmacist and inform themselves on the product range without distractions, as well as floor markings which designate the desirable distance to be kept from the customer being served.

The alphabetical storage area with drawer cabinets for storing medicines, often in modular form offering a high degree of flexibility, is close by. Two alphabetical storage units might also be considered: a unit for products which are turned over quickly, close to the sales tables, and an additional unit for other products which may be installed further to the rear. Alternatively, medicines can be stored in an automated warehouse and transported to the individual PoS by pressing a button.

Before reaching the PoS, the customer has a chance to select his own self-service products. For practical reasons, these are located in individual mobile shelf units or display gondolas, organised by product type. Shelves containing over-the-counter products are arranged behind the PoS. Where the size of the pharmacy permits, furnishings should also include a seating group and a children's play area.

INTERIOR LIGHTING | In a somewhat adapted form, the same principles applying to exterior lighting go for interior spaces. Light accompanies the customer from the moment of entry and makes orientation easier. Accent lighting points the way to specific products in the self-service and over-the-counter displays. In selecting lighting, attention should be paid to correct colour reproduction and a comfortable atmosphere which may be achieved using a warm white light. In some places, thought should be given to where the light falls, to avoid the irritating formation of shadows. Work rooms should be adequately lit, without glare or monitor reflections, aiming for low energy consumption and low heat load. The minimum requirements of DIN 5035[14] and the provisions of the German Workplace Directive must be met.

SURFACES | Visible at all times and usually also tactile, the surfaces of fittings, furniture, equipment, walls, ceilings, and floors represent the main objects of design work and challenges. For each object, the planner has to select materials from an almost unlimited range of possibilities, and decide on a certain surface structure and colour. For fittings and furniture, natural surfaces with their integral colours and man-made surfaces with a large choice of colours are available. The choice, made in keeping with the overall concept, should also give thought to resilience and easiness to clean. For hygienic reasons, equipment should be made of durable materials which are not subject to wear and tear. With regard to walls, differentiations should be made: In general, plaster which can be painted in white or colour, or papered over, is adequate; for operational and sanitary rooms, washable wall coverings should be used. Ceilings are usually suspended to make room for lighting, transport systems, ventilation, and air conditioning. Because of the easy accessibility required for servicing, they are usually made of cassettes which are often also used for improving the acoustics. To date, only a few examples of ceilings being used as a playground for imaginative design have been discovered. When choosing the right flooring, a range of aspects needs to be considered: The intended effect of the flooring in the overall design concept, the existing fitting dimensions, particularly in light of future redesigns, resilience, sound absorption, skid resistance, easiness of cleaning, and, crucially, cost-effectiveness. In laying flooring at the entrance area, one should not forget to build in a sufficiently large, floor flush doormat. With all surface objects, colours have a major impact on the overall appearance, and thus also have a psychological effect on customers and personnel. According to Frieling, colours may be calming, aggressive, oppressive, encouraging, alienating, soft, alarming, stimulating, warming, eye-catching, constricting, securing, comforting, confusing, distracting, cold, alerting, heightening, cooling, intensifying, glowing, communicative, activating, light, covering, fostering, and relaxing[15]. Well-dosed colour can therefore promote orientation in a given space, alter proportions, and furthermore stimulate certain moods. Too many different colours should be avoided as this usually only achieves a short-term effect.

ELEMENTS OF PLANNING AND
CONSTRUCTION PROCESS

Identification of basic data
Once the idea of expanding, converting or refurbishing an existing pharmacy, or of erecting a new construction has been sparked, the developer, together with other stakeholders, must compile a list of basic data, the most important of which are outlined below. These comments are only discussed briefly, as these are jobs which must anyway be performed by experienced and qualified specialist planners.

LOCATION ANALYSIS | In planning a new pharmacy, the choice of location for the future business should not be left to chance. Rather, before any further decisions are made, a thorough analysis of the relevant facts should be made. Distances from existing, competing pharmacies [pharmacy density] are important, as are proximity to medical practices, group medical practices or other medical facilities, the transport situation, location in relation to means of transport, population density, projected level of casual customers, size and purchasing power of potential target groups [families, singles, senior citizens], and plot prices. The necessary assessment and weighting of the numerous different aspects in the event of alternative locations, and the recommendation of a preferred solution, call for a great deal of experience and should be placed in the hands of a specialist, who can help to ensure that such expenditure is worthwhile.

MASTER PLANNING | Because of numerous flawed decisions with regard to construction measures, experts have for some years been advocating master planning. This works out the basics for operation and construction [and may also include the location analysis mentioned above]. In the case of redesigns, a target/actual comparison is carried out, schematic variants for reaching targets developed, and, finally, a proposal made for the chosen solution on the basis of the analysis, and with due consideration for projected investment and operating costs. Master planning should also include the pharmacy's potential future development and plan in corresponding scope for possible alterations and expansions. Supplementary sheet 4 of DIN 13080 "Division of Hospitals into Functional Areas and Functional Sections–Masterplanning for General Hospitals" contains comments which also apply to pharmacies.

Furniture production

Construction site inspection

TIME PLANNING | Alongside the estimated timing of the planning process as projected in the master plan, where a specific construction measure has been decided, precise time planning forms a major element of planning. Precise periods and completion deadlines are drawn up for all phases. An element of effective time planning is ongoing monitoring and, where required, updating. Based on the motto "time is money", punctual completion of construction can avoid the considerable extra costs which are inevitably associated with overrun deadlines. This applies in particular to refurbishments and redesigns which are carried out while the business continues its daily operations. The disturbance and unpleasantness usually suffered by customers and personnel in such situations should not continue for even a single day longer than is absolutely necessary.

SPACE ALLOCATION PLANNING | Once the basics of the pharmacy project have been established, a space allocation plan has to be drawn up. Here too, specialised planners should ideally be used, who must undertake this assignment together with the pharmacy manager and his staff right from the outset. Details of the formal breakdown and necessary details are shown in the earlier chapter "Data required for construction". It should be ensured that the space allocation plan shows each room with its required usable area. Rough sketches for critical rooms should already be prepared, which will provide an explanation of the specified usable area values. A good space allocation plan will not omit any rooms [including the caretaker's room]; developing rooms will be given a suitable reserve of space, and no unnecessary space will be demanded, as not only will its construction call for major expenditure, but it must also be operated above and beyond the pharmacy's lifetime.

COST PLANNING | The developer must consider the investment costs arising out of construction measures at all times, but particularly when commencing planning. The developer wishes to get an overview of costs at as early a stage as possible, so that he can feel secure in making fundamental decisions. If he knows the cost framework, for example, he has the opportunity to select the appropriate design option and standard of fit-out.

With regard to new constructions or expansions, there are processes which allow a rough estimation of costs immediately after the space allocation plan has been drawn up. According to the cost category process developed by the Institut für Wirtschaftliches Bauen in Bremen, an established cost can be allocated to the usable areas of the individual types of room on the basis of analyses conducted throughout Germany. In this way, a total cost can then be calculated. A rougher estimate of the projected costs can also be obtained using pharmacy-specific empirically established costs per square metre of usable area.

For redesigns and refurbishments, the above methods can only be used to a limited extent because of the wide range of variations. It is the long-standing experience of qualified planners which is required.

Planning fees

Clause 15 of the Fee Structure for Architects and Engineers [HOAI][16] specifies the fees to be paid by the client. A differentiation is made between new constructions, new plant, reconstruction, expansion, redesign, modernisation, spatial expansion, maintenance, and restoration. Based on services, the fees are broken down into nine different fee groups; important ones are briefly outlined below.

PRE-PLANNING | The most important basic services are the analysis of the basic data, the drafting of a planning concept, including the examination of alternative options, the clarification and explanation of relevant urban construction, design, functional, technical, physical construction, economic and energy-related aspects, prenegotiations with authorities, and an estimate of costs in line with DIN 276. Special services include, for example, the creation of a financing plan, or a building and operational cost-benefit analysis.

DRAFT PLANNING | Basic services largely comprise the drawing of a plan on a scale of 1:100 and, for spatial expansions on a scale of 1:50 and 1:20, negotiations with authorities and other specialists involved in planning with regard to approvability and an estimate of costs in line with DIN 276.

IMPLEMENTATION PLANNING | This phase involves the final implementation, detail and construction drawings on a scale of 1:50 to 1:1, including determination of the materials. For pharmacies, in general these plans are particularly important because they fundamentally fix the design of the details.

PROJECT MONITORING | Once the "Preparation" and "Collaboration in awarding the contract" basic services [which we have skipped here] have been provided, project monitoring has the aim of overseeing the completion of the job in keeping with the building permit and all mandatory requirements with regard to individual aspects of the design. The basic services in this phase also include the creation of a time schedule, an estimate of costs in line with DIN 276, an application for final certification from the authorities, and de-snagging. This basic service comprises 31 percent – almost one-third – of the total fees.

COMMISSIONING PLAN | Although this does not form part of the HOAI services, a detailed commissioning plan can be very practical. Particularly for building work carried out during normal business operations, a gradual, ideally interruption-free transition from the outset is desirable. An inauguration party with interesting lectures, drinks, small gifts, and special offers can be very useful in winning customer loyalty and for advertising purposes. The author of this work, anyway, very much looks forward to receiving such invitations.

Sources

1. Bundesvereinigung Deutscher Apothekerverbände: Die Apotheke. Zahlen, Daten, Fakten. Berlin 2018.
2. Apothekenbetriebsordnung of 26 September 1995. Bundesgesetzblatt I. Page 1195 and Page 1574.
3. Gesetz über das Apothekenwesen [Apothekengesetz–ApoG] of 15 Oktober 1980, Bundesgesetzblatt I. Page 874.
4. Robert-Koch-Institut: Anforderungen der Hygiene an die funktionelle und bauliche Gestaltung. Bundesgesundheitsblatt 1989.
5. Damaschke, Sabine u. a.: Apotheken. Planen, Gestalten und Einrichten. Leinfelden-Echterdingen 2000.
6. DIN 277-1: 2005-02, Grundflächen und Rauminhalte von Bauwerken im Hochbau – Part 1: Begriffe, Ermittlungsgrundlagen | DIN 277- 2; 2005- 02, Grundflächen und Rauminhalte von Bauwerken im Hochbau – Part 2: Gliederung der Netto-Grundfläche [Nutzflächen, Technische Funktionsflächen und Verkehrsflächen]
7. DIN 13080: 2016-06, Gliederung des Krankenhauses in Funktionsbereiche und Funktionsstellen
8. Spegg, Horst: Apothekenbesichtigung. Ein Handbuch zur Selbstkontrolle des Apothekenbetriebs. Stuttgart 2000.
9. DIN 12924-4: 1994-01, Laboreinrichtungen; Abzüge; Abzüge in Apotheken; Hauptmaße, Anforderungen und Prüfungen
10. Verordnung über Arbeitsstätten [Arbeitsstättenverordnung – ArbStättV], Bundesgesetzblatt I. Page 1595.
11. DIN 276-1: 2006-11, Kosten im Bauwesen – Part 1: Hochbau
12. DIN 4109-1: 2006-10, Schallschutz im Hochbau; Anforderungen und Nachweise
13. DIN 4102-1: 1998-05, Brandverhalten von Baustoffen und Bauteilen – Part 1: Baustoffe; Begriffe, Anforderungen und Prüfungen
14. DIN 5035-8: 2007-07, Beleuchtung mit künstlichem Licht – Part 8: Arbeitsplatzleuchten – Anforderungen, Empfehlungen und Prüfung
15. Frieling, Heinrich: Farbe am Arbeitsplatz. Munich 1992.
16. Honorarordnung für Architekten und Ingenieure in its current version of 10 July 2013.

Pharmacies

Zum Löwen von Aspern
ARTEC Architekten

Adler Pharmacy
Jörn Bathke

Wilhelm Pharmacy
Jörn Bathke

Alpin Pharmacy
Klaus R. Bürger

Klemensplatz Pharmacy
Klaus R. Bürger

OHM Pharmacy
Klaus R. Bürger

TRI-Haus Pharmacy
Renate Hawig

St. Anna Pharmacy [1828]
Huber Rössler

Linden Pharmacy
Ippolito Fleitz Group

Stadtklinik Pharmacy
sander.hofrichter

This unusual pharmacy, with its large exposed concrete roof, is a dispensary, event space and herb garden all in one. The two-storey building is located on a long, narrow site with streets on either side, in the centre of the Aspern district of Vienna. It has to allow as much light as possible to reach the houses on both sides, and therefore has large front and back windows and a fully glazed atrium, complete with gingko tree. The bright, transparent look is enhanced by the shelves hanging from the concrete ceiling in the sales area, which do not touch the polished asphalt floor. The slightly offset rows of shelves, which are lit from the insides and to which the ceiling strip lights form a visual continuation, add tension and rhythm to this large space. Each shelf has its own consultation area, which in turn adjoins the full-length drugs cabinet. The contrast between white furniture and grey concrete is one of the defining features of this interior. The dispensing area and a large seminar room are located at the rear of the first floor, and there is a medicinal herb garden on the roof terrace.

ARTEC ARCHITEKTEN

ZUM LÖWEN VON ASPERN
VIENNA

Pharmacies

Pharmacies

**Client	operator**	Wilhelm Schlagintweit
Design phase	10 2002–04 2003
Built	04 2003–09 2003
Gross floor area	581 sqm
Usable surface	464 sqm
Gross volume	2,500 cbm
Construction cost	1,000,000 €
Total cost	1,000,000 €

a Large windows overlook an atrium containing a gingko tree.
b The contrast between the white furniture and grey concrete is one of the interior's defining features.
c The strip lights on the suspended shelves continue to the ceiling, creating a sense of tension and rhythm in this large space.

Other key information

Catchment radius	6 km
Distance to nearest pharmacy	500 m
Nearest health facility	Hospital, medical practice
Number of items in stock	6,500
Stock turnover rate	12
Number of staff	24
Prescriptions filled per day	250
Proportion of special services	10 %
Technical facilities	Air conditioner

Nature of special services: cosmetics, skin and face analysis, yoga, qigong, fasting cures, nutrition advice, cultural program, lectures

The pharmacy is on three floors, so staff have to do a considerable amount of walking around, but there are conveniently located stairs to the upstairs office and staff lounge and basement storage area. The consultation and seminar rooms in the courtyard and the rooftop herb garden work particularly well.

Usable floor spaces

Sales	**yellow**	207 sqm	45 %
Service	**red**	67 sqm	14 %
Administration	**green**	41 sqm	9 %
Employee areas	**orange**	44 sqm	9 %
Supply	disposal	**brown**	55 sqm	12 %
Training	**purple**	50 sqm	11 %
Total		464 sqm	100 %

Diagrammatic plans, scale 1:400
Floor plans, scale 1:200

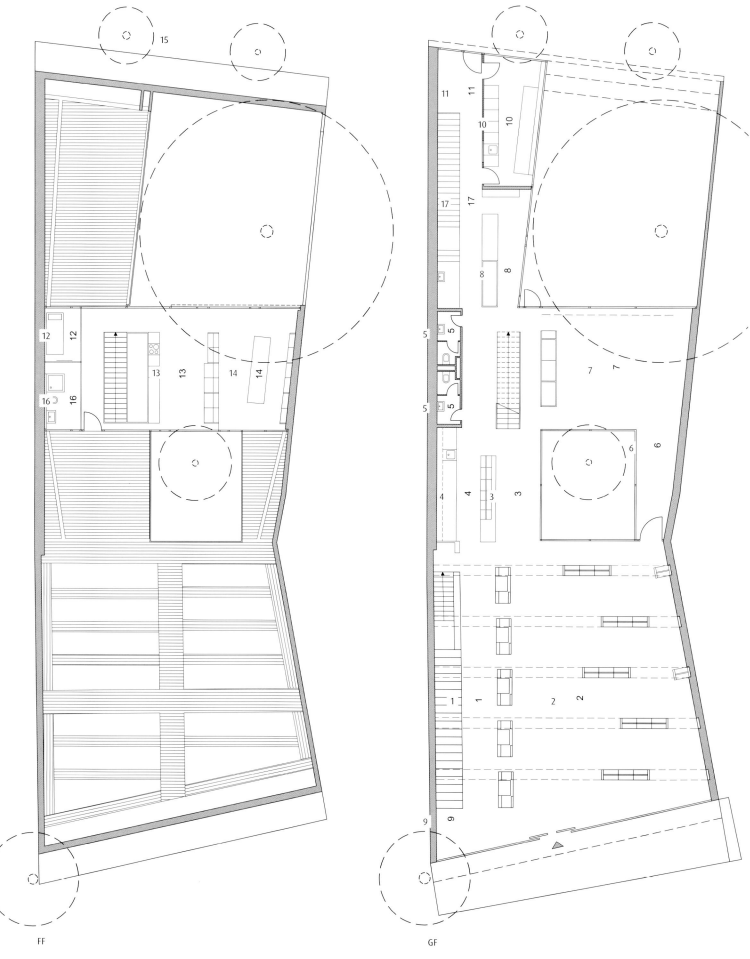

Pharmacies

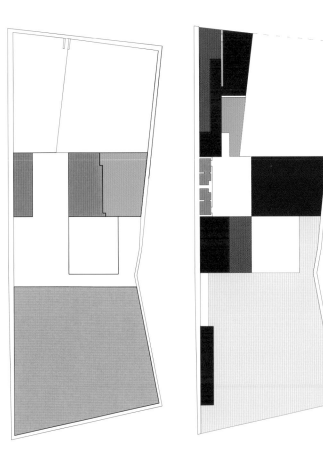
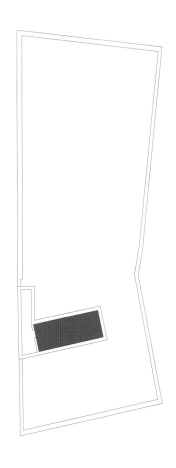

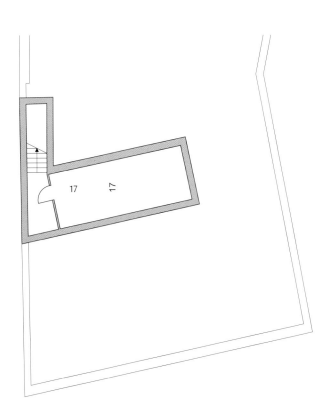

BSMT

Floor plan layout

1. Sales area
2. Self-service area
3. Drug storage
4. Prescription area
5. Bathroom
6. Consultations
7. Seminar room
8. Office
9. Alphabetic storage
10. Laboratory
11. Goods inwards
12. Night shift room
13. Staff lounge
14. Office
15. Herb garden
16. Bathroom | shower
17. Product storage

The massive curved concrete counter in the middle of the Adler Pharmacy provides an eye-catching sculptural centrepiece. Its monolithic elegance also symbolises durability and reliability, but also dynamism, movement, and kinetic energy. It is nine metres long, made from a single piece of concrete and curves diagonally through the interior with a depth of up to 120 centimetres, resembling a sinuous plesiosaur. The counter will undoubtedly remain in position for many decades, its immutability a reminder that despite the pharmacy's new location, it has a 100-year history. The space is strongly horizontal, with curving timber-clad walls, veneered cylindrical columns and glass steles in the display window. The dark Wenge wood horizontal wall cladding accentuates and frames the red-backlit shelves. The dramatic colours and lines contrast with the simple materials of the interior, which is dominated by Wenge wood, slate, and concrete. The pharmacy has two stories: the lower with the alphabetic storage area and controlled drug storage at the rear, and the upper, accessed via an internal staircase, containing the office, prescription area, and staff lounge.

JÖRN BATHKE

ADLER PHARMACY
KAMEN

Pharmacies

Pharmacies

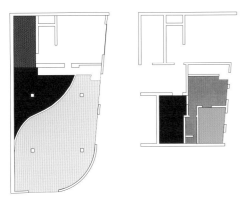
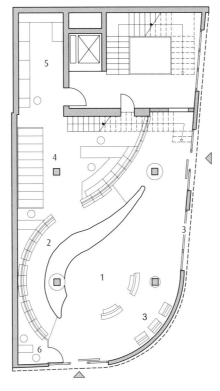
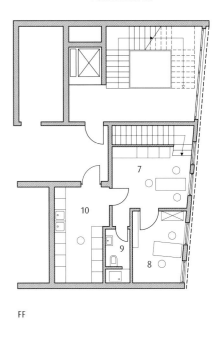

GF

FF

Client | operator	Eva Rapos	
Design phase | built	2007	
Usable surface	163 sqm	
Gross volume	584 cbm	
Total cost	168,000 €	

a The nine-metre counter, shaped from a single piece of concrete, curves diagonally though the space.
b The dark Wenge wood horizontal wall cladding accentuates the red-backlit shelves.

Diagrammatic plans, scale 1:400
Floor plans, scale 1:200

Other key information

Catchment radius	10 km
Distance to nearest pharmacy	150 m
Nearest health facility	Medical houses, medical practice
Proportion produced in-house	1 %
Drugs produced in-house	Salves, tinctures
Proportion delivered externally	15 %
Type of customers supplied	Retirement homes
Number of staff	6
Technical facilities	Air conditioner, laboratory facilities

Nature of special services: blood pressure, blood sugar, and cholesterol testing, compression stockings, baby weighing-scale, and inhaler loans, prescription pickup service, cosmetics training.

The two-storey building has some shortcomings: The combined prescription area and laboratory on the upper floor are accessible only via a staircase, though this is less important in the case of the office and lounge. Likewise, access to the office via the staff lounge is less than ideal.

Usable floor spaces

Sales	**yellow**	81 sqm	50 %
Service	**red**	40 sqm	25 %
Administration	**green**	11 sqm	7 %
Employee areas	**orange**	17 sqm	10 %
Supply | disposal	**brown**	14 sqm	8 %
Total		163 sqm	100 %

Floor plan layout

1 Sales area
2 Cash desk
3 Self-service area
4 Alphabetic storage
5 Drug storage
6 Consultations
7 Staff lounge
8 Office
9 Bathroom \| shower
10 Prescription area \| laboratory

The modern Wilhelm Pharmacy lies behind a Stalinist precast concrete-slab façade on Berlin's historic Wilhelmstraße. Specialising in traditional Chinese medicine, it also uses the Normamed® system and detailed patient case histories. The interior design is intended to bridge the gap between mainstream European medicine and complementary Asian therapies. The strict rectangular floorplan contrasts with the gently rounded counter. The counters are surrounded by colourful cylindrical columns decorated with specially made, digitally printed, and enlarged motifs from traditional Chinese medicine. Horizontal plywood sheets are fitted to the semicircular wall separating the sales area from the offices, with narrow tubular aluminium spacers being used to create an attractive light and shade effect. The counters, too, have been designed to make an impression, with small LED lights shining through the translucent green countertops. These create an enigmatic impression of depth and contrast effectively with the dark matt surfaces of the furnishings.

JÖRN BATHKE

WILHELM PHARMACY
BERLIN

Pharmacies

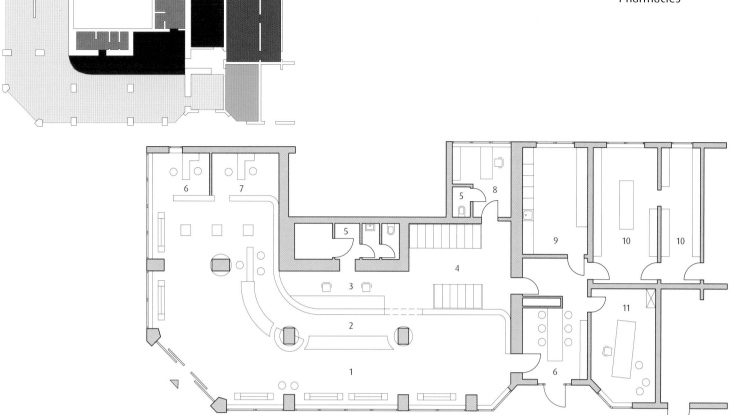

**Client	operator**	Christian Melzer
**Design phase	built**	2006
Usable surface	306 sqm
Gross volume	801 cbm
Total cost	230,000 €

a Horizontal plywood panels create an attractive chiaroscuro effect.
b Small LED lights shine through the translucent green countertops.
c Brightly coloured cylindrical columns surround the counters.

Diagrammatic plans, scale 1:400
Floor plans, scale 1:200

Other key information

Catchment radius	50 km
Distance to nearest pharmacy	300 m
Nearest health facility	Medical practice
Number of items in stock	4,000
Stock turnover rate	12
Proportion produced in-house	2 %
Drugs produced in-house	TCM
Proportion delivered externally	0 %
Number of staff	4
Proportion of special services	10 %
Nature of special services	TCM

This single-storey pharmacy has been designed for maximum user friendliness. There are three consultation rooms, one with additional outside access, but the staff lounge is too small.

Usable floor spaces

Sales	**yellow**	162 sqm	53 %
Service	**red**	59 sqm	19 %
Administration	**green**	21 sqm	7 %
Employee areas	**orange**	22 sqm	7 %
Supply	disposal	**brown**	42 sqm	14 %
Total		306 sqm	100 %

Floor plan layout

1 Sales area
2 Cash desk
3 Workstation
4 Alphabetic storage
5 Bathroom
6 Consultations
7 Private consultations
8 Night shift room
9 Prescription area | laboratory
10 Product storage
11 Office

The new Alpin Pharmacy is located in a medical centre close to a hospital in the Allgäu mountains, and a picture of the surrounding landscape is displayed above the shelves extending along the whole length of the sales counter. Local materials such as sycamore and slate have been used in an expression of the young pharmacist's love of the area and of the natural environment, which is also reflected in the name of the business. Two other fundamental themes of the design are openness and transparency. The large, open-plan sales area is visible at a glance, and the dispensary is separated from it only by a glass screen. Movable display shelves have been placed in front of the full-height windows facing the street, their small size allowing an ample view of the interior to lure passers-by inside. The sense of openness is increased by the use of simple, uniform materials: a green slate floor in the public area and sycamore parquet in the staff-only parts of the building. One unusual feature of this pharmacy, perhaps a foretaste of the future, is a drive-in counter.

KLAUS R. BÜRGER

ALPIN PHARMACY
KEMPTEN

Pharmacies

Pharmacies

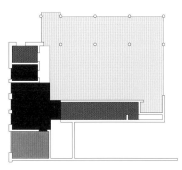

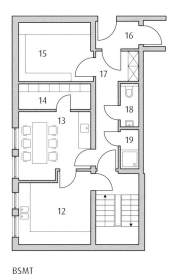

BSMT

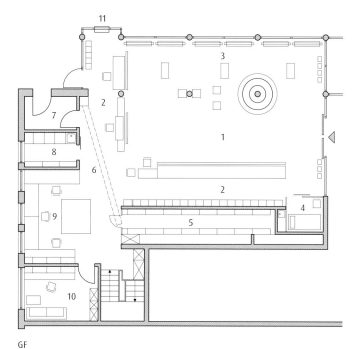

GF

Client | operator	Michael Bentz
Design phase | built	2006
Usable surface	236 sqm

a The bench in the middle of the pharmacy provides a splash of bright colour.

b The spacious, open-plan sales area makes an immediate impact on visitors.

c A picture of the surrounding mountains extends the full length of the counter.

Diagrammatic plans, scale 1:400
Floor plans, scale 1:200

Other key information

Catchment radius	20 km
Distance to nearest pharmacy	800 m
Nearest health facility	Medical practice
Number of items in stock	6,500
Drugs produced in-house	Teas, Bach flowers, vitamins
Number of staff	6 Full time, 3 Part time
Prescriptions filled per day	81
Technical facilities	Air conditioner, order picker, conveyors

Nature of special services: facial analysis, travel medicine, aromatherapy, free delivery, cosmetic studio, blister packaging, footcare practice, own brand under development

The pharmacy's very concentrated floor plan, with almost no aisles, makes it very user-friendly. The automated stockroom reduces the amount of space required, and there is even a drive-in night counter.

Usable floor spaces

Sales	**yellow**	117 sqm	50 %
Service	**red**	43 sqm	18 %
Specialist areas	**pink**	4 sqm	2 %
Employee areas	**orange**	34 sqm	14 %
Supply | disposal	**brown**	34 sqm	14 %
Plant	**blue**	4 sqm	2 %
Total		236 sqm	100 %

Floor plan layout

1	Sales area	**7**	Goods inwards	**13**	Staff lounge
2	Cash desk	**8**	Prescription area	**14**	Staff changing
3	Self-service area	**9**	Workstation	**15**	Product storage
4	Consultations	**10**	Night shift room	**16**	Sluice
5	Product storage	**11**	Night shift desk | drive-in	**17**	Server room
6	Conveyors	**12**	Laboratory	**18**	Bathroom
				19	Shower

Less is more in the new interior and exterior of the pharmacy on Klemensplatz in Düsseldorf. The futuristic, simply laid-out dispensary is almost all white, with only a few strategically positioned display surfaces. A deliberate decision has been made to have no self-service area, thus placing the entire focus on the pharmacist and her team. There are no in-your-face Buy Me signs, no overflowing shelves, and all is pure form and function. A wide counter made from dark cherrywood seems to float above its two simple metal feet, dominating the minimalist sales area. A small, simple shelf has been placed in front of the full-height display window to avoid obstructing the view from outside. At the back of the sales area a frosted glass screen, used to display the small number of behind-the-counter products, separates the tea dispensary, which is not accessible to customers. A consultation booth stands at the interface of the public and private areas, creating transparency between the two and saving space for workstations.

KLAUS R. BÜRGER

KLEMENSPLATZ PHARMACY
DÜSSELDORF

Pharmacies

Pharmacies

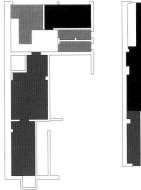
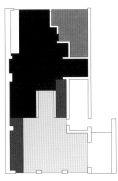
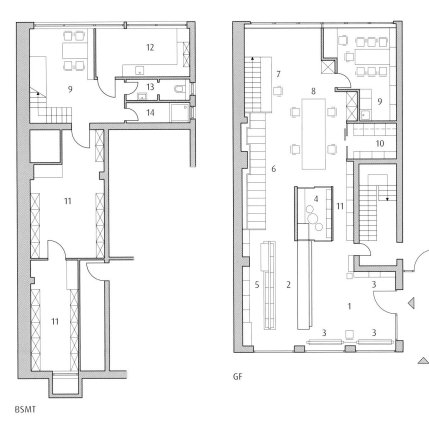

BSMT

GF

Client	operator	Karin Waldmann
Design phase	03 2003 – 02 2004
Built	02 2004 – 04 2004
Gross floor area	216 sqm
Usable surface	197 sqm
Gross volume	605 cbm
Construction cost	286,000 €
Total cost	398,000 €

a The wide, dark cherrywood counter
b The simple, futuristic sales area
c The consultation booth between the public and private areas
d Workstations at the back of the pharmacy

Diagrammatic plans, scale 1:400
Floor plans, scale 1:200

Other key information

Catchment radius	8 km
Distance to nearest pharmacy	300 m
Nearest health facility	none
Number of items in stock	17,500
Stock turnover rate	10,5 months
Proportion produced in-house	5 %
Drugs produced in-house	Capsules, salves, teas, cosmetics
Proportion delivered externally	20 %
Number of staff	7
Technical facilities	Air conditioner

The space on the first floor is designed to be simple and functional. There is a soundproofed consulting room for customers, and two rest areas for staff. The laboratory, storage, and staff rooms are in the basement, reached via a single flight of stairs.

Usable floor spaces			
Sales	**yellow**	42 sqm	21 %
Service	**red**	76 sqm	39 %
Employee areas	**orange**	35 sqm	18 %
Supply	disposal	**brown**	44 sqm	22 %
Total		197 sqm	100 %

Floor plan layout

1 Sales area	5 Tea-prescription area	10 Prescription area
2 Cash desk	6 Alphabetic storage	11 Product storage
3 Self-service area	7 Dispatch	12 Laboratory
4 Consultations	8 Workstation	13 Bathroom
	9 Staff lounge	14 Shower

Nature of special services: prescription collection, home delivery, preventive health (blood pressure, blood sugar, and cholesterol testing, body structure analysis), nutritional advice, seminars, walking courses, specialist makeup advice, compression stocking fitting, incontinence advice, vaccinations, travel medicine.

A bright orange pharmacy? Despite, or perhaps because of his low budget, the young pharmacist wanted an offbeat, spacious interior with a small number of distinctive fittings and bold shapes and colours. The spacious and impressive shop, located in a former industrial building, certainly lent itself to an unusual design. The visible joists and large shop window set the aesthetic tone for the whole pharmacy, with the open-plan, generously proportioned loft-like space being preserved by the renovation. This is underlined by the hard wearing industrial oak parquet, but the most prominent feature of the interior is the bright orange concave semicircular counter, its front decorated with pharmacy-related words. The colourful counter near the entrance, a few steps below the sales area, is equally impressive. The fittings are made using exhibition stand technology, a flexible, low-cost solution. In a further elegant detail, the countertop surface is orange leather.

KLAUS R. BÜRGER

OHM PHARMACY
ERLANGEN

Pharmacies

Pharmacies

**Client	operator**	Ingo Deinl
**Design phase	built**	2004
Gross floor area	539 sqm
Usable surface	495 sqm
Gross volume	2,156 cbm

a All of the shop fittings are made using exhibition-stand technology.
b Bright orange is the dominant colour in the spacious sales area.
c The concave semicircular counter is the most eye-catching feature of the interior.

Diagrammatic plans, scale 1:400
Floor plans, scale 1:200

Other key information

Catchment radius	Not known
Distance to nearest pharmacy	
Nearest health facility	
Number of items in stock	
Stock turnover rate	
Proportion produced in-house	
Drugs produced in-house	
Proportion delivered externally	
Type of customers supplied	
Number of staff	
Prescriptions filled per day	
Proportion of special services	
Nature of special services	

The pharmacy has plenty of space spread over two levels, and the customer area is unusually large. The equally expansive basement stockrooms are accessed from the first floor via a staircase.

Usable floor spaces			
Sales	**yellow**	221 sqm	45 %
Service	**red**	68 sqm	14 %
Administration	**green**	12 sqm	2 %
Employee areas	**orange**	14 sqm	3 %
Supply	disposal	**brown**	180 sqm	36 %
Total		495 sqm	100 %

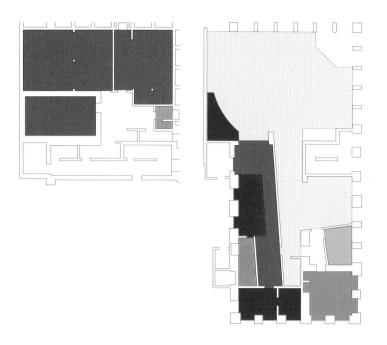

Pharmacies

Floor plan layout
1. Sales area
2. Cash desk
3. Self-service area
4. Behind-the-counter shelves
5. Workstation
6. Goods inwards
7. Product storage
8. Workstation
9. Kitchen | staff lounge
10. Product storage
11. Laboratory
12. Prescription area
13. Consultations
14. Private consultations
15. Office
16. Night shift room | office
17. Bathroom

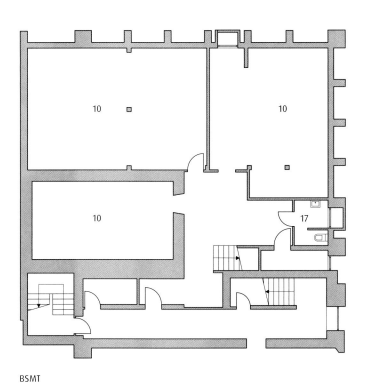

BSMT

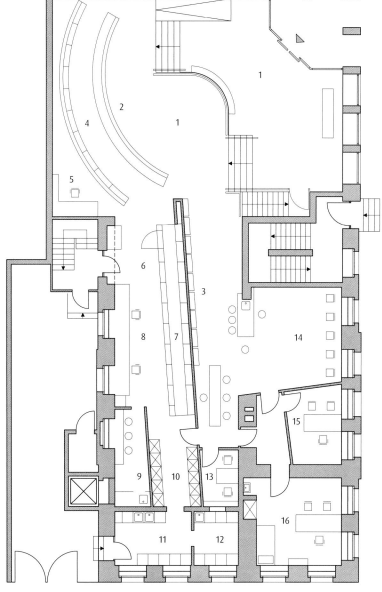

GF

The tri-Haus-Pharmacy, located in a health centre in Arnsberg-Neheim near Dortmund, is built on a triangular site. This was a difficult project, as the interior had to be designed accordingly, and the pharmacy also needed a rear entrance in addition to the front doors in one corner of the triangle. The sales area is only 57 square metres, but looks more spacious, with grey and beige stone floors enhancing the bright, friendly impression. The first thing visitors see is the amber-coloured rear walls, which also help to create a warm, harmonious environment, while a backlit storage unit of the same colour on the counter fosters a sense of continuity. The Wenge wood and lacquered glass furniture gives the space a strong sense of structure – an important element of the design which, given the triangular floorplan, was not easy to achieve. The building's energy supply is highly innovative, with geothermal probes 100 metres below the ground supplying heating and air conditioning, and a 100-square-metre solar electricity array, together reducing the pharmacy's total energy consumption to that of a 32-square-metre apartment.

RENATE HAWIG

TRI-HAUS PHARMACY
ARNSBERG-NEHEIM

Pharmacies

Pharmacies

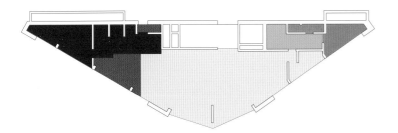

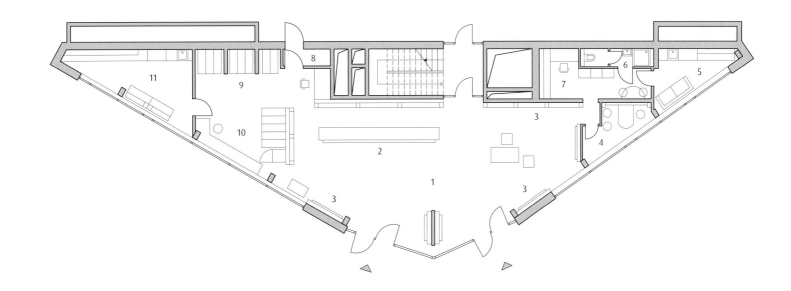

Client | operator	Elke Banz, Dietmar Riecks, Uwe Berbüße
Design phase	02 2003–11 2004
Built	12 2004–04 2005
Gross floor area	231 sqm
Usable surface	195 sqm
Gross volume	712 cbm
Total cost	150,000 €

a Grey and beige stone floors and dark Wenge wood help to create an attractive atmosphere.
b The consulting room is behind the glowing amber-coloured self-service wall to the side of the building.
c Although it is small, the sales area is bright and user-friendly.

Other key information

Catchment radius	4 km
Distance to nearest pharmacy	500 m
Nearest health facility	Medical practice
Number of items in stock	8,000
Stock turnover rate	6 Months, 2 Times a year
Proportion produced in-house	2 %
Drugs produced in-house	dermatology, salves, uvula
Number of staff	4 Full time
Prescriptions filled per day	200
Proportion of special services	5 %
Technical facilities	Heating and cooling with geothermal probe

One corner of the triangular site houses the offices and stockroom, and the other the administration and service areas. The customer area is in the middle of the building, pushing the other parts of the pharmacy out to the sides.

Usable floor spaces

Sales	yellow	112 sqm	57 %
Service	red	40 sqm	21 %
Administration	green	12 sqm	6 %
Employee areas	orange	14 sqm	7 %
Supply | disposal	brown	17 sqm	9 %
Total		195 sqm	100 %

Floor plan layout

1 Sales area
2 Cash desk
3 Self-service area
4 Consultations
5 Night shift room
6 Bathroom \| shower
7 Office
8 Goods inwards
9 Alphabetic storage
10 Stock control
11 Prescription area \| laboratory

Diagrammatic plans, scale 1:400
Floor plans, scale 1:2000

Many hurdles had to be overcome in building the St. Anna Pharmacy 1828 in central Munich, making the interior design even more spectacular. The small site, only just large enough to reach the minimum legal size for a pharmacy, extends over two floors. The owner's desire for a modern, individual design had to be reconciled with the need to protect the listed building. The interior is dominated by ceiling-height, wall-mounted shelves painted in signal-red car paint that separate the customer area from the dispensary. There is no window display, so the shelves are the first things to catch the visitor's eye, and the only external signage is an inscription on a pillar and the green-cross international pharmacy symbol in the windows. The wall-mounted shelves extend into the rear half of the interior, making space for the simple, white-painted counter. Behind this are the workroom, dispensary, and stairs to the upper floor, which houses a stockroom only seven square metres in area. This contains a fully automatic stock picker connected to the sales area by a transportation system in the timber-beamed ceiling. A small office was constructed for the pharmacist in what was previously a storage room; the specially made green glass wall and lifting device are the most instantly noticeable feature of the staircase.

HUBER RÖSSLER

ST. ANNA PHARMACY 1828
MUNICH

Pharmacies

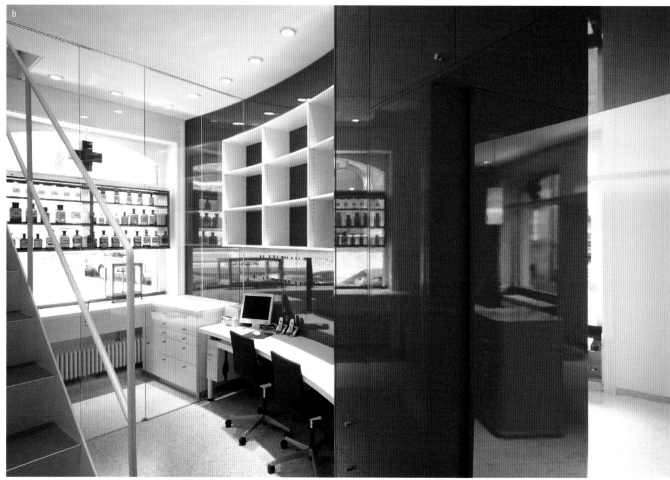

Pharmacies

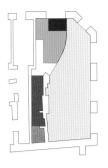
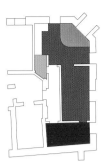

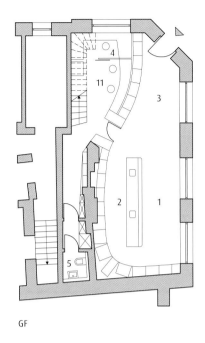

GF

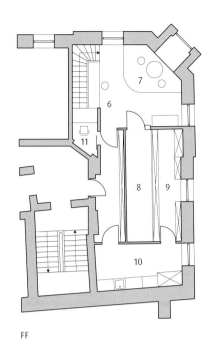

FF

Client | operator	Ludwig Gierstorfer jr.
Design phase	11 2002–02 2003
Built	02 2003–04 2003
Usable surface	124 sqm
Gross volume	604 cbm

a A deliberate decision was made not to incorporate a window display, and instead the wall of red shelves has been used to attract passers-by.

b The workroom and dispensary are visible from outside.

Diagrammatic plans, scale 1:400
Floor plans, scale 1:200

Other key information

Distance to nearest pharmacy	300 m
Nearest health facility	Medical practice
Number of items in stock	20,000
Proportion produced in-house	5 %
Drugs produced in-house	Salves, capsules, teas, uvula
Proportion delivered externally	15 %
Type of customers supplied	Medical practice, retirement homes, clinics, private clients
Number of staff	8 Full time, 5 Part time [+ messengers]
Technical facilities	Product storage, order picker

Nature of special services: skin cosmetics, oncology, inhouse production, prescription development, searches

Because the pharmacy is located in a two-storey historic building, it is not always very practically laid out. It is not easy to carry products from the first floor up the angular staircase to the second-floor stockrooms, or to access the laboratory and workstations.

Usable floor spaces

Sales	**yellow**	49 sqm	40 %
Service	**red**	14 sqm	12 %
Administration	**green**	13 sqm	10 %
Employee areas	**orange**	9 sqm	7 %
Supply | disposal	**brown**	39 sqm	31 %
Total		124 sqm	100 %

Floor plan layout

1	Sales area	4	Prescription area	8	Product storage
2	Cash desk	5	Bathroom	9	Product storage
3	Self-service area	6	Stock control	10	Laboratory
		7	Night shift room	11	Office

The granite floor of the Linden Pharmacy draws on the baroque architecture of Ludwigsburg. Full-length white shelves, some of them set into the wall, and rooms with rounded corners make the sales area look compact, with carefully designed lighting creating a strong background for the goods on display. The main focus is on the white counter jutting out on either side of the ceiling supports. Three revolving displays are used for seasonal products and their white circular bases making them stand out from the otherwise uniform cobble flooring. However, the interior's most distinctive feature is the ornamental ceiling. The strong, bold lines and pale monochrome furnishings lead the eye to the colourful fresco depicting eleven medicinal herbs in a modern interpretation of a traditional colour scheme. The pharmacist specialises in natural medicine and cosmetics, and the whole corporate design was changed to reflect this. The emphasis on herbs, the old-fashioned floor covering and the dominance of the fresco express pharmaceutical traditions in a simple form.

IPPOLITO FLEITZ GROUP

LINDEN PHARMACY
LUDWIGSBURG

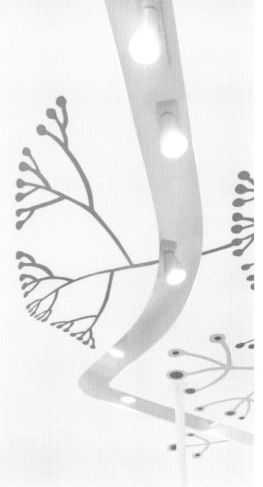

Pharmacies

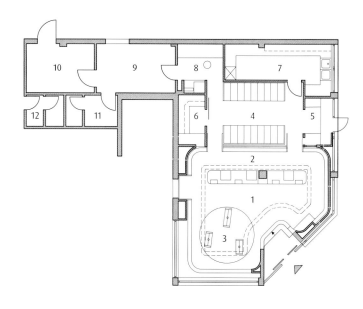

**Client	operator**	Meike Raasch		
Design phase	12 2005–06 2006		
Built	05 2006–06 2006		
Usable surface	120 sqm		
Gross volume	361 cbm		

a The granite-cobbled floor is a reference to Ludwigsburg's baroque architecture.

b White shelves set into the wall and rooms with round corners make the sales area look compact.

c The colourful plant motifs of the ceiling fresco are a modern take on a traditional theme.

d Three revolving display stands are used for seasonal products.

Diagrammatic plans, scale 1:400
Floor plans, scale 1:200

Other key information

Catchment radius	15 km
Distance to nearest pharmacy	350 m
Nearest health facility	Medical practice, medical houses
Number of items in stock	8,000
Stock turnover rate	2,5
Proportion produced in-house	5 %
Drugs produced in-house	Salves, juices, uvula, teas, eye drops
Type of customers supplied	Regular and walk-in customers
Number of staff	3
Technical facilities	Air conditioner

Every inch of the limited space available in this one-storey pharmacy has been put to good use. However, the stockroom is reached via the night service area, and the bathroom is not as accessible as it could be.

Usable floor spaces

Sales	**yellow**	50 sqm	42 %
Service	**red**	37 sqm	31 %
Administration	**green**	4 sqm	3 %
Employee areas	**orange**	15 sqm	13 %
Supply	disposal	**brown**	10 sqm	8 %
Plant	**blue**	4 sqm	3 %
Total		120 sqm	100 %

Floor plan layout

1 Sales area
2 Cash desk
3 Self-service area
4 Alphabetic storage
5 Office
6 Prescription area
7 Laboratory
8 Workstation
9 Night shift room
10 Product storage
11 Plant room
12 Bathroom

The new central pharmacy at Frankenthal municipal hospital also supplies six other hospitals in Rhineland-Palatinate and Baden-Wuerttemberg, with a total of 1,500 beds. Not all hospitals maintain their own pharmacies; increasing numbers now pool their resources with others. The pharmaceutical and medical products service centre is on the first floor of the Frankenthal hospital, a solid 1970s building. Goods can be brought in and out without an elevator, and the hospital's delivery area can also be used. The pharmacy uses a semi-automated stock picker to save space, which staff have found to be more laboratory-intensive but also more efficient than a fully automatic one. Particular attention has been given to making the pharmacy an attractive workplace, with its contemporary lighting and colour scheme and bright, friendly materials.

SANDER.HOFRICHTER

STADTKLINIK PHARMACY
FRANKENTHAL

Pharmacies

c

Client | operator	Stadt Frankenthal		
Design phase	07 2004–04 2005		
Built	05 2005–04 2006		
Gross floor area	1,300 sqm		
Usable surface	827 sqm		
Gross volume	5,148 cbm		
Construction cost	950,000 €		
Total cost	1,200,000 €		

a The corridor to the consultation room
b The rooms feature contemporary colours and lighting and bright, friendly materials.
c The pharmacy's semi-automated stock picker
d The cleanroom is used to make parenterals.

Other key information

Catchment radius	50 km
Number of items in stock	2,300
Stock turnover rate	4–5
Proportion produced in-house	20 %
Drugs produced in-house	Sterile drugs
Proportion delivered externally	80 %
Type of customers supplied	Hospital
Number of staff	9 Full time, 11 Part time
Proportion of special services	20 %
Technical facilities	Order picker, sterile laboratory, air conditioner
Nature of special services	Sterile production

The pharmaceutical and medical products service centre consists of two areas which are separated by a wide aisle: the pharmacy itself, with a semi-automated stock picker, the adjoining goods inwards, and administration areas.

Usable floor spaces			
Sales	**yellow**	11 sqm	1 %
Service	**red**	398 sqm	48 %
Specialist areas	**pink**	64 sqm	8 %
Administration	**green**	8 sqm	1 %
Employee areas	**orange**	156 sqm	19 %
Supply | disposal	**brown**	190 sqm	23 %
Total		827 sqm	100 %

Diagrammatic plans, scale 1:400
Floor plans, scale 1:200

Pharmacies

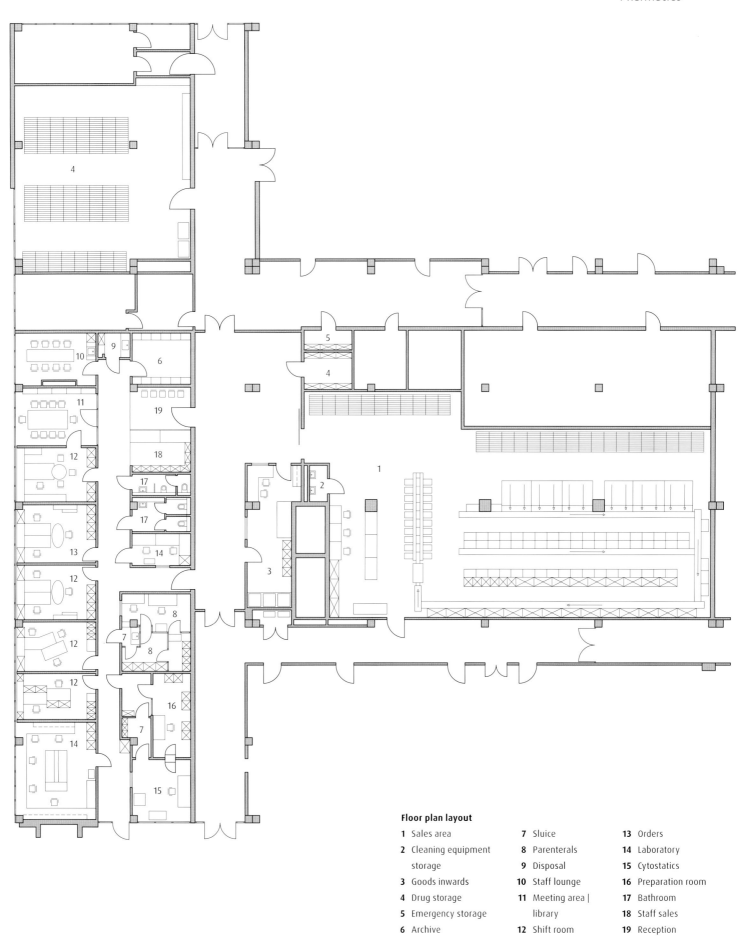

Floor plan layout
1 Sales area
2 Cleaning equipment storage
3 Goods inwards
4 Drug storage
5 Emergency storage
6 Archive
7 Sluice
8 Parenterals
9 Disposal
10 Staff lounge
11 Meeting area | library
12 Shift room
13 Orders
14 Laboratory
15 Cytostatics
16 Preparation room
17 Bathroom
18 Staff sales
19 Reception

Index of Architects and Designers

Aalto, Alvar: 020
Arcass Freie Architekten BDA: 178-183, 190-195
Arcass Planungsgesellschaft mbH: 184-189
Architekten BDA RDS Partner: 124-129
Artec Architekten: 388-393
Atelier 5: 092
Attista Alberti, Leon: 201

Bathke, Jörn: 394-401
Baumann, Thomas: 084-085
bauzeit architekten: 010-011, 035, 038-039, 041
Beeg, Geiselbrecht und Lemke: 082, 091
Berg Planungsgesellschaft mbH & Co. KG: 154-157
bhend.klammer architekten: 254-257
brandherm + krumrey: 258-261
Bürger, Klaus R.: 262-265, 357, 402-415

Cossmann_de Bruyn: 266-269
Curtis & Davis: 021

Dahmen-Ingenhoven, Regina: 270-273
Damaschke, Sabine: 367-368
Döcker, Richard: 018-020, 022
Domènech i Montaner, Lluís: 032

Eling + Novotny Mähne: 081

Geldner, Paul: 018
GRAFT: 033-034, 278-287
Grober, Julius: 082
Gruppe für Gestaltung: 274-277

Hall Black Douglas Architects: 068
Hawig, Renate: 416-419
Heinle, Wischer und Partner Freie Architekten: 144-153
Henning Larsen Architects: 040-041, 044
Hildebrand, Walter: 025
HLM Architects: 068
Hoffmann, Ludwig: 016-019
holzrausch: 288-291

IAP Isin Architekten Generalplaner GmbH: 166-171
ippolito fleitz group: 292-297, 424-427

Jencks, Charles: 027
Judtmann, Fritz: 063, 065
Junghans + Formhals: 082

Koller Heitmann Schütz: 124-129
Kromschröder, C.: 083

Labryga, Franz: 075, 077-078
landau + kindelbacher: 298-301
Le Corbusier: 024
Lundin, Stefan: 030, 031

Mayer, Walter: 084
Mebes, Paul: 018
Medplan AS: 024
Meissler, Sebastian: 023
Meuser Architekten: 247, 249, 302-307
Meuser, Philipp: 009, 015
Mikulandra-Mackat, Mateja: 308-311
Mitterberger, Gerhard: 312-317
Mocken, Franz: 021
Mullins, Michael: 027

Nickl & Partner: 025, 082
Nix Mann Perkins & Will: 023

OD 205 architectuur bv: 102

pd raumplan: 318-325
Peter Brand & Partner: 022
Philips Design: 056, 058-060
Piano, Renzo: 048
planbar 3: 201, 202
Planungsring Dr. Jüchser + Pawlik: 064, 066
Planungsring Dr. Pawlik + Co.: 136-143
Planungsring Dr. Pawlik: 083
Poyet, Bernard: 017

Rappmannsberger, Rehle und Partner: 080, 081
Rauh · Damm · Stiller · Partner: 075
Rauh, H.: 083
Riss, Egon: 063, 065
Rössler, Huber: 420-423
RRP Architekten + Ingenieure GbR: 130-135

Saalmann, Ribbert: 162-165
sander hofrichter architekten: 067, 106-109, 326-331, 428-433
Scheffer, Bernadette: 367, 368
Schirmer, Christoph: 015
Schmieden, Heinrich: 063, 064
Schröder, Jürgen: 023
Schürmann, Joachim: 190-195
Schweitzer + Partner: 102, 103
Smith Group: 023
Stefan Ludes Architekten: 091
Steffen + Peter: 162-165
Stengele+cie.: 332-335
Swensson, Earl: 022

Index of Architects and Designers

Thiede Messthaler Klösges: 172-177
Thomas Schindler Architekt BDA: 158-161
TMK Architekten Ingenieure: 110-123, 172-177

van der Rohe, Mies: 380
Voigt, Andreas: 018

Wagenknecht Architekten: 336-339
Weber, Brand & Partner: 021
Weber, Wolfgang: 022
Welsch, Max: 018
Wilhelm, Wolfgang: 084

Authors and Co-Authors

Klaus Bergdolt, Prof. Dr. med., born in Stuttgart in 1947, studied medicine in Tübingen, Vienna, and Heidelberg from 1968–1974. In 1980 and 1981, he studied history, art history, Byzantine studies, history of science, and the history of the concepts of health and medical ethics in Heidelberg and Florence. Director of the history and ethics institute of Cologne University from 1995–2014. His numerous books and other publications have been translated into several languages.

Klaus R. Bürger, born in 1948. Trained as a carpenter from 1965–1968, and then studied at Werkkunstschule Krefeld from 1969–1972. He studied interior design from 1972–1978, and philosophy and German from 1981–1985, both in Düsseldorf. He founded the interior design firm *Bürger Büro für Innenarchitektur und Design* in 1981. Based in Krefeld, Germany.

Linus Hofrichter, Prof. Dipl.-Ing., Architect BDA, born in 1959 in Ludwigshafen/Rhein. He studied civil engineering and architecture in Karlsruhe. Worked since 1987 at Walter Klumpp, where he became a partner in 1991. From 1996–2000 worked at Klumpp Sander Hofrichter, since 2001 at sander hofrichter, today a|sh architekten. Honorary professor at the Technische Hochschule Mittelhessen (University of Applied Sciences) in Giessen, Germany. Deputy chair of the board of the AKG.*

Franz Labryga, Prof. Dipl.-Ing., Architect, born in 1929 in Erfurt. Studied in Jena, Weimar, and Berlin. From 1974–1994, lecturer at the Berlin Technical University. Specialised in: design and buildings in the healthcare sector. Long-time director of the Institut für Krankenhausbau (institute for hospital building). From 1974–1977, speaker for the special research division for hospital building. Manager of several working committees (including for the German Ministry of Health and the DIN). For many years, he was a member of the AKG.*

Philipp Meuser, Prof. h. c. Dr.-Ing. Architect BDA, born 1969 in Hilden/Rhineland. Studied in Berlin and Zurich, specialising in history and the theory of architecture. International planning and construction projects in the Former Soviet Union and Western Africa. Author of numerous publications focussing on healthcare buildings and the architectural history of the Soviet Union.

Hartmut Nickel, Dipl.-Ing. Architect BDA, born 1948 in Dessau. From 1979–1984, studied in Berlin. Research associate from 1986–1988; research assistant at the institute for hospital building at the Berlin Technical University; from 1988–1994 architect in the Hannover Construction Administration dept. Member of the AKG* since 1990. Employed since 1995 and partner since 1998 at the architectural firm Schweitzer + Partner.

Lekshmy Parameswaran, co-founder of fuelfor, an innovative design consultancy specialising in healthcare, where she encourages innovation teams to challenge conventional thinking, conducts collaborative research to bring the voices of patients and care providers into the innovation process in inspiring ways and applies design thinking as a strategic tool to unlock creativity in client organisations. She holds an MEng degree from Cambridge University and an MA from the Royal College of Art in London.

Peter R. Pawlik, Dr.-Ing., Architect BDA, born 1946 in Wilhelmshaven. Studied in Oldenburg and Berlin. From 1974–1977, employed at Georgije Nedeljkov; from 1977–1982 office manager in Planungsring Dr. Jüchser (since 1983, Planungsring Dr. Jüchser + Pawlik). From 1989–2014, Planungsring Dr. Pawlik. From 2006–2015 Chair of the board of AKG.* Based in Berlin working as consultant and author.

Jeroen Raaijmakers, born in 1963 in Delft/NL. Studied Industrial Design at the Technical University in Delft, graduated in 1988. Began at Philips Design in 1996 as product designer. In 1999 he became Global Design Director Healthcare, responsible for the brand identity of Philips in healthcare. After the acquisition and integration of several healthcare companies, Raaijmakers now leads a global design team with a seat in Eindhoven/NL, Andover/USA, Seattle/USA. In 2003 he initiated the design-led innovation programme Ambient Experience Design for Healthcare which resulted in its installation in 50 hospitals worldwide.

Wolf Dirk Rauh, Dipl.-Ing. Architect BDA, born 1947 in Bochum. Studed in Berlin. From 1978–1989, in partnership with Wolfgang Rauh. Since 1989, joint owner of the Rauh• Damm• Stiller• Partner office. From 1995–2011 managing partner. Member of the AKG* and the International Hospital Federation.

Authors and Co-Authors

Christoph Schirmer, born in 1976, in Berlin. Apprenticed as a carpenter, studied interior design in Rosenheim, followed by architecture in London. From 2000–2001 Worked for Murphy/Jahn in Chicago. In 2002, freelance architect in Rosenheim and Berlin, in 2005, partner with Galandi Schirmer Architekten + Ingenieure, and managing director since 2009.

Álvaro Valera Sosa, Dipl.-Ing. Arch. MScPH, born 1973 in Caracas, Venezuela. Studied architecture at the Central University of Venezuela and Public Health at the Charité Medical University of Berlin. Since 2001 he practiced in countries such as Venezuela, Spain, and the UAE. In academia his research extends to renowned universities such as the Technical University Berlin, Chalmers University of Technology, and Polytechnic University of Milan. In early 2019 he founded HEI, the Health Environment Institute of Berlin.

Photo Credits
All photos on the pages given are by the same photographer or source, unless otherwise stated.

Albert, Andi: 279–280
Baumgarten, Ralf: 319–320, 323–324
Braun, Zooey: 293–296, 425–426
Brilo, Moritz: 417–418
Frahm, Klaus: 337–338
Hacker, Christian: 299–300
hiepler, brunier: 283–285
Huthmacher, Werner: 309–310
K+W: 289–290
Kleiner, Tom/GfG: 275–276
Knauf, Holger: 271–272
Meuser, Philipp: 303–306
Oberwalder, Zita: 313–316
Peters, Lucas: 255–256
Scheffler, Jens: 395–396, 399–400
Spiluttini, Margherita: 389–390
Spoering, Uwe: 259–260, 263–264, 267–268, 403–404, 407–408, 411–414
stengele+cie: 333–334
Traub, Markus: 412–422
Vogt, Johannes: 327–329, 429–432

* AKG = Architekten für Krankenhausbau und Gesundheitswesen e.V. (association of architects for hospital building and healthcare, with its headquarters in Berlin, Germany)

Further Readings

Here is a selection of publications that inspired the authors and co-authors while writing this book.

These titles have no claim to universal validity.

Fermand, Catherine:
les hôpitaux et les cliniques.
Architectures de la santé,
Paris 1999

This book covers French hospital buildings, their programmes and evolution. Here, the diversity of designers' solutions, influenced by societal choices are put forward thanks to a wide range of rich illustrations: over 500 floor plans, sections, drawings, and photographs.

Kjisik, Hennu:
The Power of Architecture.
Towards Better Hospital Buildings,
Helsinki 2009

This doctoral dissertation shows that future designs and concepts can improve hospital architecture, through an in-depth analysis of historical archetypes and existing hospitals. Photographs, floor plans, sections, and diagrams thoroughly illustrate the twelve chapters of this book.

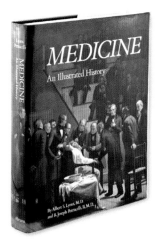

Lyons, Albert S./Petrucelli, R. Joseph:
Medicine. An Illustrated History,
New York 1978

This highly-illustrated book offers a breathtaking overview of the history of medicine through a vast collection of drawings and photographs ranging from archaeological relics to modern medical procedures. It has been translated into many languages.

Mens, Noor/Wagenaar, Cor:
Health care architecture in the Netherlands, Rotterdam 2010

This book offers a progressive survey of the architecture of healthcare buildings, from its emergence as a specific typology to the most recent care complexes. Furthermore, some fifty buildings are meticulously described and illustrated.

Further Readings

Murken, Axel Hinrich:
Vom Armenhospital zum Großklinikum.
Die Geschichte des Krankenhauses,
Cologne 1988

This book is solely based on the historical, social, and typological development of hospitals, covering the eighteenth century up to the last part of the twentieth century. Even if it is already thirty years old, this books remains a contemporary publication to which one can turn for valid inspiration.

Pawlik, Peter R.:
Moshe Zarhy.
Health Facilities in Israel,
Berlin 2014

Published when Moshe Zarhy was still alive (1923–2015) this book covers his life's extensive work, mainly dedicated to hospitals and health facilities. The architect's monograph thus also serves as a compendium of this building typology in Israel during the second half of the 20th century.

Valera Sosa, Álvaro/Matthys, Stefanie: From Concepts in Architecture to German Health Economics, Berlin 2012

In this book, light, nature, acoustics, air quality, architectural layout, interior design, and art are considered, as well as three research areas (experimental interventions, evaluation tools, and systematic reviews) to establish the guidelines for developing an amalgam of architecture and healing.

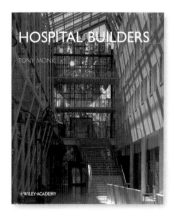

Monk, Tony:
Hospital Builders,
Chichester 2004

This book puts forth thirty-two case studies covering significant healthcare facilities around the world. Filled with photographs of buildings, models, and floor plans, this book provides inspiration from late twentieth-century architectural hospital projects.

Wischer, Robert et al.:
Zukunftsoffenes Krankenhaus,
Vienna 2007

Robert Wischer (1930–2007) focused his life's work on healthcare. He was to Germany what Moshe Zarhy was to Israel. His fundamental book is divided into seven parts – from the history of hospitals to the impact of architecture on the recovery process.

Wagenaar, Cor (ed.):
The Architecture of Hospitals,
Rotterdam 2006

Through its peppy graphics and layout, this book agilely demonstrates how architecture can contribute to better, more efficient and humane hospitals, through the themes of culture, healthcare concepts, design, technology, and new hospitals, in the Netherlands.

著作权合同登记图字：01-2019-1751 号

图书在版编目（CIP）数据

建造设计手册：医院和医疗建筑 = Construction and Design Manual：Hospitals and Medical Facilities：英文/（德）菲利普·莫伊泽编著；弗朗茨·纳布瑞加技术指导 . —北京：中国建筑工业出版社，2019.5

ISBN 978-7-112-23633-6

Ⅰ . ①建… Ⅱ . ①菲… ②弗… Ⅲ . ①医院 — 建筑设计 — 手册 — 英文 Ⅳ . ① TU246.1-62

中国版本图书馆 CIP 数据核字（2019）第 073439 号

Copyright © 2019 DOM publishers, Berlin/Germany
www.dom-publishers.com
All rights reserved.

责任编辑：姚丹宁　段　宁

Construction and Design Manual
Hospitals and Medical Facilities

建造设计手册
医院和医疗建筑

Edited By Philipp Meuser（菲利普·莫伊泽）
Scientific Advisor：Franz Labryga（弗朗茨·纳布瑞加）
*
中国建筑工业出版社出版、发行（北京海淀三里河路 9 号）
各地新华书店、建筑书店经销
北京点击世代文化传媒有限公司制版
深圳市泰和精品印刷厂印刷
*
开本：880×1230 毫米　1/16　印张：27½　字数：851 千字
2019 年 6 月第一版　2019 年 6 月第一次印刷
定价：398.00 元
ISBN 978-7-112-23633-6
（33923）

版权所有　翻印必究
如有印装质量问题，可寄本社退换
（邮政编码 100037）